Janina Krell-Rösch

Interdependence of Physical (In-) Activity, Fitness and Cognition

A Cross-Sectional Study in Young Adults

KARLSRUHER SPORTWISSENSCHAFTLICHE BEITRÄGE / BAND 5

Institut für Sport und Sportwissenschaft
Karlsruher Institut für Technologie (KIT)

Herausgeber der Schriftenreihe:
Prof. Dr. Klaus Bös
PD Dr. Michaela Knoll

Interdependence of Physical (In-) Activity, Fitness and Cognition

A Cross-Sectional Study in Young Adults

by
Janina Krell-Rösch

Dissertation, Karlsruher Institut für Technologie (KIT)
Fakultät für Geistes- und Sozialwissenschaften
Tag der mündlichen Prüfung: 16. Dezember 2013
Referenten: Prof. Dr. Klaus Bös, Prof. Dr. Alexander Woll

Impressum

 Scientific
Publishing

Karlsruher Institut für Technologie (KIT)
KIT Scientific Publishing
Straße am Forum 2
D-76131 Karlsruhe

KIT Scientific Publishing is a registered trademark of Karlsruhe
Institute of Technology. Reprint using the book cover is not allowed.

www.ksp.kit.edu

Print on Demand 2014

ISSN 1862-748X
ISBN 978-3-7315-0164-0
DOI: 10.5445/KSP/1000037914

Interdependence of Physical
(In-) Activity, Fitness and Cognition
A Cross-Sectional Study in Young Adults

Zur Erlangung des akademischen Grades eines

DOKTORS DER PHILOSOPHIE

(Dr. phil.)

von der Fakultät für Geistes- und Sozialwissenschaften

des Karlsruher Instituts für Technologie (KIT)

angenommene

DISSERTATION

von

Janina Krell-Rösch (geb. Krell)

aus Rastatt

Dekan: Prof. Dr. Andreas Böhn
Erstgutachter: Prof. Dr. Klaus Bös
Zweitgutachter: Prof. Dr. Alexander Woll
Tag der mündlichen Prüfung: 16.12.2013

Vorwort

Die Untersuchung von Persönlichkeitsmerkmalen wie der Intelligenz und der Motorik sowie die Analyse deren Zusammenhänge ist ein zentraler Bestandteil der sozialwissenschaftlichen Forschung. Auch in der Sportwissenschaft wird häufig ein positiver Zusammenhang von kognitiven und motorischen Merkmalen angenommen. Schon in den 70er Jahren warb der Deutsche Sportbund mit dem Slogan „Ein Schlauer trimmt die Ausdauer" und im Jahre 2008 formulierte Dr. Ralf Reinhardt „Laufen macht schlau" als zentrale These seiner Dissertation.
Viele Fragen sind bislang allerdings offen: Gibt es generelle Zusammenhänge zwischen Kognition und Motorik oder eher sehr spezifische Beziehungen, beispielsweise zwischen exekutiven Funktionen und aerober Ausdauerleistungsfähigkeit? Zeigen sich diese Beziehungen unabhängig von Alter und Leistungsniveau oder gibt es alters- und niveauspezifische Aspekte? Sind Verhaltenstests ein adäquater Zugang zu diesen Forschungsfragen oder sollte man konstruktnäher bei den (neuro)physiologischen Korrelaten ansetzen? Mit diesen Fragen befasst sich Frau Krell-Rösch in Ihrer Dissertation, die unter unserer Betreuung am KIT erstellt wurde.
Die Dissertation entstand in interdisziplinärer Zusammenarbeit zwischen dem Institut für Sport und Sportwissenschaft am KIT, dem Cognitive Systems Lab (CSL) am KIT unter der Leitung von Frau Prof. Dr. Tanja Schultz sowie dem Zentrum für Weltraummedizin der Charité in Berlin unter der Federführung von Dr. Alexander Stahn. Im vorliegenden Buch werden Interdisziplinarität und Komplexität der Arbeit sehr gut deutlich.

Die Ergebnisse passen insgesamt zu dem aktuellen Forschungsstand. Frau Krell-Rösch findet ebenfalls signifikante Beziehungen zwischen Aktivität, Fitness und Kognition; die Effektstärken sind jedoch durchweg relativ gering. Am stärksten sind die Zusammenhänge zwischen der „globalen" körperlich-sportlichen Aktivität und den exekutiven Funktionen.
In jedem Fall scheint es so, dass in selektiven Gruppen mit hohem kognitiven und motorischem Leistungsniveau, wie in der vorliegenden Studie mit Studierenden, nicht mit ähnlich großen Wechselbeziehungen zwischen kognitiven und motorischen Leistungsmerkmalen gerechnet werden kann wie dies bei Kindern, Senioren oder bei Personen mit eingeschränkter Leistungsfähigkeit der Fall ist. Die querschnittliche Studie lässt auch die Frage offen, ob es Interventionseffekte und kausale Beziehungen zwischen Motorik und Kognition gibt.
Es wird also weiterer, insbesondere experimenteller und längsschnittlicher Studien bedürfen, um zentrale Fragen zu den Wechselbeziehungen von kognitiven und motorischen Merkmalen zu beantworten. Die Arbeit von Frau Krell-Rösch ist ein gut gelungenes Beispiel für eine breit angelegte und fundierte Herangehensweise zu diesem komplexen Forschungsfeld.

Klaus Bös & Alexander Woll, Karlsruher Institut für Technologie (KIT)

Contents

Contents

Kurzzusammenfassung

Einleitung: Körperlicher Aktivität werden positive Einflüsse auf die Gesundheit sowie eine hohe Wirksamkeit für die Prävention verschiedener Erkrankungen zugeschrieben. Darüber hinaus gilt ein wachsendes Interesse der Erforschung der Zusammenhänge von körperlicher Aktivität, Fitness und Kognition. Regelmäßige körperliche Aktivität wird mit einer erhöhten Leistung in kognitiven Tests in Verbindung gebracht, die sich sowohl auf verhaltensbasierter als auch neurophysiologischer Ebene beispielsweise in Form von ereigniskorrelierten Potentialen (ERP) manifestieren lässt. Jedoch hat sich die Forschung bislang vor allem auf Kinder und ältere Erwachsene konzentriert und vorrangig Effekte genereller Sportpartizipation oder kardiovaskulärer Fitness aufgezeigt. Daher war es das Ziel dieser Studie, die Wechselwirkungen zwischen körperlicher (In-) Aktivität, Fitness und exekutiven Funktionen bei jungen Erwachsenen zu untersuchen.

Methodik: 152 junge Erwachsene im Alter zwischen 18 und 34 Jahren nahmen an der Querschnittsstudie teil. 119 Probanden berichteten regelmäßig Sport zu treiben, 33 Probanden waren sportlich inaktiv. Alle Studienteilnehmer bearbeiteten eine computerbasierte kognitive Testbatterie, mit der Antwortzeiten sowie die N100, P200 und P300 Komponenten (Amplituden und Latenzen) der ereigniskorrelierten Potentiale in einem auditiven Oddball sowie einem visuellen Flanker und Switching Task erhoben wurden. Zusätzlich absolvierten alle Probanden verschiedene sportmedizinische und motorische Testverfahren zur Erfassung der Leistungsfähigkeit in den Bereichen Ausdauer, Kraft und Koordination sowie zur Bestimmung der Körperkonstitution. Darüber hinaus wurden verschiedene Aspekte der habituellen körperlichen und sportlichen Aktivität wie Ausmaß, bevorzugte Sportart, Belastungsintensität, oder Häufigkeit und Dauer des Sporttreibens mittels eines Fragebogens erhoben.

Ergebnisse: Generelle Sportpartizipation (Sportler vs. Nichtsportler), die Teilnahme am Wettkampfsport sowie die Häufigkeit der Sportausübung korrelierten, verglichen mit anderen aktivitätsbezogenen Parametern wie dem Ausmaß oder der Intensität, am stärksten mit der kognitiven Leistungsfähigkeit. Bezogen auf die körperliche Fitness zeigte sich, dass die Leistung in einem Testverfahren zur Bestimmung der Koordination unter Zeitdruck die größten Zusammenhänge mit verschiedenen Kognitionsvariablen aufwies. Es konnten keine Zusammenhänge zwischen kardiovaskulärer Fitness oder der regelmäßigen Ausübung von Ausdauersport mit der kognitiven Leistung im jungen Erwachsenenalter nachgewiesen werden. Generell sind die Ergebnisse sehr heterogen und können nur für spezielle kognitive Testverfahren oder Messgrößen nachgewiesen werden. Daneben fanden sich auch geschlechtsspezifische Unterschiede.

Schlussfolgerungen: Die Ergebnisse zeigen, dass regelmäßiges Sporttreiben mit einer erhöhten kognitiven Informationsverarbeitung in Tasks zur Überprüfung der exekutiven Funktionen bei jungen Erwachsenen korreliert. Es wird vermutet, dass generelle Sportpartizipation nicht nur eine regelmäßige sportliche Aktivität der Probanden widerspiegelt, sondern auch von anderen Faktoren wie der Einstellung zu einem gesunden und aktiven Lebensstil oder der Ernährung beeinflusst wird. Diese

Faktoren könnten in der Summe die gefundenen Unterschiede zwischen Sportlern und Nichtsportlern in den exekutiven Funktionen erklären. Ebenfalls konnten positive Zusammenhänge zwischen der Häufigkeit der Sporttreibens und der Kognitionsleistung aufgezeigt werden. Zudem zeigten Sportler, die nicht an Wettkämpfen teilnehmen, verglichen mit Wettkampsportlern bessere Leistungen in einzelnen Tasks. Darüber hinaus konnte nachgewiesen werden, dass eine hohe Koordinationsleistung, die beispielsweise aus der Ausübung koordinativ-anspruchsvoller Sportarten resultieren kann, positiv mit der Kognition und Gehirnleistung verknüpft ist. Studien mit älteren Erwachsenen konnten in diesem Zusammenhang zeigen, dass Koordinationstraining funktionelle und strukturelle Veränderungen und Anpassungen in bestimmten Bereichen des Gehirns auslösen kann, was wiederum förderlich für die kognitive Leistungsfähigkeit sein könnte.

Entgegen der Erkenntnisse aus bisherigen Studien mit jungen Erwachsenen zeigten sich in der hier vorliegenden Stichprobe keine Hinweise darauf, dass die Ausübung von Ausdauersportarten oder eine gute kardiovaskuläre Fitness mit einer höheren Kognitionsleistung zusammenhängen. Jedoch ist bei der Interpretation der Ergebnisse zu beachten, dass querschnittliche Daten keine Rückschlüsse auf Kausalitätsbeziehungen ermöglichen.

Abstract

Introduction: Physical activity has been associated with improved health across the lifespan and has a beneficial effect on the prevention of several diseases. There is also growing interest in the study of the relationships between exercise, fitness and cognition. Regular physical activity has been linked to an enhanced cognitive task performance as assessed by behavioral and neurophysiologic measures such as event-related brain potentials (ERPs). However, most research has been conducted among children or elderly and had a limited focus on general sports participation and cardiovascular fitness. The purpose of this study was therefore to determine the interdependence of physical (in-) activity, physical fitness and central information processing in young adults.

Methods: A cross-sectional study involving 152 young adults between the ages of 18 and 34 years was conducted. 119 participants reported to regularly engage in sports activity and 33 participants were physically inactive. They performed a cognitive test battery assessing response times and the N100, P200 and P300 components (amplitudes and latencies) of ERPs in an auditory Oddball, visual Flanker and visual Switching paradigm. Further, various physical performance tests to measure endurance, strength, coordinative skills and body composition were performed. All participants additionally reported their physical and sports activity habits such as amount, type of sport, exercise intensity, or frequency and duration based on a questionnaire.

Results: General sports participation, participation in competitive sports and frequency of exercise rather than other aspects of physical activity such as amount or intensity revealed the strongest relationships to cognitive performance measures. With regard to physical fitness, performance in a test to determine coordination under time pressure showed the strongest associations with cognitive performance. No effects of cardiovascular fitness or endurance sport on cognition were observed. The results are very heterogeneous and could be proved only for selected tasks or cognitive measures. In addition, significant gender-specific differences were observed.

Conclusions: The results indicate that general sports participation might increase cognitive information processing during cognitive tasks in young adults. It is supposed that general sports participation not only reflects the existence of regular sport activity but may also be influenced by other factors such as the general attitude referring to a healthy and active way of life or nutritional aspects. Taken together, these factors might be responsible for explaining the difference in cognitive performance between active and inactive people. Moreover, positive relationships between exercise frequency and cognitive performance were found. Furthermore, athletes who do not regularly participate in competitive sports performed better in selected cognitive tasks than competitors. In addition, coordinative skills probably resulting from sports associated with high motor control might be particularly valuable in promoting cognition and brain health. It is hypothesized that coordination training leads to functional and structural changes and adaptation in selected brain regions which might be in turn beneficial for cognitive functioning. In contrast to findings from previous studies, no evidence for a relationship between cardiovascular fitness and cognition was found. Endurance athletes did not perform significantly better in the executive functions tasks than athletes performing other types of sport. However, for the interpretation of the results, it must be noted that this cross-sectional study design does not allow for drawing any conclusions regarding the causality of the relationships.

Chapter 1

Introduction

The investigation of the relationships between physical activity, fitness, and health is an important research field within sports science and attracts the attention of both, scientists and the general public. The preservation, promotion or recovery of physical and mental health is one of the major aims in today's societies and causes rising health care costs and medical spending all over the world (Orszag & Ellis, 2007). Accumulating evidence shows that regular physical activity has positive effects on the human organism and leads to adaptations of the cardiovascular and metabolic (Howley, 2012) and musculoskeletal system (Green, 2012). Physical activity further plays an important role in the prevention and treatment of diseases such as obesity (Ross & Janssen, 2012), diabetes mellitus (Middelbeek & Goodyear, 2012) and cancer (Lee, 2012), whereas low physical activity and fitness levels have been related to risk factors for several diseases and increased mortality (Nocon, Hiemann, Müller-Riemenschneider, Thalau, Roll et al., 2008). In addition, research supports the beneficial effects of exercise on various aspects of mental health including emotional, depressive or anxiety disorders (Raglin & Wilson, 2012). Taken together, these findings indicate the importance of regular physical activity and sport for health promotion across the lifespan (Blair, Kohl, Gordon & Paffenbarger, 1992, p. 103 ff.; Janssen & LeBlanc, 2010; Vogel, Brechat, Leprêtre, Kaltenbach, Berthel et al., 2009).

Moreover, a new field to assess the effects of physical activity and fitness on cognitive functioning has emerged in the last decade. A growing body of human and animal research has focused increasingly on the effects of physical exercise on brain function and cognitive performance. According to Hillman, Erickson and Kramer (2008, p. 58), "the roots of a mind-body connection can be traced back to at least the ancient Greek civilization". The scientific investigation of the relationship between physical exercise and cognition began in the 1930s. Burpee and Stroll (1936) found that athletes have a higher processing speed compared to adults who do not regularly engage in physical activity. In the 1970s it was observed that higher physical activity levels in elderly (but not young adults) are related to faster psychomotor speed and reaction time compared to sedentary older adults (Sherwood & Selder, 1979; Spirduso, 1975).

Nowadays, research on the relationships between physical activity, fitness and cognition can be differentiated into the investigation of acute and chronic effects. This distinction is based on the duration of physical activity (single bout of exercise vs. regular participation in physical activity) and the time lapse between the performance of exercise and the assessment of cognitive performance. Further distinctions may be related to the age of participants (e.g. children, young/middle-aged/older adults), type of fitness (e.g. cardiovascular fitness, strength, body composition), physical activity intervention (e.g. aerobic endurance, strength training), or type of cognitive tasks used in a study (e.g. executive functions vs. simple reaction time tasks).

In many studies, physical activity and fitness have proven to provide a beneficial effect on cognitive performance and brain functioning. This is indicated by improved performance in cognitive testing for higher fit and active individuals compared to lower fit and sedentary persons (e.g. Hillman et al., 2008; Smith, Blumenthal, Hoffman, Cooper, Strauman et al., 2010). This effect is most pronounced for the so-called executive functions. Executive functions are attributed to higher cognitive performances and are primarily located in the prefrontal cortex. Various cognitive processes such as attention control, action planning and monitoring, inhibition, and short-term storage of relevant information are summarized under this term (Jurado & Rosselli, 2007, p. 213 f.). Cognitive benefits of physical exercise are commonly examined using behavioral measures like response times and response accuracy or by assessing neurophysiologic measures such as event-related potentials (ERPs) in cognitive paradigms. ERPs reject patterns of voltage change in neuroelectric activity resulting from response execution or preparation (Key, Dove & Maguire, 2005, p. 183 f.). A prominent stimulus-related ERP that has been investigated in many studies to detect exercise-induced effects on cognitive performance is the P300 component, which appears about 300 ms after stimulus presentation and is considered to be an indicator for attentional processes devoted to a stimulus (Key et al., 2005, p. 195).

1.1 Problem Identification

In general, regular as well as acute physical activity have shown to positively influence or to be positively related to cognitive performance in many studies (e.g. Hillman et al., 2008). However, studies on acute effects often deal with influences of exercise intensity and have primarily focused on young adults (e.g. Coles & Tomporowski, 2008; Kamijo, Nishihira, Higashiura & Kuroiwa, 2007). There is only little evidence for older adults in this field of research. On the other hand, research on the impact of chronic physical activity on cognition is primarily conducted among elderly populations and cognitively impaired people. This may be explained by the fact that this research interest has mainly developed within the framework of cognitive aging theories (Pesce, 2009, p. 213). Little is known about the effects of regular physical activity and fitness on cognition in young adults between the ages of 18 to 35 years (Hillman et al., 2008, p. 58 ff.).

Since this age group is, with regard to the development across lifespan, on the peak of their cognitive performance, it is expectably difficult to detect exercise-induced influences or effects. However, young adults are confronted with high expectations and pressure to perform in school or business and given that a large number of adolescents take drugs to enhance their attention and concentration (Maher, 2008, p. 674; Sahakian & Morein-Zamir, 2007, p. 1158), this age group would particularly benefit from an exercise-induced improvement of cognitive performance. Furthermore, in light of evidence that physical activity interventions at workplace can beneficially influence employee health and productivity (Conn, Hafdahl, Cooper, Brown & Lusk, 2009), an exercise-related improvement of cognitive performance could be another reason for the promotion of physical activity in this setting.

Besides the fact that there is a general lack of research in young adulthood, many questions still need to be answered. It is for example not yet examined which exercise-related conditions (e.g. type of sport, amount, frequency, and intensity of exercise; social interaction during exercising) lead to optimal effects on cognitive performance. Most studies that have been conducted so far have only assessed the relationship between general sports participation and cognition (differences in cognition between athletes and nonathletes) or between cardiovascular fitness and cognitive performance. Moreover, many studies solely rely on assessing behavioral measures using simple cognitive computer-based or paper-pencil tests. Given the generally high performance of young, healthy adults in cognitive tasks and the little variance in errors or reaction times among this age group, it is not likely to examine differences between physically active and inactive persons in these measures from a methodical-statistical point of view. Investigation of exercise-induced changes in cognitive performance based on neurophysiologic or even molecular methods seems to be more promising.

Therefore, the purpose of this study was to determine the interdependences of physical (in-) activity, fitness, and cognitive performance in young adults. General sports participation as well as physical and sports activity habits such as amount, type of sport, exercise intensity, frequency and duration were assessed using a questionnaire. Physical fitness tests were performed to measure endurance and strength performance, coordinative skills and body composition. Cognition was tested based on a cognitive battery assessing response times and the N100, P200 and P300 components of ERPs in an auditory Oddball, visual Flanker and visual Switching paradigm.

1.2   Structure of the Thesis

This thesis is organized into six main chapters. In the first chapter, the reader is introduced into the topic of the thesis and is given a short identification of the problem and the aim of the thesis. The second chapter provides the theoretical foundation and the description of the background. It contains subchapters on physical activity (2.1), fitness (2.2), and cognition and executive functions (2.3) in which basic concepts and major aspects are defined. Chapter 2.4 reviews the relationship between physical activity, fitness and cognition by summarizing meta-analytic evidence across the lifespan and by providing an overview of studies among young adults. In addition, potential mechanisms underlying the relationship between regular physical activity, fitness and cognition are discussed. Chapter 3 focuses on the description of the empirical "CogniFit"-study. This includes subchapters on the study design (3.1), research questions (3.2), description of the study sample (3.3), measures and procedures (3.4) as well as data acquisition and analysis (3.5). Chapter 4 deals with the results of the study and their discussion. General trends on physical activity and fitness (4.1) as well as on cognitive performance (4.2) in the study sample are described and discussed. Then, the relationships between physical activity and cognition (4.3) and fitness and cognition (4.4) are analyzed and a summary is given. A discussion of the results is presented at the end of each subchapter. In chapter 5, the major findings are summarized (5.1) and limitations and strengths of the study are discussed (5.2). The chapter ends with recommendations for future research (5.3) and a short conclusion (5.4). References, lists of figures and tables and an appendix complete the thesis.

Chapter 2

Theoretical Foundations and Background

In this chapter, the theoretical foundations of physical activity, fitness, and cognition are basically described involving definitions and theories. Then, the current status of research on the relationships between these three factors is presented and discussed with a focus on findings in young adults. The chapter ends with a brief summary.

2.1 Physical Activity

According to Bouchard, Blair and Haskell (2012, p. 12), "physical activity comprises any bodily movement produced by the skeletal muscles that results in an increase in metabolic rate over resting energy expenditure". In research, physical activity is often distinguished between an acute bout of exercise and a regular exercise training or habitual physical activity. Acute physical activity refers to only the last few hours or days whereas habitual physical activity includes a longer time span of weeks, months or even years. Even acute bouts of exercise are known to have an impact on various functions of the physiological systems such as lipoproteins, endothelial function, insulin and glucose dynamics, blood pressure, immune function or fat oxidation (Hardman, 2012). These responses to single sessions of exercise are related to health outcomes and result in substantial changes and adaptations of the cardiovascular, metabolic and respiratory systems with positive effects on health and mental well-being when physical activity is performed regularly (Garber, Blissmer, Deschenes, Franklin, Lamonte et al., 2011, p. 1337). However, it should be noted that exercise is also related to health risks such as musculoskeletal pain which are mainly influenced by exercise type and intensity (Garber et al., 2011, p. 1348).

2.1.1 Physical Activity Levels in Adults

Across individuals, there is a wide range of physical activity behavior. It is known for example that physical activity levels decline with increasing age for both sexes and that men have a higher amount of physical activity and are more likely to engage in vigorous physical activity than women across the life span (Katzmarzyk, 2012, p. 40 ff.).

In the current "Physical Activity Guidelines" published by the World Health Organization (2010), it is recommended that adults should be physically active with a moderate intensity for at least 150 minutes per week or with a vigorous intensity for at least 75 minutes per week. An equivalent combination of both, moderate and vigorous intensity is possible. The American College of Sports Medicine recommends an engagement in moderate-intensity aerobic exercise for ≥ 30 minutes per day on ≥ 5 days per week in order to achieve 150 minutes or in vigorous-intensity exercise for ≥ 20 minutes per day on ≥ 3 days per week in order to achieve 75 minutes and provides further recommendations for other types of exercise such as resistance or flexibility exercise (Garber et al., 2011, p. 1336). Looking at the prevalence rates of physical activity, only 3.5 % of 20-59 years old adults in the United States (females 3.2 %, males 3.8 %) meet this guideline as measured by accelerometer (Troiano, Berrigan, Dodd, Mâsse, Tilert et al., 2008, p. 186). The prevalence rates for Germany are somewhat higher than compared to the United States. According to the "German Health Update 2009", 28.4 % of the 18-29 years old, 27.3 % of the 30-39 years old, 26.9 % of the 40-49 years old, and 21.1 % of the 50-59 years old adults engage in moderate physical activity for at least 30 minutes on five days per week (Lampert, Mensink & Müters, 2012, p. 105). However, these data are based on self-reports given by the participants in telephone interviews and are therefore not as objective as the accelerometer data from the study by Troiano et al. (2008). It is assumed that self-reported amounts of physical activity are higher than in reality due to social desirability, overestimation of one's own capabilities or problems to remember the correct amount of physical activity or to understand the questions (Helmerhorst, Brage, Warren, Besson & Ekelund, 2012, p. 1; Shephard, 2003, p. 199). Self-report questionnaires are though still commonly used to assess physical activity in studies due to their efficiency with regard to time, costs and effort as well as acceptance among participants. Review articles of methods to assess physical activity as well as reliability and validity of existing questionnaires are provided by several authors (e.g. Helmerhorst et al., 2012; Shephard, 2003; Woll, 2004).

2.1.2 Classification of Physical Activity

The term "physical activity" involves all daily activities performed during leisure time as well as during work (Bouchard et al., 2012, p. 12/13). Work-related physical activity comprises energy expenditure required by occupational work (timeframe of about 8 hours per day), the active travel to work by bike or foot as well as domestic work (Bouchard et al., 2012, p. 13). In today's western countries, the average energy demands of work-related physical activity are lower compared to developing countries or former times (Bouchard et al., 2012, p. 13). Finger, Tylleskär, Lampert and Mensink (2012) found in a study with more than 7,000 participants between 18 and 79 years across Germany that people with a low socio-economic status have a larger amount of occupational work and concurrently a lower participation in leisure time physical activities. This association seems to further mediate the relationship between a low level of education and low leisure time physical activity.

Physical activity during leisure time comprises all activities that lead to an increase in energy expenditure and are associated with fitness (e.g. walking the dog, cycling to the supermarket etc.). Exercise or sport can be seen as a subcategory of leisure time physical activity and is usually performed planned, structured and regularly over a longer time period and is driven by several incentives (Caspersen, Powell & Christenson, 1985, p. 128). These might for example include the improvement of fitness components, health, aesthetics, social contacts, stress relaxation, or competition (Bouchard et al., 2012, p. 12).

2.1.3 Factors of Physical Activity

There are four basic factors of physical activity described in literature: intensity, frequency, duration, and type of activity (Kenney, Wilmore & Costill, 2012, p. 508). These four patterns allow for the estimation of the exercise-induced energy expenditure.

Intensity refers to the energy expenditure in a defined period of time (usually hours or minutes). According to Shephard (2003, p. 198), it can be calculated as an absolute or relative measure. A unit which is commonly used to describe the absolute metabolic cost of any form of physical activity is the metabolic equivalent (MET). One MET is defined as 3.5 ml oxygen uptake/kg bodyweight/min or as 1 kcal/kg bodyweight/hour and represents the energy costs of sitting (Bouchard et al., 2012, p. 19). A "Compendium of Physical Activities", that provides MET values for a large number of different types of activities has been published by Ainsworth, Haskell, Herrmann, Meckes, Bassett et al. (2011). Relative measures describe the intensity in relation to individual peak performances such as maximal oxygen uptake (VO_{2max}) or maximal heart rate (HR_{max}). Depending on individual fitness levels, the same type of exercise can be a warm-up for one person but require maximal effort by another (Howley, 2001, p. 366 f.). Ratings of perceived exertion can also be used to receive subjectively rated exercise intensities by participants. A prominent scale for this purpose is the Borg RPE scale which ranges from 6 to 20 (Borg, 1982). Table 1 shows different methods to classify exercise intensity that might be used for the monitoring of exercise intensities particularly in empirical studies.

Table 1:	Classification of exercise intensity based on 20 to 60 minutes of endurance activity: comparison of three methods (Kenney, Wilmore & Costill, 2012, p. 515)

Classification of intensity	Relative intensity		Rating of perceived exertion
	HR_{max}	VO_{2max}	
Very light	< 35 %	< 30 %	< 9
Light	35 - 59 %	30 - 49 %	10 - 11
Moderate	60 - 79 %	50 - 74 %	12 - 13
Heavy	80 - 89 %	75 - 84 %	14 - 16
Very heavy	$\geq$ 90 %	$\geq$ 85 %	> 16

In order to reveal health benefits such as adaptations of the cardiovascular system, exercise intensities with at least 50 % to 60 % of VO_{2max} are commonly recommended. Exercise intensities over 80 % are not necessary to elicit positive effects on health (Kenney et al., 2012, p. 510).

Exercise frequency is characterized by the number of exercise sessions that are performed during a certain time period. Usually, this period is one week. A number of three up to five sessions per week are recommended for health benefits (Kenney et al., 2012, p. 509). It might further be important whether all activities during a day are performed in one large session or whether a person has several shorter sessions per day. According to Shephard (2003, p. 198), the second option is beneficial for persons with low fitness levels and/or who want to begin a regular exercise program. It must also be taken into account that exercise frequencies might depend on seasons and can differ extremely between summer and winter (Shephard, 2003, p. 198).

Duration describes the length of time which is spent on an exercise session and is usually given in minutes. Depending on the frequency of exercise, even short durations of 10 minutes have shown to improve selected health measures (Kenney et al., 2012, p. 510). However, it is commonly recommended to exercise with durations of at least 20 to 30 minutes since this duration is likely to pass the threshold to elicit benefits on health and fitness level. Based on the knowledge of one person's exercise frequency and duration, one can calculate the individual total amount of physical activity (minutes or hours) in a typical week (Shephard, 2003, p. 198).

The type of activity reveals some information about quality criterions such as involved or required motor abilities, complexity and the social context (Woll, 2004, p. 56). Mitchell, Haskell, Snell and Van Camp (2005, p. 1356 f.) provide a classification of sports into two broad types based on static and dynamic components during competition or performance. Sports with low static and dynamic demands are bowling or golf whereas boxing or cycling are sports that require high static and dynamic demands. According to Haskell (2012, p. 347), another classification based on the metabolic properties (aerobic versus anaerobic) is also common.

Besides physical activity, the research on physical inactivity and sedentary behavior has increased its importance in sport science in the recent past. Physical inactivity is one of the four major risk factors for global mortality besides high blood pressure, tobacco use and high blood glucose (World Health Organization, 2010, p. 10). Studies on inactivity physiology have shown that a high amount of physical inactivity is a risk factor for several metabolic diseases independently from the amount of physical activity. As stated by Hamilton and Owen (2012, p. 55), too many hours of inactivity per day might not be replaced by moderate to vigorous physical activity, even if the amount is sufficient and meets current activity guidelines. It is therefore assumed that today's western societies not only need to increase their amount of physical activity in order to promote a healthy living in future, but also need to decrease their large amount of sedentary behavior. According to Oja and Titze (2011, p. 257), recommendations on physical inactivity are thus needed in future.

2.2   Fitness

Physical fitness has been defined as "the ability to form muscular work satisfactorily" (World Health Organization, 1968; as cited by Bouchard et al., 2012, p. 14) and as "the ability to carry out daily tasks with vigor and alertness, without undue fatigue and with ample energy to enjoy leisure-time pursuits and to meet unforeseen emergencies" (President's Council on Physical Fitness and Sports, 1971; as cited by Caspersen et al., 1985, p. 128). A broader definition by Kayser (2003, p. 200) describes fitness as the "living capability of a human being as well as the current ability for intended activities".

Fitness can be differentiated between a performance-related and a health-related approach. Performance-related fitness is required for high sports performances and comprises factors like muscular strength, cardiorespiratory power, motor skills, and also motivational and nutritional aspects (Bouchard et al., 2012, p. 14). Health-related fitness results from regular physical activity and involves all body systems that are of importance for health and well-being. It is defined as "a state characterized by an ability to perform daily activities with vigor and by traits and capacities that are associated with a low risk for the development of chronic diseases and premature death" (Pate, 1988; as cited by Bouchard et al., 2012, p. 14). According to Bouchard, Shephard and Stevens (1994, p. 81), relevant components and traits of health-related fitness include a morphological component (e.g.

body composition, flexibility), a cardiorespiratory component (e.g. maximal aerobic power, heart function), a muscular component (e.g. strength, power), a motor component (e.g. balance, coordination), and a metabolic component (e.g. glucose tolerance, lipid and lipoprotein metabolism). In Germany, the concept of motor performance abilities according to Bös and Mechling (1983) is well accepted and leads to the understanding of motor performance ability as being a complex, multidimensional construct including conditioning (energetically-determined) and coordinative (information-oriented) aspects (Lämmle, Tittlbach, Oberger, Worth & Bös, 2010, p. 42).

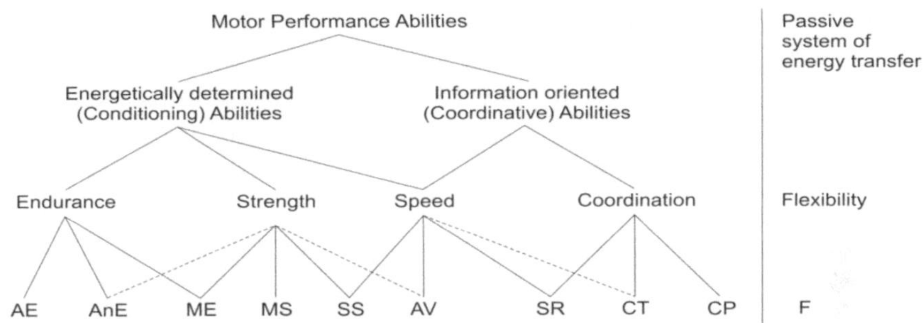

Figure 1: Differentiation of motor performance abilities (Bös, 1987, p. 94). AE = aerobic endurance, AnE = anaerobic endurance, ME = muscular endurance, MS = maximal strength, SS = speed strength, AV = action velocity, SR = speed response, CT = coordination under time pressure, CP = coordination with precision requirement, F = flexibility

The basic motor performance abilities are thus endurance, strength, speed, and coordination with flexibility being a construct that is neither energetically-determined nor information-oriented but represents the passive system of energy transfer. On a third level, these basic abilities can be further classified due to duration and intensity into 10 components (figure 1).

Since this thesis focuses on the relationships between fitness and cognition, it seems reasonable to solely address those aspects of fitness in the following that have shown to be related to cognitive performance. Primarily, these involve endurance exercise and aerobic fitness (e.g. Guiney & Machado, 2013), strength exercise (e.g. Liu-Ambrose & Donaldson, 2009), and motor fitness including balance and coordination (e.g. Voelcker-Rehage, Godde & Staudinger, 2010). Further, it is hypothesized that there is a relationship between body composition and cognition (Donnelly & Lambourne, 2011).

2.2.1 Endurance

Aerobic capacity or power is defined as the energy released by cellular metabolic processes that are dependent from the involvement and availability of oxygen and thus reflects the maximal capacity for the aerobic resynthesis of ATP (Kenney et al., 2012, p. 211). This capacity is also used as a synonym for the maximal oxygen uptake (VO_{2max}) which describes the highest rate of oxygen uptake and utilization in the body during exercise and is regarded as the most valid criterion to evaluate a person's aerobic endurance and cardiovascular fitness (Bassett & Howley, 2000, p. 70). Due to differences in energy demands dependent from body constitution, VO_{2max} is commonly expressed in relation to body weight (ml/kg/min).

The maximal oxygen uptake can be accurately assessed in the laboratory setting by a graded exercise test to exhaustion with a continuous measurement of VO_2. These tests are usually performed on a treadmill or cycle ergometer using standardized testing protocols. However, if a participant is not used to run or cycle, muscle fatigue is likely to occur before the subject has achieved its maximal oxygen uptake (VO_{2max} plateau) and test results might be biased (Kenney et al., 2012, p. 122). Further, validated submaximal or maximal field tests involving walking or running exist which allow for the estimation of aerobic capacity (e.g. Oja, Laukkanen, Pasanen, Tyry & Vuori, 1991). Although there is a discussion about several factors that could play a role in limiting VO_{2max}, evidence showed that it is primarily determined by the cardiovascular system and its ability to transport oxygen to exercising peripheral muscles (Bassett & Howley, 2000, p. 72 ff.). Among individuals, maximal oxygen uptake is influenced by both environmental factors such as endurance training and training status as well as genetic factors such as heredity and gender (Kenney et al., 2012, p. 268). VO_{2max} values for young adults range between 30 - 42 ml/kg/min for untrained females and 39 - 52 ml/kg/min for untrained males. The values are considerably higher for trained athletes and depending on type of sport. For cycling athletes for example, values can range between 47 - 57 ml/kg/min for females and 62 - 74 ml/kg/min for males (Kenney et al., 2012, p. 269). However, the concept and significance of VO_{2max} is also subject of discussion in literature (e.g. Levine, 2008).

Besides spiroergometric testing to determine VO_{2max}, lactate diagnostics is another popular method to assess a person's aerobic endurance potential (Kenney et al., 2012, p. 124). For this purpose, blood lactate is usually collected at several time points during an incremental exercise test and the resulting lactate curve generally shows an exponential relationship with the running or cycling performance (Faude, Kindermann & Meyer, 2009, p. 471). Several lactate threshold concepts from different investigators exist which can be used in order to interpret the lactate curve. These concepts can be subdivided into fixed, aerobic, and anaerobic thresholds (Faude et al., 2009, p. 475 ff.). The most common fixed lactate thresholds are the 2 mmol/l and the 4 mmol/l thresholds with the latter being first described by Mader, Liesen, Heck, Phillipi, Rost et al. (1976). However, this concept does not account for individual differences in baseline lactate or lactate accumulation. Thus, individual approaches are suggested (Faude et al., 2009, p. 473). Aerobic concepts aim at detecting the point at which blood lactate accumulates above resting values whereas anaerobic thresholds are assigned to the maximal point at which the rate of lactate accumulation is equal to the rate of lactate removal before exceeding it with increased workload (Faude et al., 2009, p. 475 f.). The individual anaerobic threshold has been established by Keul, Simon, Berg, Dickhuth, Goerttler et al. (1979) and Stegmann, Kindermann and Schnabel (1981) and is a frequently used measure for predicting endurance performance.

2.2.2 Strength

Strength can be classified into muscular (maximal) strength, muscular power (speed strength) and muscular endurance (Kenney et al., 2012, p. 210 f.). According to Güllich and Schmidtbleicher (1999, p. 223), there is a hierarchical relationship with maximal strength being superordinated. Maximal strength is defined as the maximal force that can be generated by the neuromuscular system during a maximal, voluntary muscle contraction (Güllich & Schmidtbleicher, 1999, p. 224).

The degree of voluntary maximal muscular contraction ranges from 70% for untrained persons up to 95% among highly trained individuals (Güllich & Schmidtbleicher, 1999, p. 224). Muscular power or speed strength is referred to as the "product of strength and speed of movement" (Kenney et al., 2012, p. 211) or the "ability of the neuromuscular system to create a maximal impulse of force in a given time" (Güllich & Schmidtbleicher, 1999, p. 225). Muscular strength is generally determined by the physiological cross section of a muscle, the muscle fiber composition and innervations behavior such as recruitment and frequency (Güllich & Schmidtbleicher, 1999, p. 226 ff.; Hohmann, Lames & Letzelter, 2003, p. 77). With regard to muscle contraction, one can distinguish between isometric or static and dynamic strength. Static strength is given when the muscle produces (maximal) force without a change in its length or joint movement (Kenney et al., 2012, p. 44). That is when a person tries to lift a weight that is heavier than the muscle force. A shortening of a muscle is called concentric contraction and occurs for example when the generated force is high enough to overcome a resistance. On the other hand, a lengthening of the muscle is defined as eccentric contraction (Kenney et al., 2012, p. 44). Static and dynamic strength can be measured in a biomechanical laboratory setting using specialized equipment (e.g. leg press, vertical jump tests such as counter-movement-jumps performed on force-plates) and dynamometric analysis methods.

2.2.3 Coordination

Coordination has been defined as "the interaction of the central nervous system and musculoskeletal system within a targeted movement" (Hollmann & Strüder, 2009, p. 140). Several coordinative abilities can be distinguished according to Hirtz (1985) such as the ability to respond, balance, or rhythm ability that are mainly of importance for exercise practice and training (Hohmann et al., 2003, p. 103). However, based on statistical dimension and structure analyses, Roth (1982) identified two main coordinative abilities which comprise coordination under time pressure and coordination under precision demands (Hohmann et al., 2003, p. 105). A test to assess coordination under time pressure is for example the jumping side-to-side task, coordination under precision demands can be determined using the balancing backwards task. Both tests are part of the German Motor Performance Test 6-18 (DMT 6-18; Bös, Schlenker, Büsch, Lämmle, Müller et al., 2009) and have been implemented in a large and representative study with children and adolescents in Germany (Woll, Kurth, Opper, Worth & Bös, 2011). In addition, and since the concept of the classical coordinative abilities cannot be proved based on empirical data, new concepts and approaches to define coordination and coordinate abilities have been suggested (e.g. Hirtz, 1994; Neumaier, 1999).

2.2.4 Body Composition

Body composition is the last aspect of fitness to be described in this chapter. According to Shen, St-Onge, Wang and Heymsfield (2005, p. 11) there are five levels (atomic, molecular, cellular, tissue-organ, whole-body), which can be used to describe body composition. The molecular level consists of the six components fat mass (lipid), total-body water, total body protein, bone minerals, soft tissue minerals and carbohydrates. Based on these components, various models can be created. The classical model that is widely applied in body composition research is the two compartment model involving fat and fat-free mass (Shen et al., 2005, p. 11). With regard to body anatomy, another prominent model comprises adipose tissue which is comparable but not identical to fat mass as well as skeletal muscles, bones, visceral organs and other tissues which are comparable to fat-free mass (Shen et al., 2005, p. 13).

Several methods are available to measure body composition and differ greatly regarding their underlying body composition model, operating effort, expense, and accurateness. A common and accurate method relying on the two-compartment model is densitometry. This method involves the measurement of body density which is equivalent to the ratio of body mass and body volume (Going, 2005, p. 17). In order to obtain body volume, laboratory techniques such as hydrostatic weighting and air displacement plethysmography must be used. Once predicted, body density can be converted to estimate relative body fat which is usually done using the standard equation of Siri (1956). This equation assumes that the densities of fat and fat-free mass are at constant values of 0.9007 g/cm^3 and respectively 1.100 g/cm^3 among all people (Going, 2005, p. 18). Since it is questionable that densities for fat and especially fat-free mass do not vary at all between people, this equation can be seen as the major weakness of densitometry (Kenney et al., 2012, p. 358).

Further widely used measurement methods to determine body composition are the dual-energy x-ray absorptiometry (DEXA), bioelectrical impedance analysis or spectroscopy (BIA/BIS), computed tomography and magnetic resonance imaging as well as skinfold fat thickness. Optimal body weight and particularly body composition are of great importance for sport performance (Kenney et al., 2012, p. 356). Moreover, excess adipose tissue and body fat are associated with several health risks or risk factors including diabetes, coronary heart disease and high blood pressure (Sardinha & Teixeira, 2005, p. 185).

2.3   Cognition and Executive Functions

The term cognition is derived from the Latin word "cognoscere" which means perceiving or knowing something. It represents a process that arises from awareness and perception and comprises several areas such as learning and memory, intelligence, thinking and concept formation, language, imagery, pattern and object recognition, attention and consciousness as well as sensation and perception (Frensch, 2006, p. 19; Solso, 2001, p. 7). Studies on the relationship between physical activity, fitness, and cognitive performance are primarily focusing on a prominent subdivision of cognitive performance, the so-called executive functions (Ratey & Loehr, 2011, p. 171). It is hypothesized that physical exercise selectively improves these higher cognitive processes (Davis & Lambourne, 2009, p. 251). Executive functions are often mentioned in literature, but a generally accepted or valid definition is still lacking (Jurado & Rosselli, 2007, p. 213). To date, it is further not clear whether there exists a single underlying ability that is responsible for executive functions or whether the different abilities are distinct processes that are related to each other (Jurado & Rosselli, 2007, p. 214). The term was first introduced by Baddeley and Hitch (1974) who defined it as a "central executive". Gazzaniga, Ivry and Mangun (2009, G5) describe executive functions as a "set of higher level cognitive operations that are essential for the production of goal-oriented behavior [...] and include processes such as working memory, attention, goal representation and planning, response monitoring, and error detection". Similarly, other authors such as Meyer and Kieras (1997) and Norman and Shallice (1986) note that executive functions refer to "a subset of goal-directed processes that encompass the selection, scheduling, and coordination of computational processes which are involved in perception, memory, and action" (cited by Hillman, Buck, Themanson,

Pontifex & Castelli, 2009, p. 114). A detailed description of several executive function definitions and concepts is provided by Jurado and Rosselli (2007). In conclusion, executive functions can be seen as complex higher-level cognitive functions with great importance for human adaptive behavior which involve the top-down control of abilities such as planning, initiation, preservation, and alteration of goal-directed behavior (Alvarez & Emory, 2006, p. 17 f.; Jurado & Rosselli, 2007, p. 213 f.). Executive functions typically develop during childhood, peak during young adulthood, are largely maintained among middle-aged adults and decline in old age (Craik & Bialystok, 2006; Zelazo, Craik & Booth, 2004). This course across the lifespan is mainly influenced by cortical maturation since the frontal lobes are among the last areas to mature in children and among the first which are impaired in older adults (Craik & Bialystok, 2006, p. 134). Environmental factors and learning experiences also play an important role and have the greatest influence on cognition in childhood and late adulthood (Craik & Bialystok, 2006, p. 134). The first executive function which emerges during the first year of childhood is inhibition as one part of attentional control. Planning and set shifting have shown to develop starting at an age of three years, whereas verbal fluency is the last skill which improves mainly by the age of 8 to 12 (Jurado & Rosselli, 2007, p. 221 f.). However, Hughes and Graham (2002) described several difficulties in assessing executive functions in children such as limited language ability that could lead to biased results in those studies (Jurado & Rosselli, 2007, p. 221). Another problem is that development and decline of executive functions are two distinct processes. Thus it is questionable as to which extent task performances of children and older adults can be compared since children may have an advantage over seniors regarding speed of response but a disadvantage regarding response accuracy (Jurado & Rosselli, 2007, p. 226).

The following subchapters provide an overview of brain structures related to executive functions as well as cognitive tasks and event-related brain potentials as possible assessment methods.

2.3.1 Brain Structures related to Executive Functions

The cerebrum is the largest part of the brain and is split into the left and the right hemisphere (Bear, Connors & Paradiso, 2007, p. 171). The cerebral cortex is the outer layer of the cerebrum and is often referred to as "gray matter". It consists of large sheets of layered neurons as well as blood vessels (Gazzaniga et al., 2009, p. 67). Its surface is divided into the frontal, parietal, temporal and occipital lobe (Bear et al., 2007, p. 209; see figure 2).

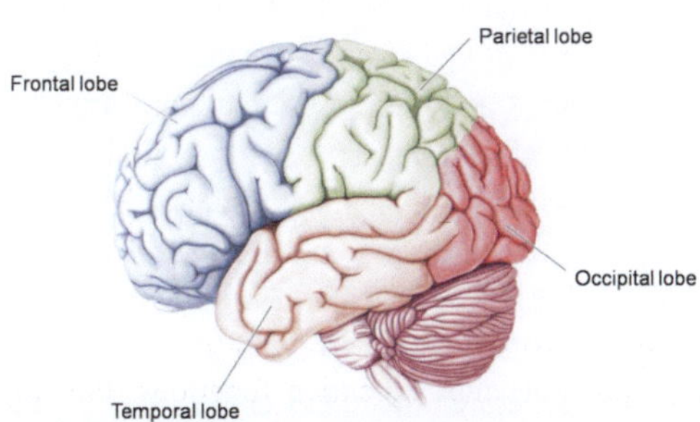

Figure 2:   Cerebral Lobes (Bear, Connors & Paradiso, 2007, p. 209)

The frontal lobe is the brain region in which the planning and execution of movements as well as cognitive control takes place. The anterior region of the frontal lobe is called the prefrontal cortex and includes the dorsolateral prefrontal cortex, the orbitofrontal cortex and the anterior cingulate and medial frontal regions (Gazzaniga et al., 2009, p. 73). The prefrontal cortex is well developed in primates compared to other mammals and it is hypothesized, that the prefrontal cortex is involved in several functions such as complex planning of behavior or problem solving which are often embraced by the term cognitive control or executive functions (Miller, 2000, p. 59). Social behavior is another function that is attributed to the prefrontal cortex (Kolb, Mychasiuk, Muhammad, Li, Frost et al., 2012, p. 17186). However, the detailed functional organization within this part of the brain has not been completely clarified yet. For example, Gläscher, Adolphs, Damasio, Bechera, Rudrauf et al. (2012) conducted a study among persons with focal frontal lesions and provided evidence for the presence of two functional-anatomical networks within the prefrontal cortex that might be distinguished from each other. They found that the dorsolateral prefrontal cortex and anterior cingulated cortex are associated with cognitive control which comprises response inhibition, conflict monitoring and switching, whereas value-based decision making was mediated by the orbitofrontal, ventromedial, and frontopolar cortex (Gläscher et al., 2012). Similar to these findings, the right dorsolateral frontal area has shown to be involved in behavior monitoring while the left dorsolateral frontal area is needed for verbal processing (Stuss, Alexander, Floden, Binns, Levine et al., 2002). In addition, Wagner and Smith (2003) found that the right inferior prefrontal cortex is activated when information manipulation is needed and that the superior frontal cortex is associated with information update and the maintenance of memory. Euston, Gruber and McNaughton (2012) further discuss the role of the medial prefrontal cortex in both, short and long-term memory and conclude that the anatomical connection to the hippocampus is of importance for the medial prefrontal cortex with regard to learning and memory consideration. A detailed review on brain correlates of executive functions is provided by Jurado and Rosselli (2007).

The development of the prefrontal cortex is influenced by several experiences and environmental events a human being is exposed to. These pre- and postnatal factors include stress, sensory stimuli, and also social experiences such as interactions between child and parents (Kolb et al., 2012, p. 17186). This prolonged development has the advantage that cognition can be formed by personal experiences but has the disadvantage that unfavorable factors can lead to lesions or neuropsychiatric disorders (Kolb et al., 2012, p. 17186). Studies in healthy adults of different ages revealed an age-related decline of gray (Zimmerman, Brickman, Paul, Grieve, Tate et al., 2006) and white (Brickman, Zimmerman, Paul, Grieve, Tate et al., 2006) brain matter especially in frontal brain regions. The results of both studies further indicate that executive functions and cognitive performance might be predicted by the interactions of age and brain matter volume.

The hippocampus is a structure that is assigned to the temporal lobe and plays a major role in the limbic system (Schuenke, Schulte & Schumacher, 2010, p. 206). Since it consists of subsidiary structures such as the dentate gyrus, it is often named hippocampal formation. The medial temporal lobe and particularly the hippocampal formation are associated with declarative memory (Squire, Stark & Clark, 2004). Declarative memory means the memory for facts and events whereas non-declarative memory represents skills and habits (Bear et al., 2007, p. 727). Most evidence on the function of the hippocampus was generated in studies among animals or human beings with hippocampal lesions. From these studies, it is hypothesized that the medial temporal lobe and the hippocampus are involved in long-term memory and also play a role in working memory (Jeneson &

Squire, 2012). However, the frontal areas are amongst others also important for short-term and working memory (Linden, 2007; Baddeley, 2003). Further, the hippocampus is associated with several types of new learning (Curlik & Shors, 2013). The hippocampus is one of three regions of the brain, for which neurogenesis and cell proliferation have been proved (Curtis, Kam & Faull, 2011, p. 171). Studies in mice showed that a larger amount of new neurons exists in the dentate gyrus of the hippocampus if the animals live in an environment full of different experiences such as toys or social contact (Kempermann, Kuhn & Gage, 1997). Since an enriched environment involves several different components that could lead to neurogenesis, Van Praag, Kempermann and Gage (1999) conducted a study among mice to test the effects of enriched environment, learning, forced swimming, voluntary running, and standard housing on neurogenesis and cell survival. They found that an enriched environment leads to a significantly increased survival of newborn cells after four weeks but not to an enhanced neurogenesis compared to the control group. However, they provided evidence that voluntary running exercise induced an increased proliferation as well as an increased survival of cells (Van Praag et al., 1999). Other authors hypothesize that a combination of mental and physical training would lead to the highest benefits compared to either form alone (Curlik & Shors, 2013). They state that both forms have an additive effect on neurogenesis since physical activity increases neurogenesis but mental training such as learning of new skills improves the survival of the cells (Curlik & Shors, 2013).

It is a widely held belief that a bigger volume of brain structures such as the frontal lobe and the hippocampus is associated with better cognitive functions. However, research on this topic has provided contradictory results. A meta-analysis conducted by Van Petten (2004) revealed a negative correlation between hippocampal volume and memory ability for adolescents and young adults. On the other hand, it was observed that a hippocampal volume decline mediated a memory decrease among older adults and that hippocampal volume decline itself was associated with reduced levels of serum BDNF (Erickson, Prakash, Voss, Chaddock, Heo et al., 2010).

It can be concluded that the frontal lobes play a major role for achieving a good performance in tasks assessing executive functions. However, recent research also demonstrates that the connections to other brain areas especially subcortical areas and the limbic system involving the hippocampus are of importance for higher cognitive functions, too (Jurado & Rosselli, 2007, p. 217).

2.3.2 Cognitive Tasks

According to the number of different cognitive abilities that have been summarized under the term "executive functions", a large number of several test paradigms exist which are used for assessing cognitive performance and executive functions such as working memory, attention and inhibition, performance monitoring, planning and decision making or cognitive flexibility (Seiferth, Thienel & Kircher, 2007, p. 267; for review on executive function tests see Chan, Shum, Toulopoulou & Chen, 2008). Nowadays, these tasks are primarily computer-based and are often used in combination with functional brain imaging. This procedure enables researchers not only to analyze behavioral measures of tasks such as response time or number of correct responses but also to detect neurophysiologic events in the brain. These result from the processing of cognitive tasks and, to some extent, might originate in those brain structures which show an increased activity during the completion of a task (Key et al., 2005, p. 185). However, it is criticized by some investigators that the validity of cognitive tasks is not always proven since many of them were developed solely to detect frontal lobe damage, or that their reliability is questionable since the ability to cope with new problems and the novelty of task stimuli is not longer given after the first test administration (Jurado & Rosselli, 2007, p. 218). In the following, four cognitive tasks are described due to their relevance for this thesis.

The n-back task is a classical test to measure working memory (Chan et al., 2008, p. 207). Working memory is referred to as a limited capacity system, which temporarily maintains and stores information and underlies human thought processes (Baddeley, 2003, p. 829). According to Baddeley and Hitch (1974), the system comprises three components which involve a control system (central executive) and two storage systems (phonological loop and visuospatial sketchpad). The phonological loop is based on language or acoustic information whereas the visuospatial sketchpad stores visual information such as colors or shapes. Both are assumed to have a limited storage capacity of three to four objects (Baddeley, 2003, p. 833). In the n-back task, subjects are faced with a continuous sequence of stimuli, generally numbers, and are instructed to press a key depending on whether they detect a repetition of an item which was presented n items ago or not (Baddeley, 2003, p. 837). For a delay of n = 2, a sequence such as 8-4-5-3-5 should evoke a positive (repetition) response for the last item. In the 3-back version, participants should press the button when they recognize a number that was shown exactly three positions before. This task requires that participants maintain and continuously update information while responding to each item (Kane, Conway, Miura & Colflesh, 2007, p. 615). Participants usually keep task-relevant information in mind by repeating it to themselves. Studies which used n-back tasks to detect working memory performance have revealed a high activity of the dorsolateral prefrontal cortex (DPFC), inferior frontal and parietal cortex (Braver, Cohen, Nystrom, Jonides, Smith et al., 1997; Cohen, Perlstein, Braver, Nystrom, Noll et al., 1997; as cited by Baddeley, 2003, p. 837). Moreover, brain activation increases as n-back task difficulty increases (Baddeley, 2003, p. 837).

The Switching task is another prominent task. It requires top-down executive control including working ability but also the ability to inhibit irrelevant information (Davidson, Amso, Anderson & Diamond, 2006, p. 2038). It can be described as a task that targets cognitive flexibility or the "ability to flexibly shift from one mindset to another" (Davidson, Amso, Anderson & Diamond, 2006, p. 2038). A Switching task generally consists of pure or homogenous conditions and mixed or heterogeneous conditions. There are several different paradigms available (Monsell, 2003, p. 135), but in a popular mode of this task, numbers from 1 to 9 are shown on the monitor screen. In the pure conditions, the subject is trained to decide whether a shown digit is e.g. odd or even or whether it is greater or less than 5. In the mixed trials, both conditions are performed within one trial and can switch suddenly and unexpected from one item to another (Monsell, 2003, p. 135). It is further possible to use mixed trials in which the task changes predictably every two or three trials. It is assumed that this method has an increased demand of working memory (Monsell, 2003, p. 136). During Switching tasks, medial and lateral areas of the prefrontal cortex have shown a higher activation during switch compared to non-switch trials (Monsell, 2003, p. 138).

The Oddball task is a commonly used task to assess attention and information processing capacity (Karch & Mulert, 2010, p. 420). Attention is defined as the ability to focus on certain sensory stimuli and can be subdivided into a voluntary or automatic attention (Herrmann & Knight, 2001, p.

465). Voluntary attention can be described as top-down influence which allows for an intentional concentration of attention. In contrast, automatic attention is a bottom-up influence and represents a stimulus-driven attention which is attracted by a sensory event (Gazzaniga et al., 2009, p. 493). Attention can either be object-selective which involves attending to a special stimulus or object while ignoring irrelevant, distracting objects or spatial-selective which comprises attending to a certain spatial location (Herrmann & Knight, 2001, p. 465). Several variations of the Oddball paradigm are available comprising visual or auditory stimulus presentation. During a three-stimulus Oddball task for example, one is faced with a sequence of three different types of stimuli. Among these are frequently presented standard stimuli, rarely presented target stimuli and rarely presented distracter stimuli. The person is instructed to ignore the standard and distracter stimuli and to react solely to the target stimulus (Polich, 2007, p. 2129). Several brain areas are likely to be involved in attentional processes such as the sensory cortices, the hippocampal formation and the prefrontal cortex (Herrmann & Knight, 2001, p. 472).

The Eriksen Flanker paradigm (Eriksen & Eriksen, 1974) is another prominent and frequently used task to investigate spatial selective attention and conflict monitoring. Thus, the paradigm ranks among the so-called conflict or congruency tasks (Davelaar, 2013, p. 1). In this task, the target stimulus is presented centrally on the monitor screen and is flanked by several distracter stimuli. Two different conditions are possible with the nearby flankers being identical to the target stimulus (congruent; < < < < <) or the nearby flankers differing from the target stimulus and pointing in the opposite direction (incongruent; < < > < <). One is required to press a key according to the pointing direction of the target stimulus. Research studies showed that response times are significantly faster for congruent trials compared to incongruent trials (Davelaar & Stevens, 2009, p. 121). This is due to the fact that performance monitoring and control processes are needed during incongruent trials to inhibit incorrect automated responses and this produces a response conflict which results in slower reaction times (Botvinick, Cohen & Carter, 2004, p. 542). Neuroimaging studies have shown that a region in the medial frontal cortex called the anterior cingulate cortex (ACC) plays an important role in monitoring response conflicts and is further linked to the error-related negativity which occurs when errors are made in a task (Botvinick et al., 2004, p. 539 ff.).

2.3.3 Event-Related Brain Potentials

The most frequently used techniques in studies of brain function and cognitive performance are methods of structural imaging such as computed tomography (CT) as well as methods of functional imaging including functional magnetic resonance imaging (fMRI) or the electroencephalography (EEG). The electroencephalography (EEG) is a method to assess the electric activity of the brain by electrodes placed on the scalp (Teplan, 2002, p. 1). EEG recordings allow for the detection of abnormalities in brain function and neurophysiologic diseases since normal EEG profiles are well established and relatively constant among healthy subjects (Gazzaniga et al., 2009, p. 148). In addition, event-related brain potentials (ERPs) have become a frequently used tool in order to assess brain activity and changes in electrophysiological signals during cognitive processing devoted to or following a stimulus event (Key et al., 2005, p. 183). ERPs need to be extracted from the EEG by means of signal averaging since they have a relatively small size (e.g. small amplitudes) compared to the normal brain activity (Key et al., 2005, p. 184). They can be characterized by their polarity which can be positive or negative, by latency, by sequence in which the peak occurs, by scalp distribution and by relation to experimental psychological measures (Duncan, Barry, Connolly, Fischer, Michie et al., 2009, p. 1884). Thus, N1 would be the first negative peak while P300 component would refer to the positive peak in the waveform that occurs 300 ms after stimulus onset (Key et al., 2005, p. 184). The latency of an ERP reflects the time course of processing activity, whereas its amplitude is correlated to the extent of neural activation due to cognitive processes (Duncan et al., 2009, p. 1884). However, brain areas located below electrodes which show maximum amplitudes in the EEG are not necessarily the brain areas with the highest activation or involvement in generation of signals (Key et al., 2005, p. 185). One advantage of the ERP method is the good temporal resolution in the range of milliseconds (Duncan et al., 2009, p. 1884; Key et al., 2005, p. 184).

Given the variety of ERP components that have been described in literature and due to their importance for this thesis, the following section primarily focuses on the P300, N100 and P200 components.

The P300 is a positive component which appears in the EEG waveform about 300 ms after a rare target or unexpected stimulus is presented in both, visual or auditory tasks (Key et al., 2005, p. 195). The P300 distributes at the centro-parietal scalp with maximal amplitudes over midline electrodes (Duncan et al., 2009, p. 1888). This ERP has gained high importance in the study of cognition and executive functions (Karch & Mulert, 2010, p. 420) and has been shown to be reduced in amplitude and increased in latency among persons with cognitive impairment (Duncan et al., 2009, p. 1891). Clinical applications of the P300 comprise mood disorders, dementia, attention-deficit disorders or schizophrenia for example (Duncan et al., 2009, p. 1891 ff.). The Oddball task is the typical test to elicit a P300, however other tasks such as the Flanker task lead also to the generation of P300 (Duncan et al., 2009, p. 1890). In order to elicit this component, a participant must pay attention and respond to stimuli (Key et al., 2005, p. 195). P300 component is assessed by measuring its amplitude and latency. According to Polich (2007, p. 2129), the amplitude is provided in µV and reflects the difference between the largest positive-going peak in the EEG within the relevant time window that ranges between 250 and 500 ms and the baseline voltage before presentation of a stimulus. P300 amplitude is typically recorded across midline electrode sites Fz, Cz and Pz and increases in magnitude from the frontal to the parietal electrodes (Polich & Kok, 1995, p. 105). It is assumed that the P300 amplitude indicates attentional resources and processes related to a task-relevant stimulus such as memory updating, active stimulus discrimination or response preparation (Key et al., 2005, p. 195 f.). P300 amplitude is further inversely associated with stimulus probability (Duncan et al., 2009, p. 1888).

The latency of the P300 is defined as the time from the onset of a stimulus to the maximum positive amplitude peak within the relevant time window (Polich, 2007, p. 2129) and is related to stimulus evaluation time with complex stimulus processing leading to longer latencies than simple stimulus processing (Duncan et al., 2009, p. 1889). A shorter latency thus means faster cognitive processing (Hillman, Castelli & Buck, 2005, p. 1968) and classification speed and can be seen as an indicator for better cognitive performance (Polich, 2007, p. 2132). P300 latencies typically range between 250 ms to 1000 ms (Duncan et al., 2009, p. 1889) and are partly related to behavioral response times depending on the cognitive task (Polich, 2007, p. 2132). The P300 component is believed to be generated in several brain regions including the medial temporal lobe with the hippocampus (Key et al., 2005, p. 196) and the ventrolateral prefrontal cortex and is affected by determinants such as circadian rhythm, fatigue and sleep deprivation, drug use, age, intelligence, personality variables, gender and genetic factors (Duncan et al., 2009, p. 1889; Polich & Kok, 1995, p. 111 ff.). Test-retest correlation coefficients for the P300 component as measured during Oddball tasks range between .50 and .80 for amplitude and between .40 and .70 for latency (Polich, 2007, p. 2132). The P300 can be subdivided into the P3a and P3b component. During a three stimulus Oddball task, a distracter stimulus elicits a P3a whereas the target stimulus leads to the generation of a P3b (Polich, 2007, p. 2133). The P3a appears shortly before the P3b in the frontal-central brain regions and is associated with top-down monitoring for the evaluation of an incoming stimulus which is probably stimulated by the anterior cingulate cortex (Polich, 2007, p. 2136). The subcomponent P3b is linked to the temporal and parietal regions of the brain and is related to memory storage and updating operations activated by attentional processes (Polich, 2007, p. 2135).

N100 component usually occurs in the EEG waveform 100 ms after the onset of a stimulus and is assumed to be associated with selective attention and intentional discrimination processing. Auditory stimuli have shown to produce larger amplitudes and shorter latencies compared to visual ones (Key et al., 2005, p. 188). An increased auditory N100 amplitude is an indicator for increased attention to a stimulus, whereas visual amplitudes are correlated to enhanced processing of an attended location. It is not affected by arousal or inhibition (Key et al., 2005, p. 189).

P200 has been linked to attention modulation of nontarget stimuli as well as stimulus classification (Key et al., 2005, p. 189). The auditory component is sensitive to physical stimulus parameters like loudness. During visual tasks, P200 amplitude increases with the complexity of stimuli (Key et al., 2005, p. 190).

2.4   Physical Activity, Fitness and Cognition

In this chapter, meta-analyses on the relationship between physical activity, fitness and cognition across lifespan will be discussed to present a general overview of the current state of human research. Further, primary studies with a focus on the effects of regular physical activity, fitness and cognitive performance as measured by electroencephalography in young adults will be reviewed. In the end of the chapter, hypothesized mechanisms underlying the relationship between physical activity and cognition will be discussed.

2.4.1 Meta-Analyses

In their prominent model, Bouchard et al. (2012, p. 18) show the complex relationship between physical activity, health-related fitness and health (figure 3).

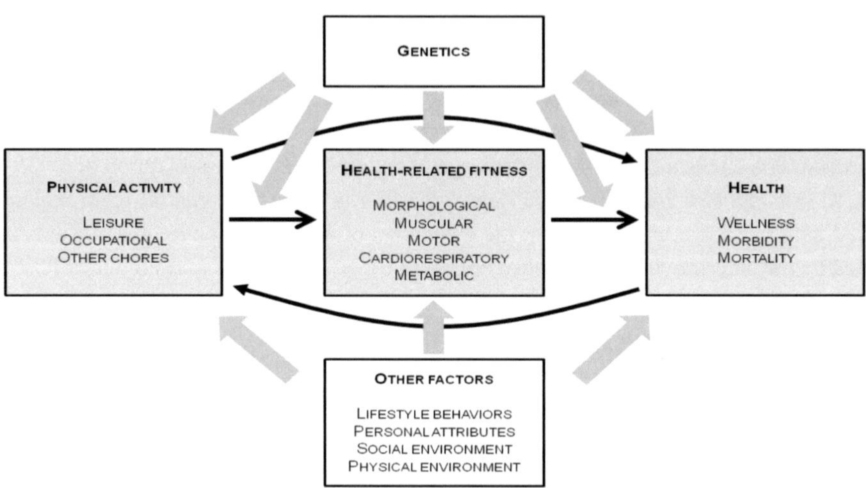

Figure 3:   Model on the relationship between physical activity, health-related fitness and health status (Bouchard, Blair & Haskell, 2012, p. 18)

Thus it appears that both, physical activity and fitness can solely impact health and that physical activity influences fitness which in turn affects health components. Further, health itself has an influence on physical activity since a bad health status might hinder an individual in exercising. In addition, several impacts of genetics and other factors relating to lifestyle behaviors, personal attributes, social or physical environment are depicted. Similar to this model, regular physical activity and fitness, especially cardiovascular fitness, are hypothesized to be associated with cognitive function across the life span.

A search for meta-analyses performed on this relationship using the PubMed, ScienceDirect and Scirus databases yielded three publications in children, one in adults, two in older adults and one across the life span. The search was conducted from January 2013 to March 2013. Key search terms included combinations of the words ("meta-analysis" OR "review") AND ("cognition" OR "cognitive performance" OR "cognitive function" OR "executive functions") AND ("(regular) physical activity" OR "exercise" OR "physical fitness" OR "sports" OR "cardiovascular fitness"). Selection criteria comprised: 1) article written in German or English; 2) study samples included only healthy participants without any known cognitive impairments; 3) information about literature search, inclusion or exclusion criteria, coding of studies/variables, data/statistical analysis and effect size calculation available in article; 4) report on effect sizes. A short description of the analyses including selected findings is provided in table 2.

The analyses constantly report a small to medium, significant overall effect of regular exercise and/or physical fitness on cognitive performance. In addition, several moderator variables such as type of exercise and cognitive outcomes, program duration and intervention unit sizes are analyzed. In children, resistance, aerobic and perceptual motor exercise as well as cardiovascular and total fitness showed the largest effects on cognition (Fedewa & Ahn, 2011; Sibley & Etnier, 2003). Small intervention group sizes with less than ten students compared to larger group sizes and a frequency of at least three days of physical activity per week also lead to higher effect sizes. Adults and older adults achieved the highest benefits on executive functions and when exercise programs had a duration of at least 6 months (Colcombe & Kramer, 2003; Smith, Blumenthal, Hoffman, Cooper, Strauman et al., 2010). Whereas Colcombe and Kramer (2003) could detect the highest effects on cognition with an exercise session duration of 31-45 minutes, Smith et al. (2010) could not find a relation between exercise duration and intensity on any cognitive outcome. Angevaren, Aufdemkampe, Verhaar, Aleman and Vanhees (2008) report small effects on few cognitive functions such as auditory and visual attention as well as cognitive speed and conclude from their analysis that aerobic exercise inducing an improved cardiovascular fitness leads to an increased cognitive function in older adults. Etnier, Nowell, Landers and Sibley (2006) demonstrated in their analysis that fitness and cognition are correlated when looking at correlational studies. But based on cross-sectional study designs or pre-post comparisons they found conflicting results, indicating that aerobic fitness might be even negatively associated with cognition in children and younger adults and positively predictive only for adults (cross-sectional) and that a gain in fitness might be a significant negative predictor of cognition (pre-post comparison). In contrast to the other meta-analyses, this publication does not support the hypothesis that (cardiovascular) fitness is correlated with a higher cognitive performance. Meta-analyses and review articles on the acute effects of exercise on cognitive performance have been published by Chang, Labban, Gapin and Etnier (2012), Lambourne and Tomporowksi (2010), McMorris, Sproule, Turner and Hale (2011) and Tomporowski (2003) for example and provide evidence for a small positive effect of acute exercise on cognition. However, the effects are dependent of moderator variables such as exercise intensity and duration, time and type of cognitive assessment, and exercise mode.

Table 2: Overview of meta-analyses on the effects of regular physical activity or physical fitness on cognition separated according to age group

Age group	Authors	Number of Studies/ Criteria	Selected Results/ Effect Sizes
Children	Fedewa & Ahn (2011)	59 studies included (1947 - 2009) *Inclusion criteria*: effect of some type of physical activity on cognition, school age (3 - 18 years), statistical data allow for effect size calculation, no replication of the same data, written in English	*Exercise on cognition*: d = 0.28 ** *Type of physical activity on cognition*: resistance/circuit training d = 0.44, perceptual motor training d = 0.15**, physical education program d = 0.20**, aerobic d = 0.35**, combined d = 0.40 *Intervention unit on cognition*: individualized d = 0.00, < 10 students d = 0.47*, 10-30 students d = 0.39**, > 30 students d = 0.31**, class d = 0.16** *Physical activity days/week on cognition*: one d = 0.16**, two d = 0.27**, three d = 0.45**, four d = 0.07, five d = 0.25* *Physical fitness on cognition*: strength d = 0.18**, flexibility d = 0.04, cardio d = 0.40**, total fitness d = 0.39**, development d = 0.47** *Exercise on cognitive outcomes*: math d = 0.44**, reading d = 0.36**, language d = 0.22**, IQ d = 0.39**, grade point average d = 0.24**, total achievement d = 0.27**, science d = 0.15 *Exercise on cognition (for gender groups)*: girls d = 0.12**, boys d = 0.07**, mixed d = 0.42** *Exercise on cognition (for cognitive status)*: normal d = 0.32**, cognitively impaired d = 0.66*, learning disabled d = 0.23*

			Exercise on cognition (for physical status): normal d = 0.30**, physically disabled d = 0.57, elite athletes d = 1.05*
	Sibley & Etnier (2003)	44 studies included (before 2002) *Inclusion criteria*: relationship between physical activity and cognition/academic performance, children, written in English	*Exercise on cognition:* 0.32* *Activity design on cognition*: chronic 0.29*, acute 0.37*, cross-sectional 0.35* *Type of activity intervention on cognition*: resistance training 0.64*, perceptual motor training 0.32*, aerobic 0.26*, physical education program 0.27* *Physical fitness on cognition (cross-sectional studies)*: overall fitness 0.34*, motor abilities 0.46* *Exercise on cognitive outcomes*: perceptual skills 0.49*, developmental level 0.39*, IQ 0.34*, math 0.20*, memory 0.03, verbal 0.17* *Exercise on cognition (for health status):* healthy 0.31*, mentally impaired 0.43*, physically disabled 0.40* *Exercise on cognition (for age):* young elementary 0.40*, old elementary 0.21*, middle school 0.48*, high school 0.24*
Adults	Smith, Blumenthal, Hoffman, Cooper, Strauman et al. (2010)	29 studies included (1982 - 2009) *Inclusion criteria*: randomized treatment allocation, mean age ≥ 18 years, treatment duration > 1 month, incorporated aerobic exercise components, supervised training, non-aerobic exercise control group	*Aerobic exercise on types of cognition*: attention and processing g = 0.16**, executive function g = 0.12*, working memory g = 0.03, memory g = 0.13* *Attention and processing*: no effects of exercise duration (r = 0.17) and intensity (r = -0.38); greater effect of combined exercise (g = 0.25*) than aerobic (g = 0.10); no effects of age (r = -0.05) *Executive function*: no effects of exercise duration (r = -0.44) and intensity (r = -0.20); no difference between combined (g = 0.16) and aerobic exercise (g = 0.11); no effects of age (r = -0.35) *Working memory*: no effects of exercise duration (r = 0.35) and intensity (r = -0.11); greater effect of combined exercise (g = 0.29*) than aerobic (g = -0.04); greater improvements in older than younger adults (r = 0.56*) *Memory*: no effects of exercise duration (r = 0.37) and intensity (r = -0.05); no effects of age (r = -0.22)
Older adults	Angevaren, Aufdemkampe, Verhaar, Aleman & Vanhees (2008)	11 studies included (1989 - 2002) *Inclusion criteria:* RCTs, studies in older people (≥ 55 years) who were not cognitively impaired, participants with age-related illnesses or disorders, Physical activity program aiming at improving cardiovascular fitness, cognitive performance: cognitive function testing (test battery or questionnaire), fitness measures (such as VO_{2max})	*Aerobic exercise vs. other intervention on cognition:* cognitive speed SMD = 0.26*, visual attention SMD = 0.26*, no significant effects for 9/11 cognitive functions *Aerobic exercise vs. no intervention on cognition:* auditory attention WMD = 0.52**, motor function WMD = 1.17*; no significant effects for 9/11 cognitive functions *Aerobic exercise vs. strength program on cognition:* no significant effects (based on only one study)

		Exclusion criteria: Studies with participants recovering from surgical treatment	
	Colcombe & Kramer (2003)	18 studies included (1966 - 2010) *Exclusion criteria*: cross-sectional design, no random assignment, unsupervised exercise program, no inclusion of aerobic fitness component, age average below 55 years	*Control vs. exercise on cognition*: 0.16* vs. 0.48** *Exercise on types of cognition*: executive function g = 0.68*, controlled tasks g = 0.46*, spatial tasks g = 0.43*, speed tasks g = 0.27* *Training type on cognition*: combined strength and aerobic = 0.59*, cardiovascular = 0.41* *Program duration on cognition*: ≤ 3 months = 0.52*, 4 - 6 months = 0.27*, > 6 months = 0.67* *Session duration on cognition*: ≤ 30 min = 0.18, 31 - 45 min = 0.61*, > 46 min = 0.47*
All ages	Etnier, Nowell, Landers & Sibley (2006)	37 studies included (1966-2004) *Inclusion criteria*: written in English, relationship between fitness or exercise program and cognition, measurement of aerobic fitness by VO_{2max}, cross-sectional or (chronic) intervention design	*Exercise/fitness on cognition*: 0.34** *Cross-sectional*: exercise/ fitness effect on cognition 0.40; aerobic fitness = significant negative predictor of cognition in children and young adults, significant positive predictor in adults, not a significant predictor for older adults *Posttest comparison between treatment and control group*: exercise/ fitness effect on cognition = 0.27; no interactions with moderator variables *Pre-post comparison for treatment group*: exercise/ fitness effect on cognition = 0.25; improvement in aerobic fitness from pre- to posttest is significant negative predictor of cognition; aerobic fitness is not a significant predictor for cognition in children, young adults and adults, but a significant negative predictor for older adults *Correlations*: fitness on cognition r = 0.29

* = p < .05, ** = p < .01; RCT = randomized clinical trial; SMD = standardized mean difference; WMD = weighted mean difference

In addition to meta-analyses, a large number of narrative review articles have been published focusing primarily on children (Chaddock, Pontifex, Hillman & Kramer, 2011; Donnelly & Lambourne, 2011; Haapala, 2013; Hillman, Kamijo & Scudder, 2011; Rasberry, Lee, Robin, Laris, Russell et al., 2011; Singh, Uijtdewilligen, Twisk, van Mechelen & Chinapaw, 2012; Tomporowski, Lambourne & Okumura, 2011) and on the life span (Burkhalter & Hillman, 2011; Guiney & Machado, 2013; Ratey & Loehr, 2011; Voelcker-Rehage & Windisch, 2013; Voss, Nagamatsu, Liu-Ambrose & Kramer, 2011). In most articles, authors conclude from their findings that there is a beneficial relation between participation in physical activity and cognitive performance in all age groups. However, effect sizes are usually smaller for young adults compared to children and particularly older adults. Interestingly, there is no meta-analysis or review available which addresses solely the exercise- or fitness-induced effects on brain health and cognition in young adulthood. Given that most articles provide evidence for the effects of aerobic and strength exercise, there is further a lack of articles focusing on cognitive benefits which might result from other types such as coordinative exercise for example.

2.4.2 Studies among Young Adults

Studies on the relationship between regular physical activity, fitness and cognition have been main-ly conducted in children, older adults or participants with mild to severe cognitive impairments. Young adults primarily participated in studies to assess cognitive effects which are induced by acute exercise bouts or served as a comparison group for older adults to provide a baseline of cognition (Hillman et al., 2008, p. 59 f.). This chapter aims at providing an overview of the studies that have been conducted in young adults. However, this overview has an emphasis on studies in which cognitive functions or performance were not only assessed by behavioral measures such as response times or accuracy, but also by measuring event-related brain potentials. Published articles were identified using PubMed, ScienceDirect and Scirus databases. The literature search was conducted from December 2012 to March 2013. Key search terms included combinations of the words ("young adults" OR "young adulthood") AND ("event-related brain potentials" OR "P300 component" OR "cognition" OR "cognitive performance" OR "executive functions") AND ("(regular) physical activity" OR "exercise" OR "physical fitness" OR "sports"). In addition, reference lists of published articles were manually searched for further relevant publications. Selection criteria were: 1) article written in German or English; 2) healthy participants without any known cognitive impairments; 3) relationship between physical activity and/or fitness and cognition; 4) mean age of participants ≥ 18 and ≤ 35 (except for children or elderly who served as comparison groups); 5) use of standardized cognitive tasks and/or EEG measurement. It must be noted though that this overview does not claim to be exhaustive.

The majority of studies (n = 18) use a cross-sectional design, but six intervention studies and one longitudinal study are also available. Table 3 provides an overview of cross sectional studies in with young adults served as a comparison group for older adults or children, table 4 summarizes cross sectional studies only performed with young adults and table 5 focuses on longitudinal and intervention studies.

Table 3:   Overview of seven cross-sectional studies (high-fit/trained/active vs. low-fit/untrained/sedentary) with young adults as comparison group for older adults or children; ordered by year of publication

Author	Participants	Exercise/ Fitness and Cognitive Testing	Selected Results (only fitness-/activity-related and age x fitness effects) and Interpretation
Pontifex, Hillman & Polich, 2009	Young: 12 higher-fit (5 males, 20 ± 1years), 13 lower-fit (4 males, 20 ± 2 years) Older: 10 higher-fit (5 males, 66 ± 4 years), 13 lower-fit (5 males, 67 ± 3 years)	Exercise/Fitness: -VO_{2max}: graded treadmill exercise test -Higher-fit: $VO_{2max} > 70^{th}$ percentile according to ACSM Guidelines, Lower-fit: $VO_{2max} < 30^{th}$ percentile according to ACSM Guidelines Cognitive: -2- and 3-stimulus Oddball task -ERP: P3a and P3b, N2	No fitness effects: Response accuracy, reaction times for 3-stimulus task, N2 amplitude and latency, P3a amplitude, P3a and P3b latency Fitness effects: Higher-fit had shorter reaction times during 2-stimulus task, larger P3b amplitude during 2- and 3-stimulus task Age x fitness effects: Fitness associated with larger P3b amplitudes during 3-stimulus task only for young adults Interpretation: -Aerobic fitness does not protect against age-related cognitive deficits and results in decreased attentional resource availability since larger P3b amplitude for difficult 3-stimulus task were only found for young adults → possible relation to locus-coeruleus-

			norepinephrine system -No effect on P3a: no fitness effects on attentional orienting processes
Hillman, Kramer, Belopolsky & Smith, 2006	<u>Young:</u> 18 active (19 ± 0.3 years), 16 sedentary (19 ± 0.2) <u>Older:</u> 17 active (64 ± 1 years), 15 sedentary (66 ± 1 years) Mixed (total: 34 males)	Exercise/Fitness: -Physical activity interview -Questionnaire -Active: > 5 hours of exercise per week, Sedentary: < 1 hour exercise per week Cognitive: -Switching task (global and local switch) -Reaction time, response accuracy -ERP: P3	<u>No fitness effects:</u> Response accuracy for global and local switch conditions <u>Fitness effects:</u> Active had faster reaction times for global and local switch conditions, larger P3 amplitudes at midline electrodes for both conditions, marginally faster P3 latencies for heterogeneous trials during global switch and faster P3 latencies for central and parietal electrodes during local switch condition <u>Age x fitness effects:</u> None <u>Interpretation:</u> -Physical activity influences perceptual/central and response-related processing (faster response times and latencies) and attention allocation in sense of memory updating (larger amplitudes) -Executive control processes are sensitive to physical activity since latencies were shorter for heterogeneous but not homogenous trials -Both, older and younger adults benefit from regular physical activity
Hillman, Motl, Pontifex, Posthuma, Stubbe et al., 2006	<u>Young:</u> 118 (58 males, 26 ± 5 years) <u>Old:</u> 123 (46 males, 50 ± 8 years)	Exercise/Fitness: -Physical activity questionnaire (days/week) -Regression analysis (no group classification) Cognitive: -Wechsler Adult Intelligence Scale -Flanker task (congruent/incongruent trials; behavioral measures)	<u>No fitness effects:</u> Wechsler Intelligence Scale, interference effect (congruent minus incongruent) of response times <u>Fitness effects:</u> Physical activity associated with faster reaction times and better response accuracy during both flanker conditions; larger interference effect for response accuracy for adults with low physical activity <u>Age x fitness effects:</u> Physical activity associated with better response accuracy and lower interference effect for response accuracy only for older adults <u>Interpretation:</u> -Physical activity is associated with performance in tasks with large but also smaller executive components: stronger relations for trials requiring greater amounts of executive control (e.g. incongruent compared to congruent trials) -Physical activity has a greater influence on cognitive performance in older adults compared to younger
Hillman, Castelli & Buck, 2005	<u>Children:</u> 12 high-fit (7 males, 9 ± 1 years), 12 low-fit (6 males, 10 ± 1 years) <u>Young adults:</u> 15 high-fit (6 males, 19 ± 1 years), 12 low-fit (7 males, 20 ± 2	Exercise/Fitness: "Fitnessgram" -PACER Aerobic endurance run -Muscle & flexibility fitness -Body composition (BMI) -For children and adult groups, the top and bottom 10% in the fitness testing of a large pool of potential	<u>No fitness effects:</u> Response accuracy <u>Fitness effects:</u> High-fit had significantly faster response times, larger P3 amplitudes and faster P3 latency along occipital sites <u>Age x fitness effects:</u> Adult groups had faster response times that children groups, but high-fit children responded faster than low-fit children; high-fit children had largest P3 amplitudes compared to other groups <u>Interpretation:</u> -Faster P3 latencies for high-fit indicate faster cogni-

	years)	participants were chosen Cognitive: -Intelligence quotient - Visual Oddball task (reaction time, response accuracy) -ERP: P3	tive processing, however they were only observed over occipital scalp region -Larger P3 amplitudes and greater cortical activation for high-fit children indicates greater allocation of attention and working memory resources -Less cortical activation in young adults might indicate that task performance is primarily dependent on cognitive efficiency
McDowell, Kerick, Santa Maria & Hatfield, 2003	Young: 21 high-active (11 males; 22 ± 1 years), 16 low-active (7 males; 23 ± 1 years) Older: 18 high-active (8 males; 66 ± 1 years), 18 low-active (7 males; 69 ± 1 years)	Exercise/Fitness: -Physical activity questionnaire (daily energy expenditure) -VO_{2max}: graded treadmill exercise test -High-active: regular physical activity with high intensity exercise and long exercise durations over 5 years preceding the study, Low-active: irregular physical activity with low-intensity exercise Cognitive: -2-stimulus Oddball task -ERP: P300	No fitness effects: P300 latency Fitness effects: Physical activity associated with P300 area under the curve (AUC) Age x fitness effects: Larger P300 area under the curve for older low-active adults than for older high-active and young adults, larger P300 amplitudes for high-active than low-active young adults at Fz, Cz (inverse effect for older adults) and Pz electrodes Interpretation: -P300 AUC reflects the extent of neural processing → low-active older adults need more neural processing for equal cognitive work and thus have a decreased efficiency compared to high-active older adults -P300 is not reliably related to physical activity history or activity x age interaction → mixed results for age groups and electrode sites, no consistent findings
Hillman, Weiss, Hagberg & Hatfield, 2002	Young: 12 fit (6 males, 22 ± 3 years), 12 sedentary (6 males, 23 ± 3 years) Older: 12 fit (6 males, 64 ± 3 years), 12 sedentary (6 males, 65 ± 3 years)	Exercise/Fitness: - Physical activity questionnaire - VO_{2max}: graded treadmill exercise test - 12 participants per age group were classified as being fit and 12 as being sedentary based on VO_{2max} and physical activity history Cognitive: S1-S2-S3-paradigm - Behavioral measures - ERPs (P3 amplitude & latency, SPN amplitude, CNV amplitude) - SPN: stimulus preceding negativity; CNV: contingent negative variation	No fitness effects: P3 to warning stimulus (S1), P3 amplitude to decision task (S2), P3 latencies to decision task at midline electrodes, SPN, CNV amplitude at lateral electrodes, behavioral measures Fitness effects: CNV amplitude at midline electrodes increased significantly in sedentary adults Age x fitness effects: P3 latencies to decision task for lateral electrodes longer for older sedentary adults than older fit and younger adults Interpretation: -Aerobic exercise does not influence memory processes associated with decision tasks (no fitness effect on P3 amplitude to S2) -Motor adaptations from physical activity might be related to benefits in cognitive domain (high-fit exhibited shorter P3 latencies) -Increased CNV amplitude in sedentary can be explained be the need to allocate more neural resources to prepare for task execution and sedentary thus have a decreased efficiency in motor preparation than fit adults -Physical activity is primarily related to adaptations of motor structures than cognition
Dustman, Emmerson, Ruhling, Shearer,	Young: 15 high-fit (males, 24 ± 3 years), 15 low-fit	Exercise/Fitness: - VO_{2max}: graded treadmill exercise test - 15 participants per age	No fitness effects: Vocabulary score, somatosensory sensitivity, ERPs except for visual-evoked potential (VEP) latencies for late waves at Oz and Cz Fitness effects: High-fit had significantly greater

| Steinhaus et al., 1990 | (males, 26 ± 3 years)
Older:
15 high-fit (males, 54 ± 3 years), 15 low-fit (males, 56 ± 3 years) | group were classified as being high-fit and 15 as being low-fit based on VO_{2max}
Cognitive:
- Vocabulary test
- Somatosensory & visual sensitivity
- Cognitive performance (factor from 4 tests)
- ERPs (including P300) | visual sensitivity, better VEP A/I (amplitude/intensity) slope (indicator for inhibition) & better cognitive performance
Age x fitness effects: P300 latencies were significantly longer for older low-fit men than older high-fit and younger men; VEP A/I slope values better for high-fit compared to low-fit old adults
Interpretation:
-Older adults have higher benefits from aerobic fitness than younger adults (aerobic fitness might postpone functional loss during aging)
-High-fit participants had better cognitive performance: lifestyle with vigorous exercise leads to a more efficient central nervous system probably due to an increased oxygen transport and utilization |

Table 4:   Overview of eleven cross-sectional studies (high-fit/trained/active vs. low-fit/untrained/sedentary) in young adults; ordered by year of publication

Author	Participants	Exercise/ Fitness and Cognitive Testing	Selected Results (only fitness-/activity-related effects) and Interpretation
Kamijo, O'Leary, Pontifex, Themanson & Hillman, 2010	Higher-fit: 32 (13 males, 20 ± 2 years) Lower-fit: 32 (12 males; 20 ± 2 years)	Exercise/Fitness: - VO_{2max}: graded treadmill exercise test → Group classification based on median split on VO_{2max} within each sex Cognitive: -Sternberg working memory task (speed/ accuracy instruction) -ERP: CNV (contingent negative variation)	Response times/accuracy: No fitness effect on response time and accuracy CNV: -Larger CNV amplitudes for lower-fit participants during speed instructions at Fz electrode (no effect for accuracy instruction) -Trend for larger CNV amplitudes for lower-fit participants at central electrode site across task instructions Interpretation: -Selective finding on frontal electrode with larger CNV amplitude for low-fit adults may be associated with the allocation of greater amounts of cognitive control to perform the task whereas higher-fit had smaller amplitudes indicating a more efficient cognitive task preparation and better top-down cognitive control -General finding on central electrode with larger amplitude for low-fit adults may be associated with the requirement of greater neural resources to prepare for motor action whereas higher-fit had smaller amplitudes indicating more efficient response preparation processes -Study indicates that aerobic fitness is related to both, cognitive and motor processes

Kamijo & Takeda, 2010	<u>Active:</u> 20 (10 males; 20 ± 0.3 years) <u>Sedentary:</u> 20 (11 males; 22 ± 0.4 years)	<u>Exercise/Fitness:</u> -Physical activity questionnaire (IPAQ) -20 participants were classified as being active and 20 as being sedentary based on physical activity history <u>Cognitive:</u> -Task switching paradigm (mixing cost, switch cost) -ERP: P3	<u>Response times:</u> -Shorter reaction times for active compared to sedentary adults during low/high trials in the mixed condition (no effect on pure condition or odd/even trials) -Larger switch effect/cost for sedentary compared to active adults <u>P300:</u> P3 amplitude during low/high trials larger for pure task compared to mixed task for sedentary adults (no effect for odd/even trials) <u>Interpretation:</u> -Sedentary adults had smaller P3 amplitudes in the mixed compared to pure task which indicates greater working memory demands and fewer available resources during the mixed task (since active adults showed no difference, this could be an indicator for better allocation of attentional resources) -Selective effects only for low/high trials might be explained by a different working memory load -Physical activity was related to lower switch cost for response times which indicates better executive control and mental flexibility
Åberg, Pedersen, Torén, Svartengren, Bäckstrand et al., 2009[#]	1,221,727 (1,221,727 males; 18 years)	<u>Fitness:</u> -Cardiovascular: Cycle ergometer test (W_{max}/kg) -Isometric muscle strength tests → Regression analysis (no group classification) <u>Cognitive:</u> Logical, verbal & visuospatial performance, technical/mechanical skills tests → global intelligence score	<u>Results:</u> -Cardiovascular fitness associated with better cognitive scores for global intelligence and subcategories; strongest association for logical and verbal intelligence -Association remained strong even after controlling for genetic factors (analyses among brother/twin pairs) -Muscular strength associated with global intelligence only for lower scores of muscular strength <u>Interpretation:</u> -Findings support cardiovascular fitness hypothesis -Association between fitness and cognition predominantly caused by environmental factors -Logical and verbal performance linked to hippocampus and frontal brain areas and showed the strongest results → regular exercise might be beneficial especially for these areas

Kamijo & Takeda, 2009	<u>Active:</u> 20 (10 males) <u>Sedentary:</u> 20 (11 males) <u>Mean age:</u> 21 ± 0.3 years	<u>Exercise/Fitness:</u> -Physical activity questionnaire (IPAQ) -20 participants were classified as being active and 20 as being sedentary based on physical activity history <u>Cognitive:</u> -Spatial priming paradigm (positive PP/ negative priming NP) -ERPs: N1, N2, P2, N2, P3a, P3b	<u>Response times:</u> -Shorter reaction times for sedentary adults in PP condition compared to control condition; larger difference in reaction times between NP and control condition for active adults -No overall difference between active and sedentary adults in reaction times <u>ERP amplitudes:</u> -Larger N1 amplitudes for active adults in control compared to PP condition at Fz; increased N2 amplitude for active adults at Fz relative to Cz and Pz in PP and control condition -Increased P3 amplitudes for active adults at Cz and Pz relative to Fz in PP, NP and control condition (in NP condition also for sedentary) -No effect on P1 and P2 amplitudes in PP condition and NP condition, no effect on N1 and N2 amplitudes in NP condition <u>ERP latencies:</u> -Shorter P3 latencies for active compared to sedentary adults for all conditions; longer P3 latencies in the NP compared to control condition for active adults -Shorter r-LRP (response-locked lateralized readiness potential) latencies for active compared to sedentary adults -No activity-related effects on P1, N1, P2, N2 latencies <u>Interpretation:</u> -Larger NP effect for response time and P3 latency in active adults might be related to higher executive control mechanisms since NP effect represents inhibitory control -Effects of physical activity on executive control are independent of task difficulty -Larger PP effect for response time and P3 latency for sedentary adults can be an indicator for worse executive control functioning -Shorter r-LRP latencies indicate faster preparation for motor responses among active adults
Scisco, Leynes & Kang, 2008	<u>High-fit:</u> 26 (12 males) <u>Low-fit:</u> 26 (7 males) <u>Mean age:</u> 20 ± 2 years	<u>Exercise/Fitness:</u> -Physical activity questionnaire -VO_{2max}: graded treadmill exercise test -High-fit: $VO_{2max} > 70^{th}$ percentile according to ACSM Guidelines, low-fit: $VO_{2max} < 30^{th}$ percentile according to ACSM Guidelines <u>Cognitive:</u> -Task switching paradigm	<u>Response times/accuracy:</u> No fitness effect on response times or accuracy <u>ERP amplitudes/latencies:</u> No fitness effect on N1, P3a amplitude, P3b amplitude, P3 peak amplitude, and P3 latency <u>Interpretation:</u> -Positive relationship between cardiovascular fitness and cognitive function emerges after early adulthood -Executive control is already highly efficient in young adults since they reached high accuracy levels -Effects in young adults only for acute exercise, no long-term effects of cardiovascular fitness

		(response time, response accuracy) -ERPs: N1, P3a, P3b, P3	
Themanson, Pontifex & Hillman, 2008	62 (23 males; 20 ± 2 years)	Exercise/Fitness: -VO$_{2max}$: graded treadmill exercise test → Regression analysis (no group classification) Cognitive: -Flanker task (congruent/ incongruent trials; accuracy/ speed conditions; behavioral data) -ERPs: ERN (error-related negativity), P$_e$ (positivity), N2	Response times/accuracy: Fitness was not related to response accuracy or response times for either trial type (congruent/ incongruent) in either instruction condition (speed/ accuracy) ERN amplitude/latency: -Fitness was related to larger ERN amplitudes (independent of sex) under accuracy but not speed instruction -Males and higher-fit adults had greater modulations in ERN amplitude across conditions Post-error behavior: -Fitness was related to greater post-error accuracy in the accuracy but not speed condition; no fitness effect on post-error response times in either condition -Higher-fit adults had greater modulations in post-error accuracy across conditions Interpretation: Larger ERN amplitudes and post-error accuracy indicate better action monitoring processes and increased flexibility and efficiency regarding cognitive control to meet task demands especially under accuracy instruction
Themanson & Hillman, 2006*	Higher-fit: 14 (7 males; 20 ± 2 years) Lower-fit: 14 (7 males; 21 ± 2 years)	Exercise/Fitness: -VO$_{2max}$: graded treadmill exercise test -High-fit: VO$_{2max}$ > 80th percentile according to ACSM Guidelines Cognitive: -Flanker task (congruent, incongruent trials; behavioral data) -ERPs: ERN, P$_e$, N2	Response times/accuracy: -No fitness effect on correct, error and matched correct trial response times -No fitness effect on response accuracy -Significant accuracy effect for higher-fit: longer response times (response slowing) for correct trials following error trials than correct following matched-correct trials; increased response slowing (longer response times) following error trials for higher-fit compared to lower-fit adults ERP amplitudes/latencies: -Higher-fit had significantly smaller ERN amplitudes during errors trials compared to lower-fit adults, no effect on ERN amplitudes on matched correct trials -Higher-fit had significantly larger P$_e$ amplitudes during errors trials compared to lower-fit adults, no effect on P$_e$ amplitudes on matched correct trials -No fitness effect on N2 amplitude Interpretation: -Smaller ERN amplitudes (index of action monitoring processes) corresponds with a relative decrease in ACC activation and thus a reduction in task-related response conflicts -Larger P$_e$ amplitudes (index of post-response error evaluation) and greater post-error response slowing indicate an increase in neural and behavioral ad-

			justments in top-down attentional control following errors to improve task performance -No effects on N2 amplitude might be explained by an already maximum performance of young adults and little room for fitness-related improvements
Iwadate, Mori, Ashizuka, Takayose & Ozawa, 2005	<u>Athletes:</u> 7 (7 males; 22 years) <u>Nonathletes:</u> 7 (7 males; 19 years)	<u>Exercise/Fitness:</u> Athletes: soccer players with participation in competitions; nonathletes: no regular participation in sports activity <u>Cognitive:</u> -Somatosensory Oddball task (electric stimuli; upper-limb and lower-limb task) -ERPs: P300, N140 (functionally analogous to N1)	<u>Behavioral measures:</u> No difference in correct answers (%) <u>ERP amplitudes/latencies:</u> -Larger P300 amplitude and shorter P300 latency in athletes during lower-limb task, not upper-limb task -More negative N140 amplitude in athletes at some electrode sites, no effect on N140 latency <u>Interpretation:</u> -Increased N140 amplitude might indicate better selective attention due to increased activation of the frontal cortex -Findings on somatosensory P300 component might be related to higher attentional resources and plasticity of somatosensory cortex and cognitive processing
Magnié, Bermon, Martin, Madany-Lounis, Suisse et al., 2000*	<u>High-fit:</u> 10 (10 males; 21 years) <u>Low-fit:</u> 10 (10 males; 23 years)	<u>Exercise/Fitness:</u> -Activity questionnaire -VO_{2max}: graded treadmill exercise test -High-fit: regular sports competitions for at least 1 year, all participants are cyclists; low-fit: regular physical training < 1 hour/week <u>Cognitive:</u> -Auditory Oddball paradigm -Task for N400 including sentences displayed word-by-word -ERPs: P300, N400, N1, P2, N2	<u>Task performance:</u> No fitness effect on task performance (error rate) <u>ERN amplitude/latency:</u> No fitness effects on P300, N1, P2, N2, N400 amplitude or latency <u>Interpretation:</u> -Aerobic fitness is not related to cognitive function in young adults -However, this study compared a group of cyclists with sedentary adults. Thus, type of sport and sport specific motor or tactic demands may play a crucial role in the relationship of regular exercise and cognition
Polich & Lardon, 1997	<u>High-exercise:</u> 11 (8 males; 30 years) <u>Low-exercise:</u> 11 (8 males; 35 years)	<u>Exercise/Fitness:</u> Exercise survey Exercise group: life-long participation in sports competitions, at least three year history of vigorous exercise, > 5 hours/week; control/low-exercise group: absence of high level sports activity, minimal engagement in exercise, < 5 hours/week <u>Cognitive:</u> -Auditory and visual Oddball task	<u>Behavioral measures:</u> No fitness effect on behavioral task performance (errors) <u>ERP amplitudes/latencies:</u> -Larger P300 amplitude for high-exercise group -Increased P300 peak latency between auditory and visual task for low-exercise group -P300 latency not affected by exercise, but tendency to decreased latency in high-exercise group for visual task -No exercise effects on N100, P200 and N200 amplitude and latency except for longer peak latencies for low-exercise group for standard stimulus data -Large individual variation in P300 for both groups <u>Interpretation:</u>

		-ERPs: P300, N100, P200, N200	-Exercise effects on ERP may result from changes in baseline EEG produced by regular aerobic exercise -EEG and ERP changes may be related to increased circulatory capacity
Lardon & Polich, 1996	Exercise: 18 (12 males; 31 ± 6 years) Control: 18 (12 males; 32 ± 7 years)	Exercise/Fitness: Exercise survey Exercise group: life-long participation in sports competitions, at least three year history of vigorous exercise; control group: absence of high level sports activity, minimal engagement in exercise Cognitive: EEG recording with eyes open/closed	Results: -Less spectral power for exercise group in the delta band, greater power in alpha and beta bands -Larger mean band frequency in the delta, theta, and beta bands for exercise group Interpretation: Exercise effects on ERP may result from changes in baseline EEG produced by regular aerobic exercise

\#: This study includes cross-sectional and longitudinal analyses as well and is therefore listed in table 4 and 5; *: This study also tested the effects of acute exercise on cognitive performance. However, only the findings on regular exercise effects are reported here

Table 5:   Overview of six longitudinal and intervention studies in young adults; ordered by year of publication

Author	Participants	Exercise/ Fitness and Cognitive Testing	Selected Results (only fitness-/activity-related effects) and Interpretation
Hopkins, Davis, Vantieghem, Whalen & Bucci, 2012*	0W-: 13 (1 male; 21 ± 1 years) 4W-: 14 (5 males; 21 ± 1 years) 4 W+: 12 (4 males; 21 ± 1 years) 0W+: 15 (4 males; 20 ± 1 years)	Fitness: - Physical activity questionnaire - Self-reported fitness level: estimation of VO_{2max} → Group classification: randomized assignment Cognitive: - Novel object recognition task - Mood and anxiety measurement Exercise training: - 4 week treadmill exercise (at least 4 sessions/week with a duration of at least 30 minutes): 4W- and 4W+ - Acute exercise on the day of posttest: 4W+ and 0W+ - No exercise intervention, maintaining physical activity level: 0W- and 0W+	Results: -Accuracy in recognition task decreased significantly in post- compared to pretest for 0W- and 4W- group -Only 4W+ group showed improvement in recognition task from pre- to posttest -Significantly increased perceived stress from pre- to posttest for 0W+ group, decreased stress for 4W+ group -Significant increase in positive mood from pre- to posttest for 4W- group -No effects on depression index, anxiety and negative mood Interpretation: -Since only the group with exercise intervention and acute exercise prior to posttest exhibited a beneficial effect on memory, there might be a gradual development in exercise related cognitive effects: an acute exercise session improves cognition dependent of regular physical activity habits -Regular exercise (4W+ and 4W-) improves positive mood and decreases anxiety

Goekint, De Pauw, Roelands, Njemini, Bautmans et al., 2010*	Experimental: 15 (12 males; 20 ± 0.4 years) Control: 8 (6 males; 22 ± 1 years)	Fitness: - Strength test: 1- repetition maximum → Group classification: randomized assignment Cognitive: - Digit span memory test (short term memory) - Picture recall task (mid-term memory) - BDNF-/IGF-1 measurement (venous blood samples) Exercise training: 10 week strength training (3 sessions/week)	Results: -Significantly increased maximal strength in experimental group after intervention period -No effect on BDNF or IGF-1 concentration -Improvement in short term memory in both groups from pre- to posttest -No intervention effect on mid-term memory Interpretation: -Resting BDNF and IGF-1 level is not influenced by strength training in young adults probably resulting from too low total work during intervention -Improvement in short term memory due to learning effect since both groups showed improvement -Memory function was not influenced by strength training since young adults already have a high level of memory function or memory test was not sensitive enough to detect small effects
Stroth, Reinhardt, Thöne, Hille, Schneider et al., 2010	Experimental: 47 (6 males; 21 ± 3 years) Control: 28 (5 males; 23 ± 5 years)	Fitness: - Incremental step test: workload at lactate threshold → Group classification: randomized assignment to five groups according to baseline fitness level Cognitive: - 2-back task (working memory) - Stroop task (attention, inhibition) - Dots-mixed task (ignoring of task-irrelevant stimuli, inhibitory control) - Positive and negative affect schedule (PANAS) - Assessment of COMT genotyping Exercise training: 50 walking/running sessions within 4 months (maximal 4 sessions/week)	Results: -Greater fitness improvement in runners compared to control group from pre- to posttest -Running group improved positive affect (not observed for control group) -No effect on negative affect -Decreased response times in 2-back task for runners carrying a Val allele; no effect for response accuracy -Val/Val runners had significantly reduced response times in congruent and neutral trials of Stroop task; no effect for incongruent trials -Better performance in dots-mixed task for runners compared to control group particularly in incongruent trials Interpretation: -COMT genotyping and dopaminergic modulation may be mediator in the influence of regular exercise on cognition -Regular exercise improves positive mood but does not influence negative affect -Results for dots-mixed task might indicate that exercise improves particularly higher cognitive functions (greatest effect for incongruent trials), however in the Stroop task there was only an effect for congruent and neutral trials
Åberg, Pedersen, Torén, Svartengren, Bäckstrand et al., 2009[#]	1,221,727 (1,221,727 males)	Fitness: - Cycle ergometer test (W_{max}/kg) at age of 18 years - Physical education grades at age of 15 years → Regression analysis (no group classification): Prediction of cognitive performance from changes in cardiovascular fitness be-	Results: -Adults with increased fitness between 15 and 18 years had significantly higher global intelligence scores than adults with increased fitness; similar results for subcategories of intelligence -Better cardiovascular fitness at the age of 18 was linked to higher educational attainment and, to a greater extent, to occupational outcome Interpretation: -Changes in cardiovascular fitness during adolescence are linked to changes in cognitive perfor-

		tween 15 and 18 years; prediction of education/ occupation between 28 and 54 years from cardiovascular fitness at the age of 18 <u>Cognitive:</u> Logical, verbal & visuospatial performance, technical/mechanical skills tests → global intelligence score	mance -Cardiovascular fitness at young adulthood predicts educational achievement in adulthood
Stroth, Hille, Spitzer & Reinhardt, 2009	<u>Experimental:</u> 14 (4 males) <u>Control:</u> 14 (5 males) <u>Mean age:</u> 20 ± 3 years	<u>Fitness:</u> - Physical activity questionnaire: participants were untrained before enrollment in study - Graded cycle ergometer test (PWC: physical work capacity) → Group classification: randomized assignment <u>Cognitive:</u> - Affect: Positive and Negative Affect Schedule (PANAS) - Visual and Verbal Memory Test (short-term visuospatial and verbal memory) - d2 Test of Attention (selective attention, concentration) <u>Exercise training:</u> 6 week running training (3 sessions/week with a duration of 30 minutes)	<u>Results:</u> -Significant increase in PWC in experimental from pre- to posttest; no effect for control group -Both groups changed in positive affect but greater increase for exercise compared to control group -No influence on negative affect -Experience group had significant increase of visuospatial memory (not observed for control group) -Both groups improved verbal memory -Both groups increased concentration performance and reduced number of errors in d2 test <u>Interpretation:</u> -Exercise leads to a benefit in well-being but does not influence negative affect probably due to the short duration of exercise intervention or floor effect in young adults -Increased performance in d2 test for both groups might be due to practice or learning effect -Exercise in young adults selectively improves visuospatial memory but not verbal memory or concentration since these tests might have been to easy → tasks that require more effortful processing are needed to detect exercise-related influences on cognition in young adults
Hansen, Johnsen, Sollers III, Stenvik & Thayer, 2004	<u>Trained:</u> 21 (21 males; 19 years) <u>Detrained:</u> 16 (16 males; 19 years)	<u>Fitness:</u> - VO_{2max}: Graded cycle ergometer test - Group classification based on application for further duty in Royal Norwegian Navy <u>Cognitive:</u> - Cognitive test battery/ Continuous performance test (simple reaction time, choice reaction time, 1-back task, addition task) - Working memory test (2-back task) <u>Exercise training:</u>	<u>Results:</u> -Detrained adults had faster response times in posttest compared to pretest on non-executive function tasks, trained had faster response times in executive function tasks -Better response accuracy (true positive responses) on executive function tasks for trained adults (not for detrained); no effect on false responses -Trained adults had higher aerobic capacity and vagally mediated cardiac control (HRV: heart-rate variability) than detrained adults after intervention <u>Interpretation:</u> -High HRV levels among trained adults are associated with better performance in executive function tasks and thus may utilize the function of the prefrontal cortex -Detrained adults had faster response times only in

		1. Both groups: 8 weeks basic training (3 hours/week) before preretest 2. Detrained: 4 weeks duty on board a navy ship; trained: 4 weeks exercise training (3 hours/week) before posttest	non-executive tasks which could be explained by a higher level of stress in the posttest and a lower HRV level

*: This study also tested the effects of acute exercise on cognitive performance. However, only the findings on regular exercise effects are reported here

The findings of the cross-sectional studies are very heterogeneous in parts. For behavioral measures, all studies except for one (Hillman, Motl et al., 2006) found no effects on response accuracy. It must be noted though, that some studies did not report any response accuracy results. With regard to response times, there are conflicting results. Whereas shorter response times for active compared to sedentary participants were observed for the Oddball task (Hillman et al., 2005; Pontifex et al., 2009), Switching paradigm (Hillman, Kramer et al., 2006; Kamijo & Takeda, 2010), and Flanker task (Hillman, Motl et al., 2006), response times were not influenced during working memory (Kamijo, O'Leary et al., 2010) and spatial priming tasks (Kamijo & Takeda, 2009). Further, no effect was found on response times for the Switching task in one study (Scisco et al., 2008) and for the Flanker task in two studies (Themanson & Hillman, 2006; Themanson et al., 2008) which is in contrast to the findings reported above.

Several different event-related brain potentials were investigated in the cross-sectional studies. For P300 amplitude, the majority of studies reported a positive relationship with physical activity behavior or fitness that is larger amplitudes for active compared to sedentary young adults. This effect has been observed in Oddball tasks (e.g. McDowell et al., 2003; Polich & Lardon, 1997), and Switching paradigms (Hillman, Kramer et al., 2006; Kamijo & Takeda, 2010). However, P300 amplitudes were not increased for active compared to sedentary adults in an S1-S2-S3 task (Hillman et al., 2002), spatial priming task (Kamijo & Takeda, 2009) and also Switching (Scisco et al., 2008) and Oddball task (Magnié et al., 2000). The results on P300 latency are also inconsistent with authors reporting effects during switching (Hillman, Kramer et al., 2006), spatial priming (Kamijo & Takeda, 2009) and Oddball tasks (Hillman et al., 2005; Iwadate et al., 2005) and others reporting no effects for Switching (Scisco et al., 2008) and Oddball tasks (Magnié et al., 2000; McDowell et al., 2003; Polich & Lardon, 1997; Pontifex et al., 2009). Due to the large number of different behavioral and neuroelectric ERP measures that have been investigated, the majority of studies find a positive relation between exercise and cognition at least specific for selected electrode sites, task conditions or trial types. However, two studies have been described in which none of the activity or fitness variables showed any association to cognitive measures (Magnié et al., 2000; Scisco et al., 2008). The authors conclude that there is no activity-induced or fitness-related beneficial effect on cognition in young adulthood and that potential activity or fitness-induced effects on cognition might emerge in older adulthood. Given that cross-sectional study designs have many advantages compared to longitudinal or intervention studies such as time efficiency and reduced dropout rate, there are nonetheless some aspects that need to be considered in terms of interpretation of results. The most important might be that cross-sectional studies do not allow for detecting any causal relationships between measures. Thus, it cannot be concluded whether a relationship between regular physical activity and cognition is caused by a large amount of physical activity resulting in higher cognitive performance or whether it is caused by a higher cognitive performance leading to a more active lifestyle.

This question can only be addressed by true experimental studies. Table 5 provides an overview of five intervention studies and one longitudinal study conducted in young adults. Unfortunately, neither study assessed ERPs. Three studies could find a beneficial effect of aerobic exercise training on positive mood and well-being whereas no effect was observed for negative affect (Hopkins et al., 2012; Stroth et al., 2009; Stroth et al., 2010). Strength training does not improve memory function (Goekint et al., 2010), whereas aerobic exercise exhibited a beneficial effect on global (Hopkins et al., 2012) and visuospatial but not verbal memory (Stroth et al., 2009). In three studies, it was concluded by the authors that exercise generally improves cognitive and especially executive functions (Hansen et al., 2004; Stroth et al., 2009; Stroth et al., 2010). The longitudinal study by Åberg et al. (2009) provides evidence for the link between an improvement in cardiovascular fitness during adolescence and a better intelligence as well as for the prediction of educational achievement in later adulthood by cardiovascular fitness at the age of 18 years.

The limitations of these studies are that they primarily focus on aerobic exercise or cardiovascular fitness. To date, there is no or only little evidence as to which extent other types of physical fitness or physical activity factors such as duration, frequency or intensity are related to cognitive performance (Kramer & Hillman, 2006, p. 55). In addition, many studies use self-reported measures to detect cardiovascular fitness and physical activity amount and further have relatively small sample sizes. Only few studies control for sex or inter-individual differences in intelligence, thus many possible moderator variables are unknown or not controlled for.

However, it can be concluded that the beneficial effects of exercise on brain function might not be limited to early childhood or the elderly population, but also seem to pertain to young adulthood. There are studies available which lead to the assumption that chronic (aerobic) exercise and a good physical fitness is positively related to cognition throughout the lifespan. Even young adults, who are on the peak of cognitive performance and have only little room for improvements might enhance their performance by regular exercise.

2.4.3 Underlying Mechanisms

There are several potential mechanisms that might mediate the effects of physical activity on cognition. The mechanisms are not mutually exclusive and it is assumed that the effects of exercise on cognition cannot be limited to only one single approach.

According to Ratey and Loehr (2011, p. 172 ff.) and in accordance with other authors (e.g. Erickson, 2012, p. 323 ff.), the effects can be located on three levels. The first level involves exercise-induced influences on the neural systems that mediate cognitive and executive functions. This hypothesis can be derived from findings in electrophysiological and neuroimaging studies showing larger P300 amplitudes and partly shorter P300 latencies during the performance of a cognitive task after short bouts of exercise compared to baseline measurements (e.g. Hillman, Snook & Jerome, 2003; Kamijo et al., 2007) and, as a result from regular physical activity, for people with a higher fitness compared to those with a lower fitness (see chapter 2.4.2). One possible underlying mechanism of this effect could be a generally higher alpha activation in the EEG for high-fit participants which in turn might influence event-related potentials (Kramer & Hillman, 2006, p. 52). Although it must be noted, that in most studies these effects on event-related brain potentials are very selective and can mainly be found for only some conditions of a cognitive task (e.g. only for incongruent but not for congruent trials).

Besides studies with electrophysiological measurements, the application of neuroimaging techniques has increased. A review on structural and functional brain changes indicated by fMRI or related techniques was written by Voelcker-Rehage and Windisch (2013) and provides a detailed overview of recent research in this field. They found that metabolic exercise or fitness is related to changes in brain activation (e.g. Voelcker-Rehage, Godde & Staudinger, 2011), cerebral blood volume (Pereira, Huddleston, Brickman, Sosunov, Hen et al., 2007) and functional connectivity of brain regions that are relevant for cognition (e.g. Burdette, Laurienti, Espeland, Morgan, Telesford et al., 2010). In addition to these findings, some studies provide evidence that physical fitness has a positive influence on brain volume in different brain regions which decline with increasing age (Voelcker-Rehage & Windisch, 2013). For example, exercise has been linked to a higher gray matter density in the frontal, temporal and parietal cortices, a greater hippocampal and gray matter volume of the prefrontal cortex, and an increased functional connectivity between prefrontal cortices (Guiney & Machado, 2013, 74 f.). However, only few studies could prove that brain volume in the frontal and temporal lobes is related to a positive impact of physical activity on cognitive performance (e.g. Weinstein, Voss, Prakash, Chaddock, Szabo et al., 2011).

On a molecular level, the influence of neurotrophins on cognition is hypothesized (Ratey & Loehr, 2011; Voss et al., 2011). Two prominent factors are discussed in this context: The brain-derived neurotrophic factor (BDNF) and the insulin-like growth factor (IGF-I). BDNF is a neurotrophic molecule that regulates the growth of neurons and promotes cell survival and synaptic plasticity (Vaynman, Ying & Gomez-Pinilla, 2003). It is produced in the brain, especially in the hippocampus, and has shown an increased release as a response to exercise. A study by Cotman and Berchtold (2002) found that hippocampal BNDF levels were significantly elevated in animals after five days of wheel running compared to sedentary controls. In addition, there was a strong positive relationship between running distance and BNDF release. Similarly, it was found in humans that the release of BDNF is increased two- to threefold during aerobic exercise (Rasmussen, Brassard, Adser, Pedersen, Leick et al., 2009). Other researchers have further investigated that an increase of BDNF in blood serum is dependent from the intensity of acute exercise with higher intensity resulting in higher BDNF changes (Ferris, Williams & Shen, 2007). Moreover, a study from Seifert, Brassard, Wissenberg, Rasmussen, Nordby et al. (2010) revealed an enhanced release of BDNF at a resting period following long-term aerobic endurance training. Furthermore, increased resting levels of BDNF initiated by a chronic 5-week aerobic training were found in young adults (Zoladz, Pilc, Majerczak, Grandys, Zapart-Bukowska et al., 2008) and therefore also provide evidence for effects of regular physical activity on BDNF. However, studies on strength training showed no effects on BDNF (Goekint et al., 2010; Schiffer, Schulte, Hollmann, Bloch & Strüder, 2009).

Only few studies exist in which BDNF is directly linked to exercise-induced influences on cognition. It was found in animals, that exercise leads to an enhanced long-term potentiation (LTP) which improves synaptic plasticity and is hypothesized to be a cellular model of memory and learning. Interestingly, BDNF was associated with this increment of LTP (Farmer, Zhao, Van Praag, Wodtke, Gage et al., 2004). Other studies in rats showed a decreased cognitive performance during exercise when BDNF receptors were blocked (Gomez-Pinilla, Vaynman & Ying, 2008; Griesbach, Hovda & Gomez-Pinilla, 2009). A comprehensive review article on exercise-induced BDNF levels in humans was provided by Knaepen, Goekint, Heyman and Meeusen (2010).

IGF-I is another neurotrophic factor which is produced in the central nervous and peripheral systems and has shown to be enhanced by aerobic exercise in rats (e.g. Trejo, Carro & Torres-Aleman, 2001). Studies in humans found that changes in IGF-1 were initiated by continuous moderate-intensity as well as intermittent high-intensity exercise (Copeland & Heggie, 2008) and that IGF-I was increased in response to strength training in older people (Cassilhas, Viana, Grassmann, Santos, Santos et al., 2007). Similar to the studies on BDNF, researchers also tried to investigate the effects of blocking IGF-I. There is evidence that blocking the IGF-I uptake in the brain results in an impaired exercise-related neurogenesis (Trejo et al., 2001).

The last level is the cellular level which is closely related to the effects on the molecular level. Research studies have supported the hypothesis that exercise-induced effects on cognitive performance might be traced back to an increased synaptic plasticity, vascular function, angiogenesis and neurogenesis (Ratey & Loehr, 2011, p. 174 ff.; Thomas, Dennis, Bandettini & Johansen-Berg, 2012; van Praag, 2006, p. 63 ff.).

Other researchers have further hypothesized that a relationship between exercise and the autonomic nervous system might explain the effects of exercise on cognition. It was found that a greater heart rate variability which is an index of parasympathetic function and vagal control was related to better executive performance and thus improved function of the prefrontal cortex (Hansen et al., 2004; Hansen, Johnsen & Thayer, 2003).

In addition, it is suggested that anxiety and other mental problems impair executive functioning and the attentional system (Eysenck, Derakshan, Santos & Calvo, 2007). Given that physical activity has been shown to improve symptoms of depression and anxiety (Meeusen, 2006, p. 130; Raglin & Wilson, 2012, p. 334 ff.), this effect might also be responsible for the relationship between physical activity and cognition.

2.5 Summary

Physical activity is referred to as any bodily movement caused by skeletal muscles and leads to increased energy expenditure. Physical activity can be subdivided into habitual, sports-related and occupational activity and is generally described based on information regarding intensity, frequency, duration and type. There is evidence that physical activity is related to mental and physical health including the cardiovascular and metabolic system. A good fitness level results from regular physical activity and comprises a set of attributes. These include energetically determined abilities such as endurance and strength, but also coordination, which is described as being information-oriented. In addition, human body composition, especially body fat percentage, is another aspect that has been assigned to the term "physical fitness". Physical activity and fitness have been shown to be related to cognition and executive functions. The latter are being defined as higher cognitive functions resulting in top-down cognitive control and comprising abilities such as attention performance, planning of goal-directed behavior or working memory. Studies have proven evidence that executive functions are primarily but not exclusively linked to the prefrontal cortex in mammals. Several computerized tasks exist that allow for the assessment of cognitive performance, especially if used in combination with functional brain imaging such as electroencephalography. This technique enables the researcher to elicit event-related brain potentials (ERPs). These potentials can be used to evaluate brain activity during the cognitive processing that is devoted to a stimulus in a cognitive paradigm. Meta-analyses on the relationship between physical activity, fitness and cognition consistently report small but positive effect sizes for children and older adults. Evidence from primary studies that have been conducted in young adults also suggests a link between exercise and cognition for this age group. However, it has not been determined yet which type of fitness or which factors related to physical activity are most effective at enhancing cognitive performance and brain function. Several potential mechanisms are discussed in literature which might underlie the association between activity, fitness and cognition. There is growing evidence that physical activity alters and improves cognition and brain functions via molecular, systems-based and emotional mechanisms. It is assumed that a combination of different processes might be responsible for exercise-induced cognitive improvements.

Chapter 3

Empirical Study

In this chapter, the aims, design, research questions and hypotheses of the CogniFit-Study are described. Further, the study sample as well as measures and procedures, which were conducted in the study, are depicted. The chapter ends with a section on data acquisition and analysis.

3.1   Study Description and Design

The Cognifit-Study was a cooperative project between the Karlsruhe Institute of Technology (KIT) and the Charité University Medicine Berlin. Involved institutes were the Institute of Sport and Sport Science (IfSS) and the Cognitive Systems Laboratory (CSL) at KIT and the Center for Space Medicine/ Institute of Physiology at Charité Berlin.

It was a prospective, quasi-experimental cross-sectional study and aimed at exploring the interdependences between physical activity, fitness and cognitive performance in young, healthy adults. The study was planned in 2009 and was audited by the ethics committee of the State Chamber of Physicians of Baden-Wuerttemberg (Germany). Recruitment of participants started in April 2010 and was conducted at the KIT and two other universities in Karlsruhe. Data acquisition lasted from May 2010 to July 2011. All participants were tested at the KIT on three separate days with a minimum of two days between sessions. Two sessions were conducted at the exercise physiology and biomechanics laboratories at the IfSS and one session was conducted at the CSL. In order to avoid or minimize acute exercise effects on physical or cognitive performance testing, participants were instructed to abstain from physical activity for at least 48 hours before they visited the laboratory. In addition, all participants were instructed to avoid food intake for the duration of two hours prior to testing day II in order to prevent errors in body composition measurement. Each session had an average duration of two to three hours. The chronological sequence of the testing days was not fixed, however most participants began with testing day I. An overview of experimental procedures of each session is given in chapter 3.4.

All participants were recruited by circular emails to student bodies of several departments, personal contacts and flyer that were distributed to students visiting large lectures, the university cafeteria and university sport courses. Moreover, posters were hung up in the university cafeteria and the involved institutes at KIT (CSL and IfSS). A call for participation in the study was further printed in the KIT's student magazine.

Interested students who were willing to participate in the study were asked to complete an online questionnaire (available via homepage http://csl.ira.uka.de/cognifit/) to report their physical activity habits as well as relevant biographic and health-related information. Based upon their answers in this questionnaire, appropriate participants were selected according to inclusion and exclusion crite-ria and invited to participate in the study.

Inclusion criteria comprised an age between 18 and 34 years, a general qualification for university entrance (undergraduate, graduate students and post-doctoral students), and at least 12 months of general sport participation or inactivity, respectively. Exclusion criteria were excessive alcohol con-sumption or smoking, acute or chronic neurological, psychiatric or medical diseases, any medica-tions affecting the central nervous system, and an inability to exercise for at least 10 minutes.

Prior to the first appointment at the IfSS or CSL, all participants were given detailed information about the aims, procedures, benefits, health risks[1], and data protection regulations of the study. Fur-ther, they received a German version of the PAR-Q (Physical Activity Readiness Questionnaire) and were instructed to bring the completed and signed questionnaire to the first session. The PAR-Q (Canadian Society for Exercise Physiology, 1992) is a self-screening tool that has been designed to identify risk factors related to physical activity and exercise. It comprises the following seven standardized questions that have to be answered with "yes" or "no".

1. Has your doctor ever said that you have a heart condition and that you should only do physical activity recommended by a doctor? (Hat Ihnen Ihr Arzt jemals gesagt, dass Ihr Herz in keinem guten Zustand ist und dass Sie Sport nur nach ärztlicher Verordnung be-treiben sollen?)

2. Do you feel pain in your chest when you do physical activity? (Haben Sie Schmerzen in der Brust, wenn Sie sich körperlich anstrengen?)

3. In the past month, have you had chest pain when you were not doing physical activity? (Hatten Sie im letzten Monat Brustschmerzen in Momenten, in denen Sie sich nicht kör-perlich anstrengten?)

4. Do you lose your balance because of dizziness or do you ever lose consciousness? (Ver-lieren Sie aufgrund von Schwindel leicht Ihr Gleichgewicht oder wurden Sie jemals ohnmächtig?

5. Do you have a bone or joint problem (for example, back, knee or hip) that could be made worse by a change in your physical activity? (Haben Sie ein Knochen- oder Gelenkprob-lem, das sich durch eine Veränderung Ihrer körperlichen Aktivität verschlimmern könn-te?)

6. Is your doctor currently prescribing drugs (e.g. beta blocker) for your blood pressure or heart condition? (Verschreibt Ihnen Ihr Arzt momentan Herzmittel oder Medikamente für Ihren Blutdruck (z. B. Betablocker)?)

7. Do you know of any other reason why you should not do physical activity? (Kennen Sie einen anderen Grund, warum Sie keinen Sport treiben sollten?)

If a participant answered "no" to all questions, he was allowed to participate in the study. In case, a participant answered "yes" to one or more questions, he was instructed to consult a physician and

---

[1] For most tests of this study, no health risks were expected. However, the test in the driving simulator is known to cause headache or dizziness in rare cases. This phenomenon is known as "simulator dizziness" or "simulator sickness". The graded cycle ergometer test requires the participant's willingness to cycle until achieving volitional exhaustion and causes an intense activation of the cardiovascular system. It might therefore lead to shortness of breath, nausea, chest pain, low or high blood pressure, and joint pain. All health risks were minimized by using standardized, approved and established test protocols and by having all tests being conducted by trained and experienced test supervisors.

submit a signed medical clearance to the investigator before taking part in the study. This applied to three participants.

All participants provided written informed consent to participate in the study on their first testing day and reported normal or corrected to normal vision. On testing day II, they further gave written informed consent to perform a graded cycle ergometer test. To facilitate the communication with participants, an email address was created especially for the study. This email account was used for making appointments with participants, answering questions about the study and even for sending a detailed overview of individual results and performances in all tests. Since participants were not paid for taking part in the study, they received this presentation of results in return.

3.2   Research Questions

The purpose of this work is to explore the relationship between physical (in-) activity, fitness and cognition in a cross-sectional study with young adults. Cognitive performance represents the dependent variable that is influenced by the independent variables physical (in-) activity and fitness. The first seven research questions address the relationship between regular physical (in-) activity and cognitive performance in young, healthy university students.

RQ 1:   Do young adults who regularly engage in physical activity have a higher cognitive performance than inactive young adults?

RQ 2:   Is the participation in competitive sports related to cognitive performance in young adulthood?

RQ 3:   Is the amount of sports and physical activity related to cognitive performance in young adulthood?

RQ 4:   Is there a difference in cognitive performance between athletes engaging in different types of sport?

RQ 5:   Is exercise intensity related to cognitive performance in young adulthood?

RQ 6:   Is exercise frequency per week related to cognitive performance in young adulthood?

RQ 7:   Is exercise duration per session related to cognitive performance in young adulthood?

The research questions eight to twelve address the relationship between several fitness parameters and cognitive performance in young, healthy university students.

RQ 8:   Is aerobic endurance related to cognitive performance in young adulthood?

RQ 9:   Is maximal strength related to cognitive performance in young adulthood?

RQ 10: Is coordinative performance related to cognitive performance in young adulthood?

RQ 11: Is body composition related to cognitive performance in young adulthood?

RQ 12: Is overall fitness related to cognitive performance in young adulthood?

3.3   Participants

In total, 163 students (50 females, 113 males) were selected for participation in the study. However, only 152 (45 females, 107 males) completed at least two out of three testing days and could therefore be included in data analysis. Mean values for participants' demographic data are provided in table 6.

Table 6:   Mean (SD) values for participant's demographic data

Measure	Females (M ± SD)	Males (M ± SD)	Total (M ± SD)
Sample size [n]	45	107	152
Age [years]	22.7 ± 2.6	23.8 ± 3.1	23.5 ± 3.0
Body height [cm]	165.5 ± 7.6	180.2 ± 6.7	175.8 ± 9,7
Body weight [kg]	60.9 ± 8.8	74.0 ± 7.4	70.1 ± 9.9
BMI [kg/m^2]	22.2 ± 2.4	22.8 ± 2.0	22.6 ± 2.2
Athletes/nonathletes ratio	29/16 (64%/36%)	90/17 (84%/16%)	119/33 (78%/22%)

26 % of the participants (had) studied natural sciences, 47 % (had) studied economics, computer sciences or humanities and social sciences and 27 % engineering sciences. Six participants stated to smoke regularly (mean: 3.3 cigarettes per day; range: 1 to 10 cigarettes). Three participants were affected by acute diseases (allergy and atopic dermatitis) and ten participants suffered from chronic diseases (allergy, asthma, thyroid hypofunction and joint problems). Further, ten participants were on a drug at the time of the study (birth control pill, allergy medication, iodine/thyroid medication). 149 participants completed testing day I, 148 completed testing day II and 146 testing day III. From 141 participants, complete data sets are available since they have completed all three testing days. However, data sets contain missing data for some participants due to technical problems that occurred during measurements.

3.4   Measures and Procedures

In this chapter, the measures and procedures of data collection are described. The following table 7 shows all procedures with relevance for this thesis and separated according to the three testing days with a chronological order. In the following text, only those procedures that are relevant for data analysis and for the results section are described in detail. Generally, the subjects were informed about the testing procedure before the beginning of the tests in every session. Participants were tested individually and every session was supervised by at least two experienced investigators.

Table 7:   Chronological sequence of measurements with relevance to this thesis and separated according to testing day[2]

Testing day I (CSL)		
General	-	Information about testing procedure
Cognition	-	Computerized cognitive testing battery: Oddball-, modified Flanker-, and Switching-Task, with 16 channel-EEG recording
Testing day II (IfSS)		
General	- -	Information about testing procedure Written informed consent to perform a graded cycle ergometer test
Physical Activity	-	Questionnaire on habitual and work-related physical activity, and recreational activities
Anthropometry & Body Comp	- -	Anthropometric measurements: height, waist and hip circumference Body composition measurement: Air-displacement plethysmography (BOD POD Gold Standard® System; Life Measurement Inc., USA)
Endurance	-	Graded cycle ergometer test to volitional exhaustion. During the test, participants were fitted with a mobile cardiopulmonary exercise testing device, two double-sensors located at the forehead and the manubrium sterni to assess body temperature via heat-flux, Polar® RS 800 heart rate monitor, and a three lead ECG
Testing day III (IfSS)		
General	-	Information about testing procedure
Intelligence	-	Intelligence test CFT 20-R (Weiß, 2006)
Coordination	-	Static and dynamic one-legged-stabilization tasks on the Posturomed® (Posturomed 202; Haider Bioswing GmbH, Germany)
Strength	- - -	Short practice on leg press and warm-up phase on treadmill Isometric lower limb maximal strength test (leg press) Dynamic lower limb strength test (Counter-movement-jump)
Coordination	-	Backward balancing and jumping side-to-side tests (Bös et al., 2009)
First session (either testing day I, II or III)		
Consents	- -	Written informed consent to participate in the study Examination of PAR-Q

---

[2] A table summarizing all measurements conducted in this study is given in the appendix

3.4.1 Physical Activity

All participants completed a standardized online questionnaire that referred to their sports-related activity habits within the twelve months prior to the study. First they declared whether they were athletes or whether they were physically inactive (Question: Do you regularly exercise?). Athletes were further asked to report

- type of preferred sport (single mention only, pre-formulated answers: Running, Swimming, Gymnastics, Tennis, Strength/Weight training, Soccer, Team handball, Basketball, others);

- frequency of exercise sessions per week (if applicable including tournaments/matches/competitions; single mention only, pre-formulated answers: less than twice, twice to three times, four to six times, daily);

- exercise duration per session (pre-formulated answers: 0.5 hours, 1 hour, 1.5 hours, 2 hours, 2.5 hours, at least 3 hours and more).

The answers to these questions were assessed separately for summer and winter season. With regard to their main sport, athletes were also asked about their average perceived exercise intensity (sweat rating while exercising; single mention only, pre-formulated answers: light activity (no sweat and shortness of breath), moderate activity (some sweat and shortness of breath), and vigorous activity (considerable sweat and shortness of breath)). Moreover, participants were asked whether they take part in competitive sport (e.g. playing tournaments) and if yes on which level (e.g. international/national/regional tournaments, league for team sports). They were also asked to report other types of sport as well as training characteristics (frequency of sessions per week and duration per session). Two questions further addressed their sport history relating to the years before two years prior to the study.

Based on the questions on the sports-related activity, the estimated energy expenditure (kcal/week) was calculated for each participant. The basis for this calculation was the definition of one MET corresponding to 1 kcal/kg bodyweight/hour (Bouchard et al., 2012, p. 19).

For this purpose, the following formula was used for both preferred type of sport as well as for other types of sport that were possibly mentioned by the participants:

Sports-related activity (kcal/week)	=	Stated hours/week (hrs) x (weeks/26) x average energy expenditure/minute (MET; Ainsworth et al., 2000) x body weight (kg)
	=	hrs x (weeks/26) x MET x kg

Stated hours per week were calculated from the frequency of exercise sessions per week and the exercise duration per session according to the following scheme (table 8).

Table 8:   Coding scheme for hours per week spent at exercising

	Pre-formulated answer	Value for calculation
Frequency	Less than twice	1
	Twice to three times	2.5
	Four to six times	5
	Daily	7
Duration	0.5 hours	0.5
	1 hour	1

1.5 hours	1.5
2 hours	2
2.5 hours	2.5
At least 3 hours and more	3

The factor for weeks was based on the assumption that several types of sport cannot be performed throughout the whole year due to weather conditions or closing times of sport clubs during holidays. Two independent investigators grouped all the types of sport that were mentioned by the participants according to the estimated number of weeks in which they can be performed. The maximal number of weeks was 26 for summer and for winter season, respectively.

Metabolic equivalent (MET) values were taken from Ainsworth et al. (2000) and are defined as 1 kcal/kg body weight/hour. The following table 9 shows the week (half-year) factor and the MET that were used for calculation.

Table 9:   Week factors and MET

Type of sport	Half-year factor Summer (Winter)	MET
Aerobic	24 (24)	7
American football, Volleyball	18 (18)	8
Badminton	22 (22)	6.5
Basketball	18 (18)	7
Beach volleyball	15 (0)	8
Bicycling	24 (24)	8
Bicycling (racing)	20 (10)	11
Dancing	20 (20)	5
Diving	6 (6)	7
Fencing	20 (20)	6
Fitness (general)	24 (24)	6.5
Golf	15 (2)	4.5
Handball, Rugby	18 (18)	10
Kung Fu, Thai boxing, Judo, Taekwondo	24 (24)	10
Mountain biking	15 (10)	8.5
Rock climbing	6 (6)	8
Rowing	20 (20)	12
Running	24 (22)	8.5
Skiing (cross-country)	0 (2)	8
Soccer	18 (15)	10
Swimming, gymnastics, strength/weight training	24 (24)	7
Tennis	18 (15)	7
Track and field	20 (18)	6
Track and field (sprint)	20 (18)	9
Trampoline	22 (22)	3.5
Triathlon	24 (24)	11
Underwater rugby	18 (18)	9

The following example (table 10) shows the procedure of calculating the sports-related energy expenditure for a fictitious participant (body weight: 80 kg).

Table 10:   Example for questionnaire data of a fictitious participant

Question	Answer
Preferred type of sport? S (W)*	Tennis (Tennis)
Frequency of sessions/week? S (W)	Twice to three times (Less than twice)
Exercise duration/session? S (W)	1.5 hours (1.5 hours)
Another type of sport	Swimming
Frequency of sessions/week?	Once
Exercise duration per session?	1 hour

*S (W)=Answers seperated for summer and winter season

Calculation of the whole year sports-related energy expenditure (EE) per week was made as shown in table 11.

Table 11:   Calculation of the sports-related energy expenditure per week

Sports-related EE (preferred sport; Summer)	=	3.75 hrs x (18/26) x 7 METS x 80 kg
	=	1453.8 kcal/week
Sports-related EE (preferred sport; Winter)	=	1.5 hrs x (15/26) x 7 METS x 80kg
	=	484.6 kcal/week
Sports-related EE (other sport)	=	1 hr x (24/26) x 7 METS x 80 kg
	=	516.9 kcal/week
Sports-related EE (total)	=	(1453.8 + 484.6)/2 + 516.9
	=	1486.1 kcal/week

Participants also completed a questionnaire on habitual and work-related physical activity. They were asked about the time they spend for walking or cycling during an average week. Based on this answer, the weekly habitual activity energy expenditure (kcal) was estimated by the formula (hrs per week x (52/52) x 3.5 MET x body weight (kg)).

3.4.2 Fitness

To assess physical fitness, several tests were performed by the participants. Maximal aerobic power (VO_{2max}; maximal oxygen uptake) was obtained during a graded cycle ergometer test (figure 4). Prior to the test, all participants were seated in an upright position and the position of the seat and handle bar was individually adjusted to each participant's preferred sitting position. For the test, a standardized 50-25-2 protocol was used for all participants. The protocol involved cycling on an ergometer (medical 8i®; daum electronic, Germany) at a constant cadence of 70 to 90 revolutions per minute. The test started at a workload of 50 watts with an increment of 25 watts every two minutes until volitional exhaustion. Vocal encouragement was provided by the test supervisors during the last stages. Volitional exhaustion was defined as inability to continue cycling due to fatigue or if problems such as chest pain or nausea occurred (subjective evaluation by participant and test supervisor). During the test, gas exchange values such as VO_2 and VCO_2, respiratory quotient (VCO_2/ VO_2), oxygen pulse, and ventilatory equivalent were continuously recorded by a mobile cardiopulmonary exercise testing device (Metamax® 3B; Cortex Biophysik GmbH, Germany). Prior to each test, the device was calibrated according to the manufacturer's recommendations. VO_{2max} was determined by the software Metasoft® (Cortex Biophysik GmbH, Germany) marking the highest VO_2 peak during the test. Blood lactate was collected from the right earlobe before the beginning of the test (baseline measure) and every two minutes (with increment of watts after each stage) until the end of the test. To document recovery phase, blood lactate and heart rate were obtained one and three minutes after the end of the test at a reduced workload of 50 watts. Blood lactate values were analyzed using the Biosen® C-Line Sport device (EKF-diagnostic GmbH, Germany) and the Ergonizer® software for performance diagnostics in sports medicine and exercise physiology to predict the individual anaerobic threshold (Ergonizer Software, Freiburg, Germany; www.ergonizer.de). Heart rate was continuously recorded during the test using a heart rate monitor (Polar® RS 800; Polar Electro Oy, Finland).

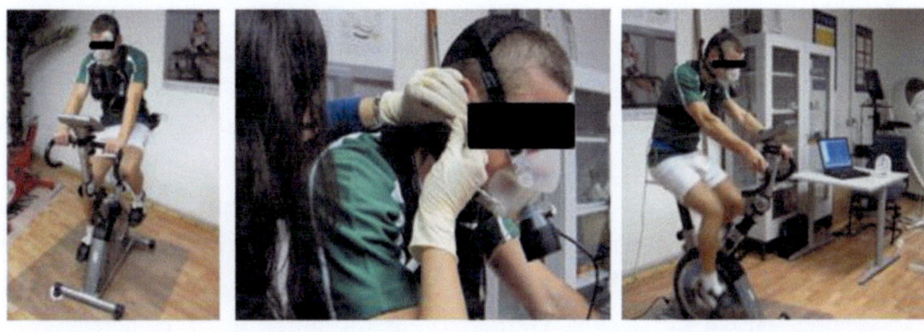

Figure 4:   Male participant during spiroergometric and blood lactate testing on a cycle ergometer

Further, all participants performed an isometric and a dynamic lower limb strength test after a five minutes warm-up phase with a speed of 5 km/h on a treadmill. For the isometric condition, the participants were positioned on an instrumented leg press equipped with two separated force plates (self-construction, BioMotion Center, Karlsruhe Institute of Technology, 1000 Hz). Each foot was placed on one plate with a 120° knee angle as measured with a knee goniometer. Subjects were asked to press with maximum strength and with both legs for the duration of five seconds. This test was performed three times with a resting time of one minute between each trial.

To assess dynamic lower limb strength, the subjects performed three counter-movement-jumps (CMJ) with arms akimbo on two force platforms embedded in the laboratory's floor (self-construction, BioMotion Center, Karlsruhe Institute of Technology, 1000 Hz). They started from an upright standing position and were instructed to jump as high as possible. The correct movement sequence (starting in an upright position, then flexion of knees and hips to initiate moving downwards and finally immediate extension of knees to vertically jump up off the ground with both feet simultaneously) was demonstrated by the test supervisor. A rest period of one minute was set between the jumps. The relevant parameters for the isometric lower limb strength are the maximum force (F_{max}, in N) and for the CMJ the maximum jump height (h, in m) as calculated by the impulse. All data of the strength tests were analyzed using the software Templo® (Contemplas GmbH, Germany) regardless of leg laterality. The best out of three trials in isometric and dynamic testing was chosen for further analysis. The test setup is shown in figure 5.

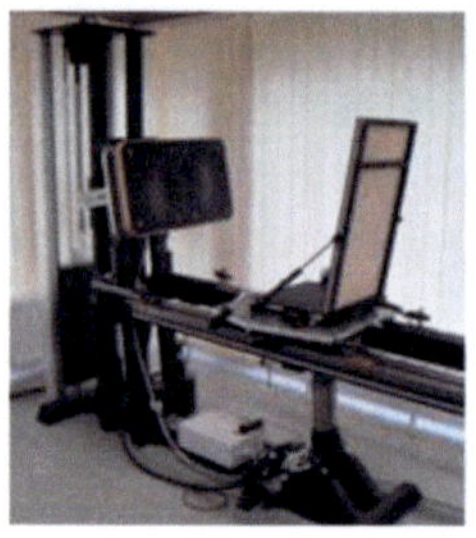

Figure 5:   Female participants during isometric (leg press) and dynamic (counter-movement-jump) lower limb strength testing

To assess coordinative skills, all participants performed six static and afterwards six dynamic one-legged stabilization (balance) tasks on the Posturomed® 202 (Haider Bioswing GmbH, Germany). Within each condition, they were asked to perform three trials per leg alternately. The Posturomed is a station which comprises an unstable and swinging quadratic platform that shifts freely in all directions of the transversal plane. During the trials, the subjects had their eyes opened and wore shoes that were standardized for all participants. In the static condition, the subjects' task was to keep the platform in balance. In the dynamic condition, a perturbation impulse was unexpectedly applied by eliciting a fixed provocation unit and subjects were instructed to re-stabilize platform movements as quickly as possible. The movement of the platform was recorded by the software Microswing 5 (Haider Bioswing GmbH, Germany) for the duration of 12 seconds. Performance scores between 0 and 1,000 points per trial (the faster the re-stabilization the higher the score) could be reached. For data analysis, the scores of the six static and the six dynamic trials were averaged to one mean score for static and dynamic condition, respectively.

Participants also performed two coordination tests from a standardized physical fitness test battery (Deutscher Motorik-Test 6-18/ German Motor Performance Test 6-18; Bös et al., 2009). The first gross motor test performed in the study is called "backward balancing" and assesses coordination under precision constraint. Participants were instructed to balance backwards on a 6, 4.5 and 3 cm wide and 3 m long beam without touching the ground with their feet. The subjects had two trials per beam with the numbers of steps balanced on each beam added. The maximal number of steps per beam was set at 8; the maximal score for the complete test is thus 48. To determine coordination under time pressure, the "jumping side-to-side" test was performed. The participants' aim was to jump sideways as quickly as possible during two 15 second intervals with a resting period of one minute between. The jumps had to be performed in a square affixed on the ground that was divided by a middle line. Participants were not allowed to touch the middle line during jumping. The relevant parameter for this test is the average number of jumps in 15 seconds. The setups of the coordinative skills tests are given in figure 6.

Figure 6:   Female participant during coordinative performance testing

To determine BMI (body mass index; kg/m^2) and waist to hip ratio, participants had their height (cm), waist (cm) and hip size (cm) measured using a calibrated stadiometer and a plastic tape measure. For this measurement, they were barefooted and wore form fitting underwear or swimsuits. Afterwards, weight and body fat measurements were conducted via whole body densitometry using a BOD POD Gold Standard® System (Model 2007; Life Measurement Inc., USA; now: COSMED, Italy).

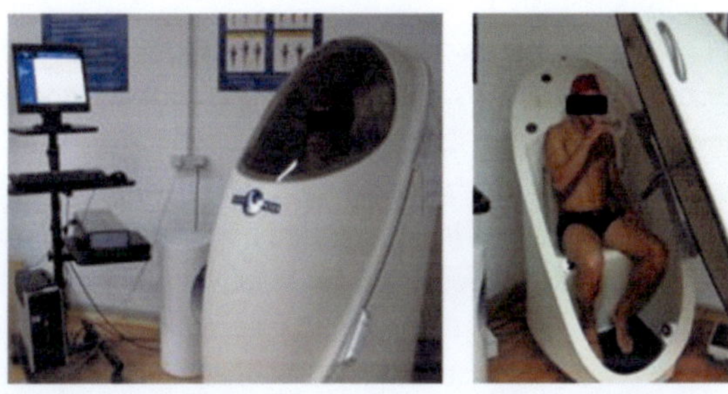

Figure 7:   Male participant in the BOD POD

The BOD POD allows for a non-invasive and quick estimation of fat (% and kg) and fat free mass (% and kg) using whole body densitometric principles. Thus, the measurement is based on the two compartment model of body composition. The system assesses body mass by using an electronic scale as well as body volume based on the principle of air-displacement plethysmography. For the measurement of body volume, the participant enters the test chamber of the BOD POD and is instructed to relax, breathe normally and limit any movement while sitting (figure 7). Two body volume measurements with durations of approximately 50 seconds are conducted followed by a thoracic gas volume measurement. Based on body mass (M_B) and volume (V_B), body density (D_B) can be calculated: $D_B=M_B/V_B$. Since fat and fat free mass have different densities, the relative amounts of a participant's fat and fat free mass can be estimated using equations (Dempster & Aitkens, 1995). The BOD POD has been shown to be a reliable and valid technique for estimating body fat and fat free mass in adults if compared with reference devices such as hydrostatic weighing or dual x-ray absorptiometry. However, it slightly (2-3 %) underestimates body fat compared with multicompartment models (Fields, Goran & McCrory, 2002).

3.4.3 Cognition

In the Cognitive Systems Lab at KIT, participants performed a cognitive testing battery which included three standardized tasks: an auditory Oddball paradigm, a visual modified Flanker task and a visual Switching paradigm. Between the tasks, they had a rest period of at least 2 minutes. The tests took place in a separate room with a low acoustic level. Participants were seated 0.5 m in front of a table. On the table, the monitor screen as well as the keyboard was fixed. The keyboard was adapted to suit testing. Thus, only task relevant buttons (space bar, left and right arrow) were visible to the participants. To minimize the number of signal artifacts, participants were instructed to avoid any body movements except of finger movements to press task relevant keyboard buttons during the tests and also to minimize head movements.

In this chapter, the three tests as well as the recording setup and technique are described. The description is based on the diploma thesis by Jeremias Engelmann (2012, chapter 3 and 4) who was a student helper of the study at the CSL and analyzed the data recorded during the cognitive tests.

The Oddball task is commonly used to investigate selective attention, direction of attention and information processing capacity (Karch & Mulert, 2010, p. 420). For the CogniFit-study, the setup of a three-stimulus Oddball task was as follows: Participants were sitting in front of a computer with a blank screen wearing headphones. They listened to a sequence of tones, consisting of three different frequencies: a high tone (2 KHz), a middle tone (1 KHz) and low tone (200 Hz). These tones were played with probabilities of 12.5 %, 75 %, and 12.5 % respectively and were equally distributed across 200 trials. After the first 100 trials, there was a short resting period. Each tone lasted for 200 ms with constant volume. After the presentation of each tone, there was a response window of two seconds. Only the high (= target) tone had to be attended by pressing a button on the keyboard with the right index finger whereas the two other irrelevant (= distractor) tones had to be ignored. After giving a response window of two seconds, the next tone was immediately played. Before the testing session started, participants were familiarized with the different tones and performed a training session consisting of 20 trials. Relevant parameters for statistical analysis of the study's research questions were response times, response accuracies and event-related brain potentials (P300 for target stimulus = P3b, N100, P200).

The Eriksen Flanker paradigm (Eriksen & Eriksen, 1974) is another prominent test to investigate attentional control processes, especially spatial selective attention. In the present study, participants were sitting 0.5 m in front of the monitor screen. In the middle of the black screen, fixation crosses were first presented to the participant followed by lines consisting of five white arrows with the same height of 4 cm. The participant's task was to ignore the two left and two right arrows (= distractor stimuli) and to only focus on the middle arrow (= target stimulus). According to the direction in which the middle arrows pointed, a left or a right button on the computer keyboard had to be pressed. There were two different conditions: The nearby flankers were identical to the target stimulus (congruent: < < < < <, > > > > >) or the nearby flankers differed from the target stimulus and pointed in the opposite direction (incongruent: < < > < <, > > < > >). The test comprised two blocks, each of it with a randomized sequence of 48 congruent and 48 incongruent trials (total: 96 trials per block). During the first block, participants were instructed to respond as precisely as possible and in the second block as quickly as possible. The appearance probabilities for congruent and incongruent trials were equally distributed (50 % to 50 %) and all trials were presented with a time lag of 2 seconds and for the duration of 80 ms. Between both blocks, participants were given a two minute resting period. Prior to the beginning of the test, participants performed a sequence of 20 trials to get familiarized with the test procedure. Relevant parameters for statistical analysis of the study's research questions were response times, response accuracies and event-related brain potentials (P300, N100, P200). It was shown in several research studies that response times are faster and response accuracies are higher for congruent compared to incongruent trials (e.g. Davelaar & Stevens, 2009, p. 121). During incongruent trials, stimulus processing is hampered and attentional control processes are necessary to inhibit incorrect responses. This response conflict results in slower reaction times and is termed 'Flanker effect'.

The Switching task allows for the analysis of cognitive flexibility in consideration of working memory and inhibition. Participants were sitting 0.5 m in front of the monitor screen. In the middle of the monitor screen, 4 cm high numbers were presented for the duration of 200 ms. The numbers were surrounded by either a dashed or a continuous line marking a quadrate. In case of a continuous line, the participants had to decide whether the shown number was greater or less than 5. In case of a dashed line, they had to decide whether the number was even or odd. Participants performed three blocks of trials. In the first block (Switching task I), 128 trials surrounded with a continuous line and in the second block (Switching task II), 128 trials surrounded with a dashed line were presented. Thus, block I and II consisted of homogenous trials. In the third block (Switching task III), both conditions were randomly mixed to 256 heterogeneous trials. The numbers 1 to 4 and 6 to 9 were equally distributed across all trials. An example of the Switching task is given in figure 8.

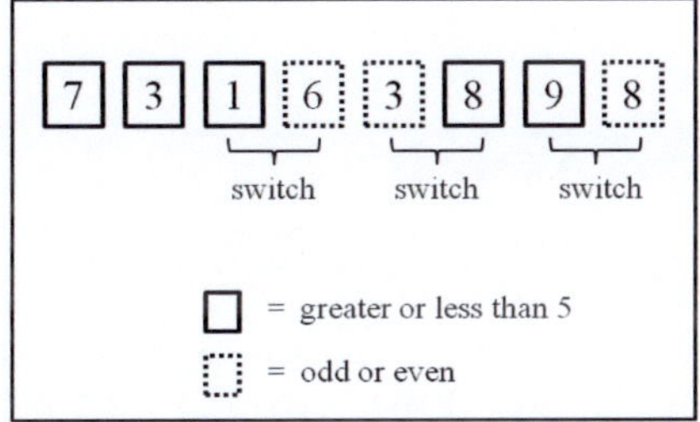

Figure 8:   Example of Switching task III comprising switch and non-switch trials

Before beginning the test, participant performed 20 trials to get familiarized. Relevant parameters for statistical analysis were response times, response accuracies and event-related brain potentials (P300, P200).

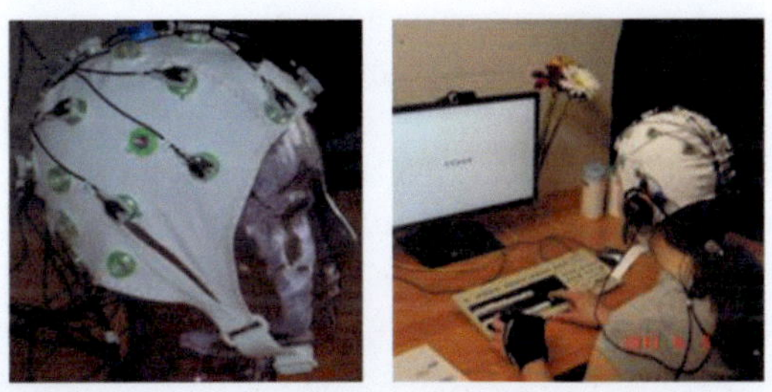

Figure 9:   Female participant during cognitive testing on the computer

For all three cognitive tasks, EEG was continuously recorded at 16 different electrode sites (FP1, FP2, F3, F4, F7, F8, T3, T4, C3, Cz, C4, P3, Pz, P4, O1, O2) with reference to the left mastoid electrode and according to the international 10-20 system. For this purpose, an active EEG-cap with Ag/AgCI sensors (BrainProducts actiCap®, Germany) as well as a self-developed recording software BiosignalStudio (Cognitive Systems Lab, KIT, Germany) were used. Amplification and analog-digital conversion were performed using a 16 channel VarioPort biosignals recording system (Becker Meditec, Germany) with a sampling rate of 256 Hz. Electrode impedances were kept below 20 kΩ. Raw EEG signals were filtered using an offline 40 Hz low-pass filter. Data were analyzed using the Matlab® (The MathWorks, Inc.) and the EEGLAB® software (http://sccn.ucsd.edu/eeglab/). Eye movements and artifacts caused by muscle activity were corrected and rejected based on a detector for extreme values with a rejection level of ±45 µV.

Peak amplitudes and latencies were determined for all participants and across all trials by averaging stimulus-devoted EEG signals which appeared 200 ms before and until 1000 ms after stimulus onset. Averages of one person were rejected if the number of correct trials with no artifacts was less than half the maximum number of trials. Peak latencies were discarded when absolute peak height was not higher than two times the standard deviation of the period 200 ms before the stimulus. A peak was generally defined as the highest amplitude within the relevant time window. The N100 component was defined as the largest negative-going peak between 50 ms and 200 ms after stimulus. The P200 and P300 components were defined as the largest positive-going peaks between 100 ms and 275 ms, and 275 ms and 700 ms, respectively. Amplitude was measured as a change from the pre-stimulus baseline and peak latency was defined as the time point of the peak amplitude.

3.5   Data Acquisition and Analysis

Statistical analysis was performed using SPSS Statistics 19.0. Charts or graphs were created using Microsoft Office Excel 2007. Analyses were conducted separately for behavioral measures (response times, response accuracy) and ERPs (P300, N100 and P200 amplitudes and latencies). For statistical analysis of the Flanker and Oddball task, P300 amplitudes and latencies were taken from the midline electrode sites Fz, Cz and Pz, whereas analyses on the Switching task were conducted to Cz, Oz and Fp2 electrodes. Analyses on N100 and P200 amplitudes and latencies were conducted to Cz electrode site. P300 component for the Oddball task was taken from the target stimulus. Therefore P300 amplitude is equivalent to P3b. P3a was not analyzed in this study.

To investigate general trends on physical activity and fitness (chapter 4.1), univariate ANOVAs (analysis of variance) were conducted with gender and general sports participation as between-subject factors. In addition, t-tests for independent samples were conducted. For general trends on cognitive performance (chapter 4.2), 2 (condition) x 2 (trials) repeated measures ANOVAs with between-subject factor gender were performed for behavioral measures and 2 (condition) x 2 (trials) x 3 (electrode) repeated measures ANOVAs with between-subject factor gender were conducted for ERPs. Repeated measures ANOVAs were further performed to determine differences in measures between conditions, type of trials or electrode sites.

To assess the relationship between physical activity (chapter 4.3) and fitness measures (chapter 4.4) and cognitive performance, 2 (condition) x 2 (trials) (x 3 (electrode)) repeated measures ANOVAs (Greenhouse Geisser-correction) with between-subject factors (e.g. athletes vs. nonathletes to determine differences in group means) were performed. For post-hoc testing, Bonferroni corrected t-tests were used. Additional correlations were conducted using Pearson's correlation coefficient. All analyses on physical activity and fitness relationships to cognition were performed separately for males and females and for the total sample, too. In general, fitness variables were divided into two groups to detect mean differences in cognitive performance. A division in three groups was not possible in most cases due to the limited sample size. Especially for gender-specific analyses, there would have been too less participants in some groups. A description of the groups used for analysis is provided in the beginning of each subchapter.

The level of significance was set at $p < .05$ for all analyses. If not mentioned otherwise, mean values and standard deviations (SD) or confidence intervals (CI) are given. In figures, error bars represent 95% confidence intervals or standard deviations. Since there are missing data for several cognitive measures, the number of included participants varies between the different analyses. The following tables 12 and 13 provide an overview of the dependent and independent variables used for statistical analysis in this thesis.

Table 12:   Overview of all dependent variables used for analysis (Cognition Tasks)

Task	Type of Measure	Conditions	Trials	Electrodes	Number of variables
Flanker	Response time [ms]	Accuracy, Speed	Congruent, incongruent	-	4
	P300 amplitude [µV]	Accuracy, Speed	Congruent, incongruent	Fz, Cz, Pz	12
	P300 latency [ms]	Accuracy, Speed	Congruent, incongruent	Fz, Cz, Pz	12
	N100 amplitude [µV]	Accuracy, Speed	Congruent, incongruent	Cz	4
	N100 latency [ms]	Accuracy, Speed	Congruent, incongruent	Cz	4
	P200 amplitude [µV]	Accuracy, Speed	Congruent, incongruent	Cz	4
	P200 latency [ms]	Accuracy, Speed	Congruent, incongruent	Cz	4
Switching	Response time [ms]	Homogenous, Heterogeneous	</> 5, odd/even	-	4
	Errors [no.]	Homogenous, Heterogeneous	</> 5, odd/even	-	4
	P300 amplitude [µV]	Homogenous, Heterogeneous	</> 5, odd/even	Cz, Oz, Fp2	12
	P300 latency [ms]	Homogenous, Heterogeneous	</> 5, odd/even	Cz, Oz, Fp2	12
	P200 amplitude [µV]	Homogenous, Heterogeneous	</> 5, odd/even	Cz	4
	P200 latency [ms]	Homogenous, Heterogeneous	</> 5, odd/even	Cz	4
Oddball	Response time [ms]	-	-	-	1
	P300 amplitude [µV]	-	-	Fz, Cz, Pz	3
	P300 latency [ms]	-	-	Fz, Cz, Pz	3
	N100 amplitude [µV]	-	-	Cz	1

N100 latency [ms]	-	-	Cz	1
P200 amplitude [µV]	-	-	Cz	1
P200 latency [ms]	-	-	Cz	1

Table 13:   Overview of all independent variables used for analysis (Physical activity and fitness testing)

Test		Measures
Physical activity	Questionnaire	- General sports participation (athletes vs. nonathletes) - Participation in competitive sports - Type of preferred sport - Total amount of sports activity and habitual physical activity (kcal/week) - Exercise intensity - Frequency of exercise sessions per week - Exercise duration per session
Endurance	Graded cycle ergometer test	- VO_{2max} (ml/kg/min) - Performance at individual anaerobic threshold (Watts)
Strength	Leg press Counter-movement-jump	- Max. isometric strength (N) - Max. dynamic strength (jump height in m)
Coordination	Posturomed Balancing backwards Jumping side-to-side	- Static performance (score from 0 to 1,000) - Dynamic performance (score from 0 to 1,000) - Number of steps - Number of jumps
Body composition	Anthropometry BOD POD®	- BMI (kg/m^2) - Body fat (%)

Given the large number of variables included in analyses, not all findings from the repeated measures ANOVAs can be presented and discussed in this thesis. Only main effects and two-way interactions (e.g. condition x general sports participation) that involve physical activity or fitness measures are presented in the results section. The main effects are given for the total sample as well as separated for males and females subsamples, whereas two-way interactions are reported only for the total sample. Main effects or interactions solely for within-subject factors (condition, trials, and electrode) or three-way interactions are not reported. All results are presented in tables and figures.

Chapter 4 _______________

Results and Discussion

In this chapter, the results with regard to general trends on physical activity, fitness and cognitive performance in the study sample and the relationships between physical activity aspects and cognition on the one hand and fitness measures and cognition on the other hand are described. At the end of each subchapter, a discussion of the results is provided.

4.1 Physical Activity and Fitness in the Study Sample

4.1.1 General Trends

Since all tested participants were in a limited age range between 18 and 34, age does not have a significant effect on physical activity or fitness measures in this study. Therefore, only the effects of gender were tested.

As table 14 shows, 2 (gender) x 2 (general sports participation) between-subject factorial ANOVA revealed no significant effect on BMI. According to the WHO classification of BMI, the mean BMI for males and females in this study was in a normal range (18.5 - 25.0). For body fat percentage, there was a main effect of gender (F [1, 144] = 116.33, p = .00, η^2 = .447) and general sports participation (F [1, 144] = 8.22, p < .01, η^2 = .054) indicating larger body fat percentages for females compared to males and for nonathletes compared to athletes. There was also an effect of general sports participation on intelligence quotient (F [1, 141] = 6.68, p = .01, η^2 = .045) with higher quotients for nonathletes than for athletes.

2 (gender) x 2 (general sports participation) between-subject factorial ANOVA showed significant effects of gender (F [1, 137] = 41.63, p = .00, η^2 = .233) and general sports participation (F [1, 137] = 14.51, p = .00, η^2 = .096) on VO_{2max} with larger maximal oxygen uptake for males compared to females and for athletes compared to nonathletes (table 15). Further, there were significant effects of gender (F [1, 144] = 37.77, p = .00, η^2 = .208) and general sports participation (F [1, 144] = 14.76, p = .00, η^2 = .093) on the individual anaerobic threshold indicating larger performances for males than females and athletes than nonathletes (table 15).
Regarding maximal isometric strength, there was only a gender effect (F [1, 134] = 42.31, p = .00, η^2 = .240) with a larger isometric strength for males than females. Analyses also revealed a gender effect for maximal dynamic strength (F [1, 142] = 138.51, p = .00, η^2 = .494) indicating a larger performance for males compared to females and a significant gender x general sports participation interaction (F [1, 142] = 5.01, p = .03, η^2 = .034) indicating a larger dynamic strength for female

athletes compared to female nonathletes. No difference was observed between male athletes and male nonathletes. For the jumping side-to-side task, there was a general sports participation effect (F [1, 142] = 10.77, p < .01, η^2 = .070) with better coordination under time pressure for athletes than for nonathletes. For both Posturomed measures as well as the balancing task, no significant effects could be detected. Further, there were no effects of gender or general sports participation on habitual physical activity (kcal/week). Also, no gender effect could be observed for sports activity (kcal/week) but males tend to have a larger amount of sports activity compared to females (table 15).

Table 14: Effects of gender and general sports participation on body composition and intelligence (M ± SD)

		Athletes n	Athletes M ± SD	Nonathletes n	Nonathletes M ± SD	Total n	Total M ± SD	t	p
Age [years]	♂	90	23.9 ± 3.2	17	23.6 ± 2.5	107	23.8 ± 3.1	-.37	.71
	♀	29	22.8 ± 2.6	16	22.6 ± 2.6	45	22.7 ± 2.6	-.17	.87
	Total	119	23.6 ± 3.1	33	23.1 ± 2.6	152	23.5 ± 3.0	-.84	.40
BMI [kg/m²]	♂	88	22.8 ± 1.8	16	22.8 ± 3.0	104	22.8 ± 2.0	-.09	.93
	♀	29	21.9 ± 2.3	15	22.6 ± 2.7	44	22.2 ± 2.4	.93	.36
	Total	117	22.6 ± 2.0	31	22.7 ± 2.8	148	22.6 ± 2.2	.18	.86
Body Fat [%]	♂	88	14.8 ± 5.2	16	17.1 ± 6.6	104	15.2 ± 5.5	1.57	.12
	♀	29	26.7 ± 6.3	15	31.3 ± 7.0	44	28.3 ± 6.8	2.22	.03
	Total	117	17.7 ± 7.5	31	24.0 ± 9.8	148	19.0 ± 8.4	3.29	<.01
IQ	♂	86	118.0 ± 12.1	16	121.6 ± 12.2	102	118.5 ± 12.1	1.12	.27
	♀	28	115.7 ± 15.2	15	125.9 ± 13.1	43	119.3 ± 15.1	2.21	.03
	Total	114	117.4 ± 12.9	31	123.7 ± 12.6	145	118.7 ± 13.0	2.44	.02

Note: Body Fat % = measured with BOD POD Gold Standard, IQ = intelligence quotient; measured with CFT 20-R (German version of Culture Fair Intelligence Test); independent samples t-test: difference between athletes and nonathletes

Table 15: Effects of gender and general sports participation on fitness and physical activity measures (M ± SD)

		Athletes n	Athletes M ± SD	Nonathletes n	Nonathletes M ± SD	Total n	Total M ± SD	t	p
VO$_{2max}$ [ml/kg/min]	♂	84	51.8 ± 9.1	15	47.0 ± 10.3	99	51.0 ± 9.4	-1.84	.07
	♀	28	42.0 ± 8.3	14	32.4 ± 5.3	42	38.8 ± 8.7	-3.93	.00
	Total	112	49.3 ± 9.8	29	39.9 ± 11.0	141	47.5 ± 10.6	-4.48	.00
IAT [W]	♂	88	189.7 ± 43.5	16	147.4 ± 29.4	104	183.2 ± 44.3	-3.73	.00
	♀	29	128.9 ± 26.7	15	109.8 ± 30.5	44	122.4 ± 29.2	-2.15	.04
	Total	117	174.7 ± 47.8	31	129.2 ± 35.1	148	165.1 ± 49.0	-4.95	.00
MIS [N]	♂	83	4313 ± 1036	16	4112 ± 930	99	4280 ± 1018	-.72	.47
	♀	24	3044 ± 747	15	2738 ± 711	39	2926 ± 739	-1.27	.21
	Total	107	4028 ± 1111	31	3447 ± 1075	138	3898 ± 1126	-2.58	.01
MDS [m]	♂	87	.36 ± .06	16	.37 ± .06	103	.37 ± .06	.66	.51
	♀	28	.25 ± .06	15	.20 ± .03	43	.23 ± .05	-2.74	<.01
	Total	115	.34 ± .08	31	.29 ± .10	146	.33 ± .09	-2.28	.03

Stat. Postur.	♂	74	412 ± 280	13	328 ± 244	87	400 ± 275	-1.03	.31
	♀	26	465 ± 289	15	394 ± 308	41	439 ± 294	-.74	.46
	Total	100	426 ± 282	28	363 ± 277	128	412 ± 281	-1.05	.30
Dyn. Postur.	♂	74	317 ± 222	13	242 ± 230	87	305 ± 224	-1.11	.27
	♀	26	275 ± 219	15	267 ± 195	41	272 ± 208	-.12	.91
	Total	100	306 ± 221	28	256 ± 209	128	295 ± 219	-1.08	.28
Bal. [no. of steps]	♂	87	41.7 ± 7.0	16	40.0 ± 8.1	103	41.5 ± 7.1	-.90	.37
	♀	28	44.2 ± 4.2	15	40.9 ± 7.8	43	43.1 ± 5.8	-1.78	.08
	Total	115	42.3 ± 6.5	31	40.5 ± 7.8	146	42.1 ± 6.5	-1.37	.17
Jump [no. of jumps]	♂	87	44.6 ± 5.9	16	41.3 ± 8.2	103	44.1 ± 6.4	-1.94	.06
	♀	28	43.9 ± 5.3	15	39.0 ± 4.1	43	42.2 ± 5.4	-3.09	<.01
	Total	115	44.4 ± 5.8	31	40.2 ± 6.5	146	43.5 ± 6.2	-3.53	<.01
SA [kcal/week]	♂	88	2997 ± 1812	16	0 ± 0	104	2536 ± 1988		
	♀	29	2429 ± 1261	15	0 ± 0	44	1601 ± 1546		
	Total	117	2856 ± 1705	31	0 ± 0	148	2258 ± 1911		
HA [kcal/week]	♂	88	646 ± 472	16	498 ± 356	104	623 ± 457	-1.19	.24
	♀	29	662 ± 417	15	603 ± 300	44	642 ± 378	-.49	.63
	Total	117	650 ± 457	31	549 ± 329	148	629 ± 434	-1.15	.25

Note: IAT = individual anaerobic threshold, MIS = maximal isometric strength; MDS = maximal dynamic strength; Bal = balancing backwards; Jump = jumping side-to-side; SA = sports activity; HA = habitual activity; independent samples t-test: difference between athletes and nonathletes

Tables 16 and 17 show the results on body composition, intelligence, fitness and physical activity separated for athletes participating in competitive sports and those who are not participating. 2 (gender) x 2 (competitive sports participation) between-subject factorial ANOVA revealed no significant effect on BMI. For body fat percentage, there was a main effect of gender (F [1, 113] = 94.43, p = .00, η^2 = .455) and participation in competitive sports (F [1, 113] = 6.18, p = .01, η^2 = .052) indicating larger body fat percentages for females compared to males and for athletes not participating in competitive sports compared to those who participate. No effect could be found on intelligence. There was further a significant main effect of gender (F [1, 108] = 23.12, p = .00, η^2 = .176) and participation in competitive sports (F [1, 108] = 6.02, p = .02, η^2 = .053) on VO_{2max} with larger maximal oxygen uptake for males and for athletes participating in competitive sports. For the individual anaerobic threshold, there was a gender effect (F [1, 113] = 47.71, p = .00, η^2 = .297) with a larger performance for males and a marginal effect of competitive sports participation (F [1, 113] = 3.68, p = .06, η^2 = .032). Regarding maximal isometric strength, a gender effect was observed (F [1, 103] = 32.22, p = .00, η^2 = .238) with a larger isometric strength for males than females as well as a gender x competitive sports interaction (F [1, 103] = 4.23, p = .04, η^2 = .039). The interaction results from the fact that male participants in competitive sports revealed a larger maximal isometric strength than those who do not participate, whereas female athletes who do not participate in competitive sports had a larger isometric strength than participating ones. Analyses also revealed a gender effect for maximal dynamic strength (F [1, 111] = 58.98, p = .00, η^2 = .347) indicating a larger performance for males compared to females. For the static Posturomed performance, there was an effect of competitive sports participation (F [1, 96] = 3.84, p = .05, η^2 = .038) indicating that athletes participating in competitive sports had a higher performance than non-competitive athletes. In addition, there was a significant effect of participation in competitive sports on sports activity (kcal/week; F [1, 113] = 6.83, p = .01, η^2 = .057) and habitual physical activity (kcal/week; F [1, 113] = 4.09, p = .05, η^2 = .035) with a larger amount of sports and habitual physical activities for competitors. No significant effects could be found for the dynamic Posturomed performance, the balancing backward and jumping side-to-side task.

Table 16: Effects of gender and participation in competitive sports on body composition and intelligence (M ± SD)

		Competitive sports		No competitive sports		Total			
		n	M ± SD	n	M ± SD	n	M ± SD	t	p
Age [years]	♂	42	24.3 ± 3.6	48	23.5 ± 2.8	90	23.9 ± 3.2	-1.09	.28
	♀	9	22.4 ± 3.2	20	22.9 ± 2.3	29	22.8 ± 2.6	.43	.67
	Total	51	24.0 ± 3.6	68	23.4 ± 2.6	119	23.6 ± 3.1	-1.03	.31
BMI [kg/m^2]	♂	41	22.9 ± 2.1	47	22.7 ± 1.6	88	22.8 ± 1.8	-.49	.62
	♀	9	22.0 ± 2.0	20	21.9 ± 2.4	29	21.9 ± 2.3	-.17	.87
	Total	50	22.8 ± 2.1	67	22.5 ± 1.9	117	22.6 ± 2.0	-.78	.44
Body Fat [%]	♂	41	12.8 ± 4.7	47	16.6 ± 5.0	88	14.8 ± 5.2	3.64	.00
	♀	9	25.2 ± 8.9	20	27.4 ± 4.8	29	26.7 ± 6.3	.68	.51
	Total	50	15.0 ± 7.4	67	19.8 ± 7.0	117	17.7 ± 7.5	3.57	<.01
IQ	♂	39	116.3 ± 10.7	47	119.3 ± 13.0	86	118.0 ± 12.1	1.14	.26
	♀	8	117.1 ± 15.1	20	115.1 ± 15.6	28	115.7 ± 15.2	-.31	.76
	Total	47	116.5 ± 11.4	67	118.0 ± 13.8	114	117.4 ± 12.9	.67	.51

Note: Body Fat % = measured with BOD POD Gold Standard, IQ = intelligence quotient; measured with CFT 20-R (German version of Culture Fair Intelligence Test); independent samples t-test: difference between athletes participating and not participating in competitive sports

Table 17: Effects of gender and participation in competitive sports on fitness and physical activity measures (M ± SD)

		Competitive Sports		No competitive sports		Total			
		n	M ± SD	n	M ± SD	n	M ± SD	t	p
VO$_{2max}$ [ml/kg/min]	♂	39	55.1 ± 7.8	45	48.8 ± 9.2	84	51.8 ± 9.1	-3.34	<.01
	♀	9	44.2 ± 10.2	19	40.9 ± 7.3	28	42.0 ± 8.3	-.99	.33
	Total	48	53.1 ± 9.2	64	46.5 ± 9.3	112	49.3 ± 9.8	-3.70	.00
IAT [W]	♂	41	204.4 ± 39.8	47	177.0 ± 43.0	88	189.7 ± 43.5	-3.09	<.01
	♀	9	133.2 ± 21.0	20	127.0 ± 29.2	29	128.9 ± 26.7	-.57	.57
	Total	50	191.6 ± 46.2	67	162.1 ± 45.4	117	174.7 ± 47.8	-3.45	<.01
MIS [N]	♂	38	4619 ± 1168	45	4054 ± 840	83	4313 ± 1036	-2.48	.02
	♀	6	2703 ± 735	18	3157 ± 735	24	3044 ± 747	1.31	.20
	Total	44	4357 ± 1296	63	3798 ± 903	107	4028 ± 1111	-2.47	.02
MDS [m]	♂	39	.37 ± .06	48	.36 ± .07	87	.36 ± .06	-.63	.53
	♀	8	.27 ± .09	20	.24 ± .04	28	.25 ± .06	-1.02	.34
	Total	47	.35 ± .07	68	.32 ± .08	115	.34 ± .08	-1.87	.06

Stat. Postur.	♂	31	498 ± 294	43	351 ± 256	74	412 ± 280	-2.28	.03
	♀	8	546 ± 267	18	429 ± 298	26	465 ± 289	-.95	.35
	Total	39	507 ± 286	61	374 ± 269	100	426 ± 282	-2.36	.02
Dyn. Postur.	♂	31	401 ± 238	43	255 ± 191	74	317 ± 222	-2.93	<.01
	♀	8	305 ± 222	18	262 ± 223	26	275 ± 219	-.46	.65
	Total	39	382 ± 235	61	257 ± 199	100	306 ± 221	-2.84	<.01
Bal.	♂	39	43.2 ± 5.3	48	40.6 ± 7.9	87	41.7 ± 7.0	-1.84	.07
[no. of	♀	8	44.6 ± 4.1	20	44.0 ± 4.4	28	44.2 ± 4.2	-.35	.73
steps]	Total	47	43.4 ± 5.1	68	41.6 ± 7.2	115	42.3 ± 6.5	-1.52	.13
Jump	♂	39	46.9 ± 5.8	48	42.7 ± 5.4	87	44.6 ± 5.9	-3.45	<.01
[no. of	♀	8	44.1 ± 4.4	20	43.8 ± 5.7	28	43.9 ± 5.3	-.14	.89
jumps]	Total	47	46.4 ± 5.6	68	43.0 ± 5.5	115	44.4 ± 5.8	-3.21	<.01
SA	♂	41	3644 ± 2002	47	2432 ± 1424	88	2997 ± 1812	-3.31	<.01
[kcal/week]	♀	9	2917 ± 1447	20	2209 ± 1139	29	2429 ± 1261	-1.42	.17
	Total	50	3513 ± 1921	67	2365 ± 1340	117	2856 ± 1705	-3.81	.00
HA	♂	41	688 ± 550	47	609 ± 393	88	646 ± 472	-.78	.44
[kcal/week]	♀	9	896 ± 660	20	557 ± 187	29	662 ± 417	-1.51	.17
	Total	50	725 ± 570	67	593 ± 344	117	650 ± 457	-1.45	.15

Note: IAT = individual anaerobic threshold, MIS = maximal isometric strength; MDS = maximal dynamic strength; Bal = balancing backwards; Jump = jumping side-to-side; SA = sports activity; HA = habitual activity; independent samples t-test: difference between athletes participating and not participating in competitive sports

To test the differences between athletes participating in competitive sports and nonathletes, 2 (gender) x 2 (competitive sports vs. nonathletes) between-subject factorial ANOVA were conducted (tables 18 and 19). There was no effect on BMI, but significant gender (F [1, 77] = 71.95, p = .00, η^2 = .483) and competitive sports vs. nonathletes effects (F [1, 77] = 11.07, p < .01, η^2 = .126) on body fat with larger body fat percentages for females and nonathletes. For intelligence, a significant competitive sports vs. nonathletes effect was found (F [1, 74] = 4.97, p = .03, η^2 = .063) with higher intelligence quotients for nonathletes. For VO_{2max} and individual anaerobic threshold, significant gender (VO_{2max}: F [1, 73] = 34.45, p = .00, η^2 = .321; IAT: F [1, 77] = 36.98, p = .00, η^2 = .324) and competitive sports vs. nonathletes effects (VO_{2max}: F [1, 73] = 21.26, p = .00, η^2 = .226; IAT: F [1, 77] = 20.23, p = .00, η^2 = .208) were observed. They detect larger aerobic endurance performances for males compared to females and for competitors compared to nonathletes. For maximal isometric strength there was only a gender effect (F [1, 71] = 32.56, p = .00, η^2 = .314), whereas for dynamic strength, a gender (F [1, 74] = 82.13, p = .00, η^2 = .526), competitive sports vs. nonathletes effect (F [1, 74] = 4.12, p = .05, η^2 = .053) as well as a significant interaction (F [1, 74] = 6.03, p = .02, η^2 = .075) was found. This indicates that males had a higher isometric and dynamic maximal strength and that overall, competitors had a larger dynamic strength than nonathletes. However, the interaction shows that there was no difference between male competitors and male nonathletes but only a difference for females.

For the static Posturomed performance, there was a competitive sports vs. nonathletes effect (F [1, 63] = 4.23, p = .04, η^2 = .063), whereas no significant effect was found for the dynamic performance. For the balancing backwards and jumping side to side tasks, analyses revealed competitive sports vs. nonathletes effects (Balancing: F [1, 74] = 4.16, p = .05, η^2 = .053; Jumping: F [1, 74] =

11.49, p < .01, η^2 = .134) with better performances for competitors compared to nonathletes. No effect was found for habitual physical activity.

Table 18:   Effects of gender and participation in competitive sports vs. being a nonathlete on body composition and intelligence (M ± SD)

		Competitive Sports		Nonathletes		Total			
		n	M ± SD	n	M ± SD	n	M ± SD	t	p
Age [years]	♂	42	24.3 ± 3.6	17	23.6 ± 2.5	59	24.1 ± 3.3	.85	.40
	♀	9	22.4 ± 3.2	16	22.6 ± 2.6	25	22.6 ± 2.8	-.15	.88
	Total	51	24.0 ± 3.6	33	23.1 ± 2.6	84	23.6 ± 3.2	1.25	.22
BMI [kg/m²]	♂	41	22.9 ± 2.1	16	22.8 ± 3.0	57	22.9 ± 2.4	.24	.81
	♀	9	22.0 ± 2.0	15	22.6 ± 2.7	24	22.4 ± 2.4	-.58	.57
	Total	50	22.8 ± 2.1	31	22.7 ± 2.8	81	22.7 ± 2.4	.13	.90
Body Fat [%]	♂	41	12.8 ± 4.7	16	17.1 ± 6.6	57	14.0 ± 5.6	-2.40	.03
	♀	9	25.2 ± 8.9	15	31.3 ± 7.0	24	29.0 ± 8.2	-1.86	.08
	Total	50	15.0 ± 7.4	31	24.0 ± 9.8	81	18.4 ± 9.4	-4.38	.00
IQ	♂	39	116.3 ± 10.7	16	121.6 ± 12.2	55	117.9 ± 11.3	-1.60	.12
	♀	8	117.1 ± 15.1	15	125.9 ± 13.1	23	122.9 ± 14.1	-1.46	.16
	Total	47	116.5 ± 11.4	31	123.7 ± 12.6	78	119.4 ± 12.3	-2.63	.01

Note: Body Fat % = measured with BOD POD Gold Standard, IQ = intelligence quotient; measured with CFT 20-R (German version of Culture Fair Intelligence Test); independent samples t-test: difference between athletes participating in competitive sports and nonathletes

Table 19:   Effects of gender and participation in competitive sports vs. being a nonathlete on fitness and physical activity measures (M ± SD)

		Competitive Sports		Nonathletes		Total			
		n	M ± SD	n	M ± SD	n	M ± SD	t	p
VO$_{2max}$ [ml/kg/min]	♂	39	55.1 ± 7.8	15	47.0 ± 10.3	54	52.8 ± 9.2	-3.13	<.01
	♀	9	44.2 ± 10.2	14	32.4 ± 5.3	23	37.0 ± 9.5	-3.68	<.01
	Total	48	53.1 ± 9.2	29	39.9 ± 11.0	77	48.1 ± 11.8	-5.62	.00
IAT [W]	♂	41	204.4 ± 39.8	16	147.4 ± 29.4	57	188.4 ± 45.1	-5.19	.00
	♀	9	133.2 ± 21.0	15	109.8 ± 30.5	24	118.6 ± 29.2	-2.02	.06
	Total	50	191.6 ± 46.2	31	129.2 ± 35.1	81	167.7 ± 51.9	-6.45	.00
MIS [N]	♂	38	4619 ± 1168	16	4112 ± 930	54	4468 ± 1119	-1.54	.13
	♀	6	2703 ± 735	15	2738 ± 711	21	2728 ± 699	.10	.92
	Total	44	4357 ± 1296	31	3447 ± 1075	75	3981 ± 1284	-3.21	<.01
MDS [m]	♂	39	.37 ± .06	16	.37 ± .06	55	.37 ± .06	.38	.71
	♀	8	.27 ± .09	15	.20 ± .03	23	.23 ± .06	-2.13	.07
	Total	47	.35 ± .07	31	.29 ± .10	78	.33 ± .09	-2.92	<.01
Stat. Postur.	♂	31	498 ± 294	13	328 ± 244	44	447 ± 288	-1.83	.07
	♀	8	546 ± 267	15	394 ± 308	23	447 ± 297	-1.18	.25
	Total	39	507 ± 286	28	363 ± 277	67	447 ± 289	-2.07	.04
Dyn. Postur.	♂	31	401 ± 238	13	242 ± 230	44	354 ± 244	-2.05	.05
	♀	8	305 ± 222	15	267 ± 195	23	281 ± 201	-.42	.68

	Total	39	382 ± 235	28	256 ± 209	67	329 ± 231	-2.27	.03
Bal. [no. of steps]	♂	39	43.2 ± 5.3	16	40.0 ± 8.1	55	42.3 ± 6.3	-1.73	.09
	♀	8	44.6 ± 4.1	15	40.9 ± 7.8	23	42.2 ± 6.9	-1.24	.23
	Total	47	43.4 ± 5.1	31	40.5 ± 7.8	78	42.2 ± 6.4	-2.04	.05
Jump [no. of jumps]	♂	39	46.9 ± 5.8	16	41.3 ± 8.2	55	45.2 ± 7.0	-2.90	<.01
	♀	8	44.1 ± 4.4	15	39.0 ± 4.1	23	40.8 ± 4.8	-2.75	.01
	Total	47	46.4 ± 5.6	31	40.2 ± 6.5	78	43.9 ± 6.7	-4.49	.00
SA [kcal/week]	♂	41	3644 ± 2002	16	0 ± 0	57	2621 ± 2364		
	♀	9	2917 ± 1447	15	0 ± 0	24	1094 ± 1676		
	Total	50	3513 ± 1921	31	0 ± 0	81	2169 ± 2283		
HA [kcal/week]	♂	41	688 ± 550	16	498 ± 356	57	635 ± 508	-1.27	.21
	♀	9	896 ± 660	15	603 ± 300	24	713 ± 477	-1.50	.15
	Total	50	725 ± 570	31	549 ± 329	81	658 ± 497	-1.57	.12

Note: IAT = individual anaerobic threshold, MIS = maximal isometric strength; MDS = maximal dynamic strength; Bal = balancing backwards; Jump = jumping side-to-side; SA = sports activity; HA = habitual activity; independent samples t-test: difference between athletes participating in competitive sports and nonathletes

4.1.2 Summary and Discussion

The following tables 20, 21 and 22 summarize all effects that were found on the dependent variables. They further provide effect sizes which are given in the units of the measures and as percentage differences to reference groups.

Table 20:   Athletes vs. nonathletes - Summary of effects and mean difference (unit of the measures and %) to reference group on fitness and physical activity measures

Task	Gender effect[1]		Sports participation effect[2]		Interaction
	unit	%	unit	%	
BMI [kg/m^2]	+.6	+2.7	-.1	-.4	-
Body Fat [%]	**-13.1***	**-46.3***	**-6.3***	**-26.3***	-
IQ	-.8	-.7	**-6.3***	**-5.1***	-
VO$_{2max}$ [ml/kg/min]	**+12.2***	**+31.4***	**+9.4***	**+23.6***	-
IAT [W]	**+60.8***	**+49.7***	**+45.5***	**+35.2***	-
MIS [N]	**+1354***	**+46.3***	+581	+16.9	-
MDS [m]	**+.14***	**+60.9***	+.05	+17.2	*
Stat. Postur.	-39	-8.9	+63	+17.4	-
Dyn. Postur.	+33	+12.1	+50	+19.5	-
Bal. [no. of steps]	-1.6	-3.7	+1.8	+4.4	-
Jump. [no. of jumps]	+1.9	+4.5	**+4.2***	**+10.4***	-
SA [kcal/week]	+935	+58.4	/////////////	/////////////	
HA [kcal/week]	-19	-3.0	+101	+18.4	-

Note: Bold and * = significant effect (p <.05); * = significant interaction (p <.05); [1] = reference group are females, [2] = reference group are nonathletes

Table 21: Athletes participating in competitive sports vs. athletes not participating in competitive sports - Summary of effects and mean difference (unit of the measures and %) to reference group on fitness and physical activity measures

Task	Gender effect[1]		Competitive sports effect[2]		Interaction
	unit	%	unit	%	
BMI [kg/m^2]	+.9	+4.1	+.3	+1.3	-
Body Fat [%]	**-11.9***	**-44.6***	**-4.8***	**-24.2***	-
IQ	+2.3	+2.0	-1.5	-1.3	-
VO$_{2max}$ [ml/kg/min]	**+9.8***	**+23.3***	**+6.6***	**+14.2***	-
IAT [W]	**+60.8***	**+47.2***	+29.5	+18.2	-
MIS [N]	**+1269***	**+41.7***	+559	+14.7	*
MDS [m]	**+.11***	**+44.0***	+.03	+9.4	-
Stat. Postur.	-53	-11.4	**+133***	**+35.6***	-
Dyn. Postur.	+42	+15.3	+125	+48.6	-
Bal. [no. of steps]	-2.5	-5.7	+1.8	+4.3	-
Jump. [no. of jumps]	+.7	+1.6	+3.4	+7.9	-
SA [kcal/week]	+568	+23.4	**+1148***	**+48.5***	-
HA [kcal/week]	-16	-2.4	**+132***	**+22.3***	-

Note: Bold and * = significant effect (p <.05); * = significant interaction (p <.05); [1] = reference group are females, [2] = reference group are athletes not participating in competitive sports

Table 22: Athletes participating in competitive sports vs. nonathletes - Summary of effects and mean difference (unit of the measures and %) to reference group on fitness and physical activity measures

Task	Gender effect[1]		Competitive sports vs. nonathletes effect[2]		Interaction
	unit	%	unit	%	
BMI [kg/m^2]	+.5	+2.2	+.1	+.4	-
Body Fat [%]	**-15.0***	**-51.7***	**-9.0***	**-37.5***	-
IQ	-5.0	-4.1	**-7.2***	**-5.8***	-
VO$_{2max}$ [ml/kg/min]	**+15.8***	**+42.7***	**+13.2***	**+33.1***	-
IAT [W]	**+69.8***	**+58.9***	**+62.4***	**+48.3***	-
MIS [N]	**+1740***	**+63.8***	+910	+26.4	-
MDS [m]	**+.14***	**+60.9***	**+.06***	**+20.7***	*
Stat. Postur.	±0	±0	**+144***	**+39.7***	-
Dyn. Postur.	+73	+26.0	+126	+49.2	-
Bal. [no. of steps]	+.1	+.2	**+2.9***	**+7.2***	-
Jump. [no. of jumps]	+4.4	+10.8	**+6.2***	**+15.4***	-
SA [kcal/week]	+1527	+139.6	///////////	///////////	///////////
HA [kcal/week]	-78	-10.9	+176	+32.1	-

Note: Bold and * = significant effect (p <.05); * = significant interaction (p <.05); [1] = reference group are females, [2] = reference group are nonathletes

Data show that BMI was unaffected by gender and general sports as well as competitive sports participation. Since BMI equals a person's weight divided by height squared, a lower body fat mass for athletes is usually compensated by a larger muscle mass. This could explain why no differences between athletes and nonathletes and between competitive sports groups were found. BMI was marginally higher for males but females had a greater variation as indicated by a larger standard deviation. This finding is in line with previous results from a large study in young adults (Schousboe, Willemsen, Kyvik, Mortensen, Boomsma et al., 2003, p. 413).

There was a gender and general sports participation effect on body fat percentage. Both findings are not surprising since males are known to have a lower body fat percentage than females caused by a larger fat-free mass, total body bone mineral content and muscle mass (Malina, 2005, p. 297). In addition, as showed by a study among college students from the United States, young athletes also had a lower body fat percentage as measured by the BOD POD than nonathletes (Ode, Pivarnik, Reeves & Knous, 2007, p. 404 f.). A comparison between competitors and non-competitive athletes showed that competitive athletes had a significantly lower body fat percentage. This could be explained by a generally higher importance of sports, higher efforts for practicing and healthy nutrition for those athletes who regularly participate in competitions.

Interestingly, intelligence quotients were significantly lower for athletes (overall and competitive only) compared to nonathletes with the difference being greater by trend for females than compared to males. This finding confirms a study from 1964 which compared high-school athletes and nonathletes and also found higher quotients for nonathletes (Slusher, 1964). However, the longitudinal study by Aberg et al. (2009) revealed a relationship between cardiovascular fitness which is positively influenced by regular aerobic physical activity and intelligence in young men whereas Hillman, Motl et al. (2006, p. 681) found no associations between physical activity and intelligence. Therefore, it is uncertain why nonathletes in this study sample exhibited larger intelligence quotients than athletes. Yet, it must be noted that intelligence quotients in this homogenous sample of young and well-educated adults were on a very high level compared to the average of the German adult population. A recent study by Rindermann, Baumeister and Gröper (2013) among engineering students at a German university reported a mean IQ (as measured by the Cognitive Abilities Test) of 116.2 for the total sample (n = 30) and of 116.7 for males (n = 23) and 114.4 for females (n = 7), respectively. This result is somewhat lower than in the current sample but also emphasizes the above average intelligence quotients of university students compared to the assumed IQ-norm of 100.

Aerobic capacity as assessed by VO_{2max} and performance at individual anaerobic threshold was higher for males compared to females and for athletes (overall and competitive) compared to nonathletes. This result was expected since there is much evidence that athletes have a larger aerobic capacity compared to nonathletes resulting from the adaptation of the cardiovascular system to their regular exercise. The difference between females and males might be explained by the larger body fat carried by women and their generally lower hemoglobin concentration (Kenney et al., 2012, p. 477 f.). In addition, a significant difference in VO_{2max} but not performance at IAT was observed between competitive and non-competitive athletes. The obtained VO_{2max} values of this study sample are further within the ranges for characteristic gender- and athletic-specific values (see chapter 2.2.1).

It was further not surprising that males had a higher maximal isometric and dynamic strength than females which can be explained by the larger cross-sectional areas and mass of the skeletal muscles (Kenney et al., 2012, p. 474). No effects of general sports participation were detected except for isometric strength between female athletes and nonathletes. However, male athletes also tend to have larger isometric strength compared to male nonathletes. In addition, competitive athletes had a significantly higher dynamic strength than nonathletes.

For the Posturomed performance, there was a tendency for a better performance of females compared to males in the static condition and vice versa for the dynamic condition. Further, female and male athletes tended to reveal higher scores and thus better performances than nonathletes for both conditions. However, due to the considerably large standard deviations, no significant differences were found. For the static Posturomed test, significantly higher performances were found for competitive athletes compared to non-competitive athletes and nonathletes. For the dynamic condition, the same tendency was observed but was not significant due to high standard deviations.

In the balancing backwards task, females compared to males and athletes compared to nonathletes tended to have a larger number of steps, however these differences became significant only for the competitive athletes. A big problem of this test was the maximum score of 48 steps. 31 % of all participants reached this maximum value and 40 % of all participants had a score $\geq$ 45. Thus, the range in this measurement was very low within the study sample and did not allow for detecting any differences that could be resulting from gender or general sports participation.

In the jumping side-to-side test, there was a significant difference between athletes (overall and competitive) and nonathletes. This result indicates that general sports participation might be beneficial for coordination under time pressure and confirms a finding from the German Motoric Module-study in which highly active adolescents between the ages of 14 and 17 exhibited a significantly better performance in this test than inactive youth (Opper, Oberger, Worth, Woll & Bös, 2008, p. 69).

Regarding sports activity, a tendency with males being more active than females was observed that became not significant. Previous studies from Germany have also reported higher amounts of sports activity for males than females in young adulthood (Krug, Jordan, Mensink, Müters, Finger et al., 2013, p. 768; Rütten, Abu-Omar, Lampert & Ziese, 2005, p. 9). In general, the amount of sports activity was very high among this study sample. This might be due to the fact, that study participants were young University students who are well educated and know of the health importance of being physically active. In addition, young adults with a high amount of physical activity might be more likely to participate in a study like the current one which included several fitness tests and a physical activity questionnaire. Interestingly, significantly higher amounts of sports activity were found for athletes participating in competitions compared to non-competitive athletes.

Since the amount of sports activity was calculated based on frequency, duration of exercise and METs, it is important to look on gender differences in these variables. It can be observed that there is a similar distribution between females and males with most participants performing 2-3 sessions per week and a comparable number of participants performing less than 2 and 4 or more sessions (table 23). Regarding exercise duration, females tend to larger session durations, whereas males tend to shorter sessions (table 24). There is no significant gender effect on either frequency (p=.92) or duration (p=.18) of exercise sessions.

Table 23:   Number of participants in exercise frequency groups

	< 2 sessions n (%)	2 - 3 sessions n (%)	≥ 4 sessions n (%)	Total
♂	14 (15.5)	61 (67.8)	15 (16.7)	90
♀	2 (6.9)	25 (86.2)	2 (6.9)	29
Total	16 (13.4)	86 (72.3)	17 (14.3)	119

Table 24:   Number of participants in exercise duration/session groups

	≤ 1 hour/session n (%)	1.5 hours/session n (%)	≥ 2 hours/session n (%)	Total
♂	35 (38.9)	37 (41.1)	18 (20.0)	90
♀	9 (31.0)	10 (34.5)	10 (34.5)	29
Total	44 (37.0)	47 (39.5)	28 (23.5)	119

However, males had larger MET means compared to females (table 25; Summer: p = .00; Winter: p = .00; Other: p = .76), which in sum leads to a higher amount of sports activity by trend.

Table 25:   METs (Mean ± SD) for males and females

	MET Summer		MET Winter		MET Other	
	n	M ± SD	n	M ± SD	n	M ± SD
♂	88	8.46 ± 1.38	88	8.34 ± 1.30	54	7.87 ± 1.28
♀	29	7.38 ± 1.46	29	7.21 ± 1.70	22	7.77 ± 1.27
Total	117	8.19 ± 1.47	117	8.06 ± 1.49	76	7.84 ± 1.27

Note: MET Summer and Winter: preferred type of sport; MET other: mean MET for other types of sport mentioned in the physical activity questionnaire

Habitual activity was higher by trend for athletes (overall and competitive) compared to nonathletes which might be explained by a generally more active lifestyle accompanying general sports participation.

4.2 Cognitive Performance in the Study Sample

4.2.1 General Trends

In this subchapter, the effects of gender on cognitive performance in the study sample are analyzed. Age does not have an influence on cognitive measures in this study. Further, general trends within the cognitive tasks regarding differences between task conditions or trials will be described. Means that are not presented in tables 26-28 are given in the text.

4.2.1.1 Flanker Task

Behavioral Measures
There was a significant gender effect on response times (F [1, 130] = 6.74, p = .01, η^2 = .049) with shorter response times for males than for females. Further, significant effects for condition (F [1, 131] = 237.50, p = .00, η^2 = .645) and congruency (F [1, 131] = 1716.18, p = .00, η^2 = .929) with faster response times for speed than accuracy condition and faster times for congruent than incongruent trials could be observed. An analysis of response accuracy was not possible since too little errors were made by the participants in the Flanker task.
P300

There was no significant gender effect on P300 amplitudes (F [1, 122] = 1.20, p = .28, η^2 = .010) but a significant electrode x gender interaction (F [2, 121] = 7.02, p = <.01, η^2 = .054) with larger amplitudes for females (3.55 µV, 95 % CI 3.10 - 3.99) than males (3.40 µV, 95 % CI 3.13 - 3.67) at Fz electrode site and smaller amplitudes for females at Cz (5.06 µV, 95 % CI 4.46 - 5.66) and Pz (4.58 µV, 95 % CI 4.08 - 5.08) electrode sites compared to males (Cz: 5.64 µV, 95 % CI 5.28 - 6.01; Pz: 5.07 µV, 95 % CI 4.77 - 5.37). Moreover, significant effects were detected of condition (F [1, 123] = 40.32, p = .00, η^2 = .247), congruency (F [1, 123] = 16.71, p = .00, η^2 = .120) and electrode site (F [2, 122] = 244.02, p = .00, η^2 = .665) with larger amplitudes for speed than accuracy condition, larger amplitudes for congruent than incongruent trials and for Cz (5.49 µV, 95 % CI 5.18 - 5.80) and Pz (4.94 µV, 95 % CI 4.68 - 5.20) electrode sites compared to Fz (3.44 µV, 95 % CI 3.21 - 3.67). There were further significant condition x electrode (F [2, 122] = 9.88, p = .00, η^2 = .074), congruency x electrode (F [2, 122] = 22.06, p = .00, η^2 = .152) and condition x congruency x electrode interactions (F [2, 122] = 5.84, p = .00, η^2 = .045).

There was no significant gender effect on P300 latencies (F [1, 100] = .02, p = .88, η^2 = .000) but a significant electrode x gender interaction (F [2, 99] = 3.77, p = .04, η^2 = .036) with larger latencies for females (415.37 ms, 95 % CI 393.22 - 437.53) than males (400.48 ms, 95 % CI 386.85 - 414.11) at Fz electrode site and smaller amplitudes for females at Cz (387.12 ms, 95 % CI 370.04 - 404.21) and Pz (369.37 ms, 95 % CI 356.50 - 382.24) electrode sites compared to males (Cz: 392.64 ms, 95 % CI 382.13 - 403.15; Pz: 382.76 ms, 95 % CI 374.84 - 390.67). Analyses further revealed significant effects of congruency (F [1, 101] = 135.59, p = .00, η^2 = .573) and electrode site (F [2, 100] = 14.08, p = .00, η^2 = .122) with shorter latencies for congruent than incongruent trials and longer latencies for Fz (404.57 ms, 95 % CI 392.95 - 416.19) than for Cz (391.13 ms, 95 % CI 382.21 - 400.05) and Pz (379.08 ms, 95 % CI 372.27 - 385.89) electrodes. There was further a significant condition x electrode (F [2, 100] = 8.32, p = .00, η^2 = .076) and condition x congruency interaction (F [1, 101] = 7.35, p = .01, η^2 = .068).

N100

There was no significant gender effect on N100 amplitudes (F [1, 125] = 1.34, p = .25, η^2 = .011) and latencies (F [1, 44] = .03, p = .86, η^2 = .001) in the Flanker task. No interactions could be observed. Analyses further revealed no effects of condition or congruency on N100 amplitudes or latencies in the Flanker task.

P200

There was a significant gender effect on P200 amplitudes (F [1, 125] = 6.63, p = .01, η^2 = .050) with larger amplitudes for females than males. There was no gender effect on P200 latencies (F [1, 94] = .16, p = .69, η^2 = .002). Analyses revealed a significant effect of condition for amplitudes (F [1, 126] = 27.14, p = .00, η^2 = .177) with larger amplitudes for the speed than accuracy condition.

4.2.1.2 Switching Task

Behavioral Measures

There was no significant gender effect on response times (F [1, 126] = .00, p = .97, η^2 = .000), but significant effects for condition (F [1, 127] = 1588.08, p = .00, η^2 = .926) and type of trials (F [1, 127] = 10.17, p = <.01, η^2 = .074) with faster response times for homogenous than heterogeneous condition and faster times for odd/even than </> 5 trials. These effects were overlaid by a condition x type interaction (F [1, 127] = 132.85, p = .00, η^2 = .511) with faster response times for odd/even trials (498.49 ms, 95 % CI 486.69 - 510.30) compared to </> 5 ones (558.43 ms, 95 % CI 544.99 - 571.87) under the homogenous condition but longer times for odd/even trials (895.66 ms, 95 % CI 872.62 - 918.69) compared to </> 5 trials (858.64 ms, 95 % CI 836.55 - 880.73) under the heterogeneous condition. There was no gender effect on number of errors (F [1, 130] = .22, p = .64, η^2 = .002), but a significant effect of condition (F [1, 131] = 102.08, p = .00, η^2 = .438) with less errors for the homogenous compared to the heterogeneous condition, type of trials (F [1, 131] = 36.62, p = .00, η^2 = .218) with less errors for odd/even than </> 5 trials and condition x type interaction (F [1, 131] = 8.09, p = <.01, η^2 = .058).

P300

There was no gender effect on P300 amplitudes (F [1, 82] = 3.34, p = .07, η^2 = .039). Significant effects could be detected of condition (F [1, 83] = 14.02, p = .00, η^2 = .144) and electrode site (F [2, 82] = 8.34, p = .00, η^2 = .091) with larger amplitudes for homogenous than heterogeneous condition and larger amplitudes for Cz (4.40 µV, 95 % CI 4.10 - 4.70) compared to Oz (3.59 µV, 95 % CI 3.32 - 3.86) and Fp2 (3.78 µV, 95 % CI 3.38 - 4.17) electrode sites. These main effects were superseded by a significant condition x electrode interaction (F [2, 82] = 10.71, p = .00, η^2 = .114) and condition x type x electrode interaction (F [2, 82] = 5.82, p = <.01, η^2 = .066). There was no gender effect on P300 latencies (F [1, 49] = .01, p = .92, η^2 = .000). A significant effect was only detected of electrode site (F [2, 49] = 73.95, p = .00, η^2 = .597) with longer latencies for Cz (406.48 ms, 95 % CI 390.84 - 422.13) and Fp2 (464.85 ms, 95 % CI 434.23 - 495.46) compared to Oz (283.48 ms, 95 % CI 266.54 - 300.41) electrode site.

P200

There was no gender effect on P200 amplitudes (F [1, 85] = 2.22, p = .14, η^2 = .026) and latencies (F [1, 78] = .14, p = .71, η^2 = .002) in the Switching task. There was a significant condition x type interaction (F [1, 86] = 12.15, p = <.01, η^2 = .124) with larger amplitudes for </> 5 trials (4.34 µV, 95 % CI 3.93 - 4.75) than odd/even trials (3.92 µV, 95 % CI 3.56 - 4.28) during the homogenous condition and larger amplitudes for odd/even trials (4.32 µV, 95 % CI 3.92 - 4.71) than </> 5 trials (3.91 µV, 95 % CI 3.56 - 4.26) during the heterogeneous condition. There was further a significant condition effect on P200 latencies (F [1, 79] = 7.17, p = <.01, η^2 = .083) with faster latencies for the homogenous condition than heterogeneous condition.

4.2.1.3 Oddball Task

Behavioral Measures
There was no significant gender effect on response times in the Oddball task (F [1, 131] = 2.83, p = .10, η^2 = .021). Since too little errors were made by the participants in this task, it was not possible to analyze response accuracy.

P300
No gender effect on P300 amplitudes could be observed (F [1, 101] = 1.78, p = .19, η^2 = .017), but an electrode effect (F [2, 101] = 15.89, p = .00, η^2 = .135) with larger amplitudes for Cz (6.40 µV, 95 % CI 5.84 - 6.95) and Pz (5.93 µV, 95 % CI 5.47 - 6.38) compared to Fz electrode (5.41 µV, 95 % CI 4.89 - 5.93). A gender effect was found for P300 latencies (F [1, 93] = 6.99, p = .01, η^2 =

.070) with shorter latencies for females than for males as well as an electrode effect (F [2, 93] = 16.87, p = .00, η^2 = .152) with longer latency for Fz electrode (456.37 ms, 95 % CI 429.32 - 483.43) compared to Cz (416.12 ms, 95 % CI 399.31 - 432.93) and Pz electrode (394.94 ms, 95 % CI 383.36 - 406.53).

N100

There was no significant gender effect on N100 amplitudes (F [1, 102] = 3.56, p = .06, η^2 = .034) and latencies (F [1, 92] = 3.71, p = .06, η^2 = .039).

P200

There was no significant gender effect on P200 amplitudes (F [1, 102] = .57, p = .45, η^2 = .006) and latencies (F [1, 71] = .06, p = .82, η^2 = .001).

Table 26:   Effects of gender on cognitive measures (means with 95% CI over all task conditions)

		Females		Males		
		n	M (95% CI)	n	M (95% CI)	p
Flanker	Resp. time [ms]	37	513.87 (499.73-528.01)	95	492.00 (483.18-500.82)	.01
	P300 ampl. [μV]	33	4.40 (3.92-4.87)	91	4.71 (4.42-4.99)	.28
	P300 lat. [ms]	28	390.62 (376.18-405.07)	74	391.96 (383.07-400.85)	.88
	N100 ampl. [μV]	34	-1.88 (-2.18 - -1.58)	93	-1.67 (-1.86 - -1.49)	.25
	N100 lat. [ms]	14	140.82 (133.53-148.11)	32	140.03 (135.20-144.85)	.86
	P200 ampl. [μV]	34	3.79 (3.26-4.32)	93	2.99 (2.67-3.31)	.01
	P200 lat. [ms]	25	225.00 (214.15-235.85)	71	222.44 (216.00-228.88)	.69
Switching	Resp. time [ms]	38	703.17 (675.24-731.09)	90	702.65 (684.51-720.80)	.98
	Errors [no.]	39	4.06 (3.33-4.80)	93	3.86 (3.38-4.33)	.64
	P300 ampl. [μV]	23	4.25 (3.83-4.67)	61	3.80 (3.54-4.06)	.07
	P300 lat. [ms]	16	385.96 (361.90-410.02)	35	384.47 (368.20-400.74)	.92
	P200 ampl. [μV]	23	4.53 (3.90-5.16)	64	3.98 (3.60-4.36)	.14
	P200 lat. [ms]	23	190.63 (182.47-198.78)	57	188.83 (183.65-194.01)	.71
Oddball	Resp. time [ms]	40	531.38 (502.32-560.44)	93	501.83 (482.77-520.89)	.10
	P300 ampl. [μV]	29	6.41 (5.53-7.29)	74	5.72 (5.16-6.27)	.19
	P300 lat. [ms]	26	390.5 (362.31-418.68)	69	434.53 (417.23-451.83)	.01
	N100 ampl. [μV]	30	-5.48 (-6.26 - -4.71)	74	-4.61 (-5.10 - -4.12)	.06
	N100 lat. [ms]	28	176.40 (170.54-182.25)	66	169.61 (165.80-173.43)	.06
	P200 ampl. [μV]	30	2.73 (1.97-3.48)	74	3.07 (2.59-3.55)	.45
	P200 lat. [ms]	16	260.69 (250.68-270.70)	57	259.36 (254.06-264.67)	.82

Note: p = taken from ANOVA

Table 27:   Difference between task conditions (Flanker: accuracy vs. speed; Switching: homogenous vs. heterogeneous)

		Accuracy/ Homogenous		Speed/ Heterogeneous		
		n	M (95% CI)	n	M (95% CI)	p
Flanker	Resp. time [ms]	132	533.54 (522.78-544.30)	132	462.72 (456.21-469.23)	.00
	P300 ampl. [μV]	124	4.36 (4.12-4.60)	124	4.89 (4.61-5.17)	.00

		n	M (95% CI)	n	M (95% CI)	p
	P300 lat. [ms]	102	391.23 (382.67-399.79)	102	391.96 (384.39-399.52)	.81
	N100 ampl.[µV]	127	-1.71 (-1.87 - -1.54)	127	-1.75 (-1.93 - -1.58)	.50
	N100 lat. [ms]	46	139.72 (134.58 - 144.86)	46	140.82 (135.75 - 145.89)	.73
	P200 ampl. [µV]	127	2.95 (2.68-3.23)	127	3.45 (3.14-3.77)	.00
	P200 lat. [ms]	96	221.12 (215.03 - 227.22)	96	225.09 (219.51 - 230.67)	.05
Switching	Resp. time [ms]	128	528.46 (516.88 - 540.04)	128	877.15 (855.35 - 898.95)	.00
	Errors [no.]	132	2.72 (2.39 - 3.05)	132	5.12 (4.55 - 5.68)	.00
	P300 ampl. [µV]	84	4.14 (3.87 - 4.40)	84	3.71 (3.47 - 3.94)	.00
	P300 lat. [ms]	51	391.64 (377.68 - 405.61)	51	378.23 (361.61 - 394.85)	.08
	P200 ampl. [µV]	87	4.13 (3.77 - 4.49)	87	4.11 (3.78 - 4.45)	.88
	P200 lat. [ms]	80	185.62 (181.01 - 190.23)	80	193.07 (187.42 - 198.71)	<.01

Note: p = taken from ANOVA

Table 28: Difference between task trials (Flanker: congruent vs. incongruent; Switching: odd/even vs. <>5)

		Congruent/ odd-even		Incongruent/ <>5		
		n	M (95% CI)	n	M (95% CI)	p
Flanker	Response time [ms]	132	462.42 (454.81 - 470.02)	132	533.85 (525.79 - 541.91)	.00
	P300 amplitude [µV]	124	4.80 (4.52 - 5.08)	124	4.45 (4.21 - 4.69)	.00
	P300 latency [ms]	102	369.99 (362.57 - 377.40)	102	413.20 (403.95 - 422.45)	.00
	N100 amplitude [µV]	127	-1.72 (-1.89 - -1.54)	127	-1.74 (-1.90 - -1.58)	.73
	N100 latency [ms]	46	139.59 (134.67 - 144.51)	46	140.95 (136.44 - 145.46)	.59
	P200 amplitude [µV]	127	3.22 (2.92 - 3.52)	127	3.19 (2.91 - 3.46)	.69
	P200 latency [ms]	96	224.97 (218.76 - 231.18)	96	221.24 (215.46 - 227.03)	.12
Switching	Response time [ms]	128	697.07 (681.65 - 712.50)	128	708.54 (692.83 - 724.24)	<.01
	Errors [no.]	132	3.33 (2.95 - 3.72)	132	4.50 (4.02 - 4.99)	.00
	P300 amplitude [µV]	84	3.96 (3.72 - 4.20)	84	3.89 (3.65 - 4.12)	.33
	P300 latency [ms]	51	382.21 (366.97 - 397.45)	51	387.66 (373.64 - 401.68)	.37

P200 amplitude [µV]	87	4.12 (3.78 - 4.46)	87	4.12 (3.79 - 4.46)	.98
P200 latency [ms]	80	190.19 (185.09 - 195.28)	80	188.50 (183.65 - 193.36)	.49

Note: p = taken from ANOVA

4.2.2 Summary and Discussion

The following tables 29 and 30 summarize all effects that were found on the dependent cognition variables. They further provide effect sizes which are presented as mean differences to reference groups (females, accuracy/ homogenous condition, congruent/ odd/even trials). No effect sizes are given for the electrode effects since this included a comparison of three groups and for interactions.

Table 29: Summary of gender and condition effects and mean difference to reference group (unit of the measures and %) on cognitive performance measures

Task	Type of Measure	Gender effect[1]		Condition effect[2]	
		unit	%	unit	%
Flanker	Resp. time [ms]	**-21.87***	**-4.3***	**-70.82***	**-13.3***
	P300 ampl. [µV]	+.31	+7.0	**+.53***	**+12.2***
	P300 lat. [ms]	+1.34	+.3	+.73	+.2
	N100 ampl. [µV]	-.21	-11.2	+.04	+2.3
	N100 lat. [ms]	-.79	-.6	+1.10	+.8
	P200 ampl. [µV]	**-.80***	**-21.1***	**+.50***	**+16.9***
	P200 lat. [ms]	-2.56	-1.1	+3.97	+1.8
Switching	Resp. time [ms]	-.52	-.1	**+348.69***	**+66.0***
	Errors [no.]	-.20	-4.9	**+2.40***	**+88.2***
	P300 ampl. [µV]	-.45	-10.6	**-.43***	**-10.4***
	P300 lat. [ms]	-1.49	-.4	-13.41	-3.4
	P200 ampl. [µV]	-.55	-12.1	-.02	-.5
	P200 lat. [ms]	-1.80	-.9	**-7.45***	**+4.0***
Oddball	Resp. time [ms]	-29.55	-5.6		
	P300 ampl. [µV]	-.69	-10.8		
	P300 lat. [ms]	**+44.03***	**+11.3***		
	N100 ampl. [µV]	-.87	-15.9		
	N100 lat. [ms]	-6.79	-3.8		
	P200 ampl. [µV]	+.34	+12.5		
	P200 lat. [ms]	-1.33	-.5		

Note: Bold and * = significant effect (p <.05); * = significant interaction (p <.05); [1] =reference group are females, [2] = references are accuracy/ homogenous condition

Table 30: Summary of trial and electrode effects and mean difference to reference group (unit of the measures and %) on cognitive performance measures

Task	Type of Measure	Trial effect[3]		Electrode effect	Interactions
		unit	%		
Flanker	Resp. time [ms]	**+71.43***	**+15.4***		-
	P300 ampl. [µV]	**-.35***	**-7.3***	*	*
	P300 lat. [ms]	**+43.21***	**+11.7***	*	*
	N100 ampl. [µV]	+.02	+1.2	-	-

Switching	N100 lat. [ms]	+1.36	+1.0	-	-
	P200 ampl. [µV]	-.03	-.9	-	-
	P200 lat. [ms]	-3.73	-1.7	-	-
	Resp. time [ms]	**+11.47***	**+1.6***		*
	Errors [no.]	**+1.17***	**+35.1***		-
	P300 ampl. [µV]	+.07	+1.8	*	*
	P300 lat. [ms]	-5.45	-1.4	*	-
Oddball	P200 ampl. [µV]	±0	±0	-	*
	P200 lat. [ms]	+1.69	+.9	-	-
	Resp. time [ms]				
	P300 ampl. [µV]			*	-
	P300 lat. [ms]			*	-
	N100 ampl. [µV]				
	N100 lat. [ms]				
	P200 ampl. [µV]				
	P200 lat. [ms]				

Note: Bold and * = significant effect (p <.05); * = significant interaction (p <.05; see description in text); [3] = references are congruent/ odd/even trials

For response times during the Flanker task, significant gender, condition, and trial effects were found. It was not surprising that a condition effect with larger response times for the accuracy compared to the speed condition has been observed since it was the participants' task to respond as quickly as possible during the speed condition. Thus, this finding proves that participants considered this demand and it is also in line with reports from previous studies (Themanson et al., 2008). The result that response times were faster during congruent compared to incongruent trials also confirms the findings from existing studies (Hillman, Motl et al., 2006; Themanson et al., 2006; Themanson et al., 2008) and indicates a smaller response conflict that leads to faster responses during the congruent trials. Females responded faster during the Flanker task than males, but the origins of this effect are uncertain. Since most studies on the relationship between physical activity, fitness and cognition have been performed with a small sample size only few studies provide a gender-specific analysis of data. A study by Conroy and Polich (2007, p. 28 f.) among 120 university students observed faster response times for males compared to females during a three-stimulus Oddball task and stated that different rates at which target processing occurs across genders might be an explaining factor. However, it is unknown whether this is transferable to the Flanker task. Response times during the Oddball task were not significantly influenced by gender in this study sample, but males tended to elicit faster response times compared to females. A result, that is similar to the study by Conroy and Polich (2007). Response times during the Switching task were unaffected by gender but influenced by type of condition and type of trial. Faster times were found for the homogenous compared to the heterogeneous condition and for odd/even compared to </> 5 trials. Descriptive data by Kamijo and Takeda (2010, p. 307) also reveal faster response times in young adults for homogenous compared to heterogeneous Switching task conditions. However, participants in their study responded faster during </> 5 trials compared to odd/even trials. It is uncertain why there is a difference between the two studies since the setup of the paradigm was very similar. Error rate during the Switching task was affected by condition and trials with a higher rate for the heterogeneous condition and for </> 5 trials. Again, Kamijo and Takeda (2010, p. 307) reveal the same effect for condition but found more errors during odd/even trials. Scisco et al. (2008,

p. 56) performed the same task with young adults and found no difference between task condition and response accuracy.

P300 amplitude during the Flanker task was larger during the speed condition, during congruent trials and at parietal and central compared to frontal electrode sites. P300 amplitude during the Switching task was larger during the homogenous condition and at central compared to frontal and occipital electrode. For the Oddball task, P300 was also affected by electrode position and larger for parietal and central compared to the frontal electrode. It is known from several studies that P300 amplitude has the largest peak at parietal and the smallest at frontal electrode sites (e.g. Kamijo & Takeda, 2010, p. 307; Polich & Lardon, 1997, p. 495; Magnié et al., 2000, p. 373). Hillman, Kramer et al. (2006, p. 36) could also find larger P300 amplitudes for homogenous compared to heterogenous Switching task conditions. Since larger P300 amplitudes are an indicator for a greater amount of attention devoted to a stimulus it is arguable why larger amplitudes were observed during the congruent trials of the Flanker and the homogenous condition of the Switching task. The congruent Flanker trials as well as the homogenous switching conditions are assumed to require less amounts of executive control and thus one would expect smaller amplitudes. However, young adults are on the peak of their cognitive performance and might compensate differences in task difficulty by an increased cognitive efficiency or other processes that are not observable in P300 amplitude. P300 latencies during the Flanker task were faster for congruent trials and at parietal and central compared to frontal electrode. The same effect of electrode site was found for the Oddball and Switching task. These results are consistent with previously reported ones (Scisco et al., 200, p. 56). Since faster latencies mean faster cognitive processing devoted to a stimulus, is was expected that congruent trials would be related to faster latencies. However, no effects of type of condition or trial were observed in the Switching task. N100 amplitude and latency which was assessed in the Flanker and Oddball task was unaffected by gender or task-specific variables in the study sample. P200 amplitudes were enhanced for females and during the speed condition of the Flanker task. P200 latency was faster during the homogenous conditions of the Switching task. This might indicate that the speed (Flanker) and heterogeneous (Switching) condition required a greater attention modulation. It is further known that visual P200 amplitude increases with enhanced stimulus complexity as given in the speed condition of the Flanker task compared to the accuracy condition.

4.3 Relationships between Physical Activity and Cognition

4.3.1 General Sports Participation

In this chapter, the differences between athletes and nonathletes concerning their cognitive performance are described. The independent variable is general sports participation (athletes vs. nonathletes).

> *RQ 1*: Do young adults who regularly engage in physical activity have a higher cognitive performance than inactive young adults?

4.3.1.1 Results

Behavioral Measures

For the Flanker and Switching tasks, a 2 (condition) x 2 (congruency) repeated measures ANOVA revealed no significant effect of general sports participation on response times. However, numerical trends detect faster response times for athletes compared to nonathletes for both genders. It was found that athletes responded significantly faster than nonathletes during the Oddball task (F [1, 131] = 5.46, p = .02, η^2 = .046; males: F [1, 91] = .73, p = .40, η^2 = .008; females: F [1, 38] = 4.94, p = .03, η^2 = .115). The mean response times for the cognitive tasks are presented in table 31 and figure 10.

Table 31:   Mean response times for athletes and nonathletes in the Flanker, Switching and Oddball task

		Athletes		Nonathletes			
		n	ms (95 % CI)	n	ms (95 % CI)	p	η^2
Flanker	♂	80	490.65 (481.30 - 499.99)	15	499.22 (477.65 - 520.80)	.47	.006
	♀	25	511.21 (491.86 - 530.55)	12	519.41 (491.50 - 547.33)	.63	.007
	Total	105	495.54 (486.99 - 504.09)	27	508.20 (491.34 - 525.06)	.19	.013
Switching	♂	74	698.50 (679.76 - 717.24)	16	721.86 (681.56 - 762.15)	.30	.012
	♀	25	698.34 (657.45 - 739.22)	13	712.45 (655.76 - 769.15)	.69	.005
	Total	99	698.46 (681.23 - 715.69)	29	717.64 (685.81 - 749.47)	.30	.009
Oddball	♂	77	498.17 (477.63 - 518.71)	16	519.43 (474.36 - 564.50)	.40	.008
	♀	27	508.58 (472.14 - 545.01)	13	578.74 (526.23 - 631.25)	.03	.115
	Total	104	500.87 (483.02 - 518.72)	29	546.02 (512.21 - 579.82)	.02	.040

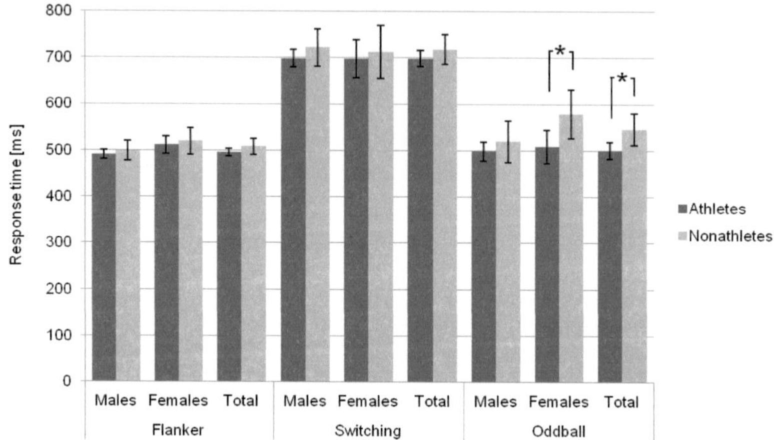

Figure 10: Mean response times during Flanker, Switching and Oddball task for athletes and nonathletes; error bars represent 95% confidence intervals

No significant effect of general sports participation on errors in the Switching task could be observed (table 32 & figure 11), but there was a marginally significant condition x general sports participation interaction (F [1, 130] = 3.54, p = .06, η^2 = .027) with more errors for athletes under the homogenous condition (4.53, 95 % CI 3.98 - 5.09; nonathletes: 4.39, 95 % CI 3.33 - 5.46) and more errors for nonathletes under the heterogeneous condition (3.50, 95 % CI 2.66 - 4.34; athletes: 3.29, 95 % CI 2.85 - 3.73).

Table 32: Errors in the Switching task for athletes and nonathletes

		Athletes		Nonathletes			
		n	No. (95 % CI)	n	No. (95 % CI)	p	η^2
Errors	♂	77	3.87 (3.38 - 4.36)	16	3.81 (2.74 - 4.88)	.93	.000
	♀	27	4.04 (2.98 - 5.09)	12	4.13 (2.55 - 5.70)	.93	.000
	Total	104	3.91 (3.46 - 4.36)	28	3.95 (3.08 - 4.81)	.94	.000

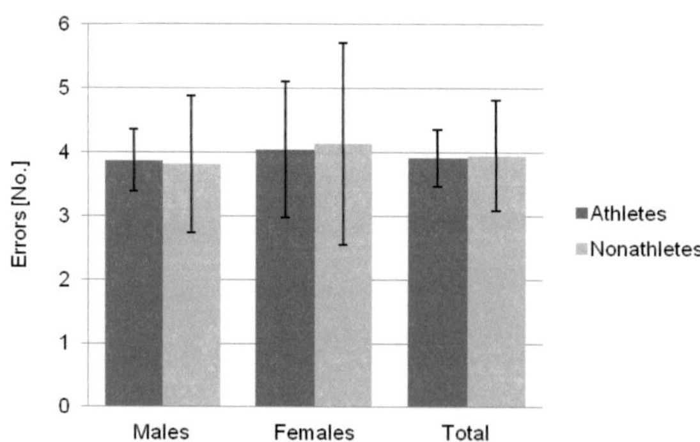

Figure 11: Errors during the Switching task for athletes and nonathletes; error bars represent 95% confidence intervals

P300

It appears that athletes tended to have larger amplitudes throughout all conditions of the Flanker task, but no significant overall effect could be observed. However, there was an electrode x general sports participation interaction (F [2, 121] = 4.07, p = .02, η^2 = .032) indicating larger P300 amplitudes for athletes compared to nonathletes at Cz (5.64 µV, 95 % CI 5.29 - 5.99 vs. 4.89 µV, 95 % CI 4.20 - 5.58) and Pz (5.09 µV, 95 % CI 4.80 - 5.37 vs. 4.37 µV, 95 % CI 3.80 - 4.94) electrode sites. No effects could be detected for P300 amplitude in the Switching task. Regarding the Oddball task, there was a significant effect of general sports participation on P300 amplitude with larger amplitudes for athletes compared to nonathletes (F [1, 101] = 4.94, p = .03, η^2 = .047; males: F [1, 72] = 4.07, p = .05, η^2 = .054; females: F [1, 27] = 1.44, p = .24, η^2 = .050). The mean P300 amplitudes for the cognitive tasks are presented in table 33 and figure 12.

Table 33: Mean P300 amplitudes for athletes and nonathletes in the Flanker, Switching and Oddball task

		Athletes		Nonathletes			
		n	µV (95 % CI)	n	µV (95 % CI)	p	η^2
Flanker	♂	76	4.80 (4.49 - 5.10)	15	4.26 (3.56 - 4.95)	.16	.022
	♀	23	4.53 (3.91 - 5.15)	10	4.10 (3.16 - 5.04)	.44	.019
	Total	99	4.73 (4.46 - 5.01)	25	4.19 (3.65 - 4.74)	.08	.025
Switching	♂	48	3.82 (3.55 - 4.09)	13	3.72 (3.21 - 4.24)	.75	.002
	♀	14	4.25 (3.56 - 4.94)	9	4.26 (3.40 - 5.12)	.99	.000
	Total	62	3.92 (3.66 - 4.18)	22	3.94 (3.50 - 4.38)	.92	.000
Oddball	♂	61	5.97 (5.36 - 6.59)	13	4.50 (3.17 - 5.82)	.05	.054
	♀	22	6.69 (5.72 - 7.66)	7	5.54 (3.83 - 7.26)	.24	.050
	Total	83	6.16 (5.65 - 6.68)	20	4.86 (3.82 - 5.91)	.03	.047

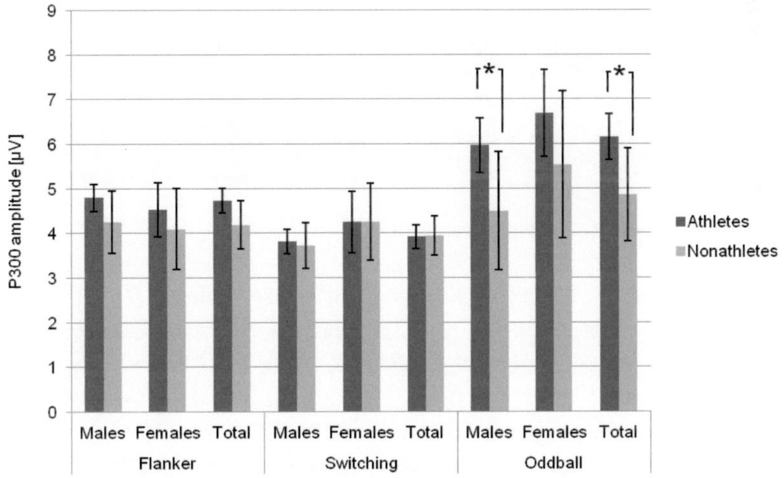

Figure 12: Mean P300 amplitudes during Flanker, Switching and Oddball task for athletes and nonathletes; error bars represent 95% confidence intervals

The grand average for all participants for P300 amplitude at Pz electrode during the Oddball task is given in figure 13. It is clearly noticeable that athletes have a larger P300 amplitude peak than nonathletes. Topoplots for athletes and nonathletes during the Oddball tasks are presented in figure 14. This figure also shows a more intense brain activation (color: dark red) for athletes compared to nonathletes (color: orange-red) in the relevant time range (300-400 ms) after the stimulus was presented.

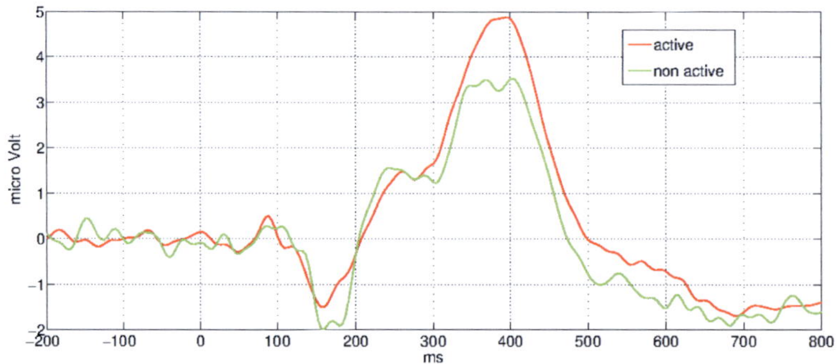

Figure 13:   Grand average for Oddball P300 amplitude at Pz electrode (red: athletes, green: nonathletes)

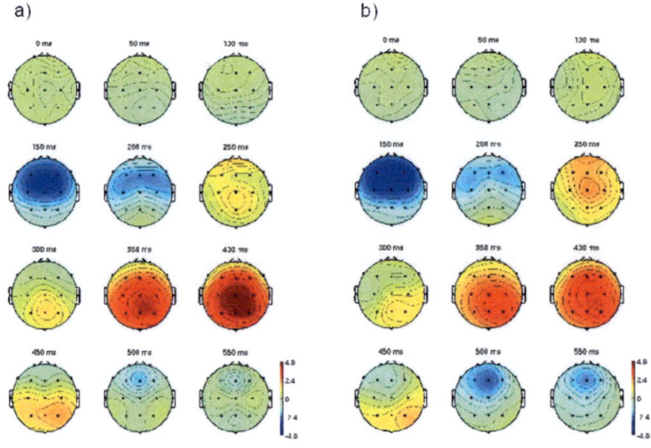

Figure 14:   Topoplots for athletes (a) and nonathletes (b)

A significant overall effect (F [1, 100] = 5.88, p = .02, η^2 = .056; males: F [1, 72] = .06, p = .82, η^2 = .001; females: F [1, 26] = 9.44, p = < .01, η^2 = .266) was observed with faster latencies for athletes compared to nonathletes in the Flanker task (table 34 & figure 15). There was further a congruency x general sports participation interaction (F [1, 100] = 4.94, p = .03, η^2 = .047) indicating faster P300 latencies for athletes for congruent (362.95 ms, 95 % CI 355.01 - 370.90) but not incongruent trials (410.53 ms, 95 % CI 400.03 - 421.03) compared to nonathletes (congruent: 394.15 ms, 95 % CI 379.43 - 408.87; incongruent: 422.37 ms, 95 % CI 402.90 - 441.83). No effect was found on P300 latencies in the Switching task as well as Oddball task.

Table 34: Mean P300 latencies for athletes and nonathletes in the Flanker, Switching and Oddball task

		Athletes		Nonathletes			
		n	ms (95 % CI)	n	ms (95 % CI)	p	η^2
Flanker	♂	60	391.56 (383.68 - 399.44)	14	393.68 (377.37 - 409.99)	.82	.001
	♀	19	371.53 (349.00 - 394.06)	9	430.93 (398.20 - 463.67)	<.01	.266
	Total	79	386.74 (378.38 - 395.10)	23	408.26 (392.77 - 423.75)	.02	.056
Swit-ching	♂	26	390.85 (371.87 - 409.82)	9	366.05 (333.79 - 398.30)	.19	.052
	♀	9	377.76 (343.44 - 412.08)	7	396.49 (357.58 - 435.41)	.45	.041
	Total	35	387.48 (371.26 - 403.70)	16	379.37 (355.38 - 403.35)	.58	.006
Oddball	♂	58	437.55 (416.93 - 458.17)	11	418.61 (371.26 - 465.96)	.47	.008
	♀	21	387.80 (363.79 - 411.80)	5	401.82 (352.63 - 451.02)	.60	.012
	Total	79	424.32 (407.59 - 441.06)	16	413.36 (376.17 - 450.56)	.60	.003

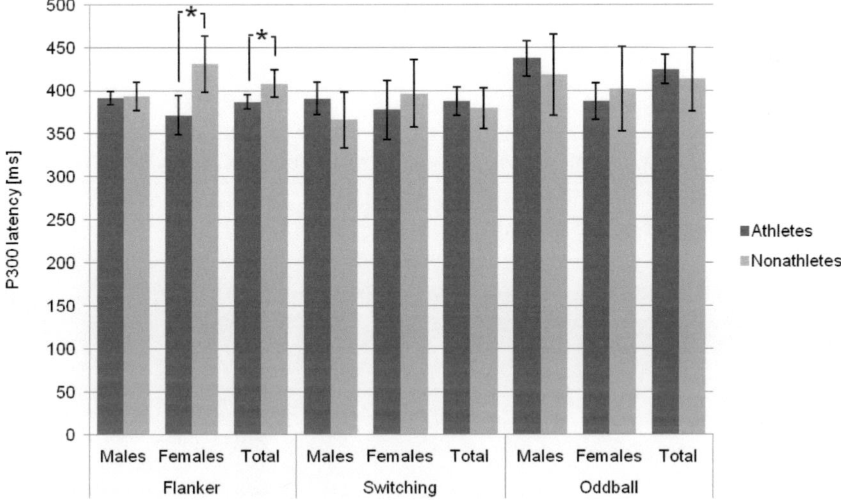

Figure 15: Mean P300 latencies during Flanker, Switching and Oddball task for athletes and nonathletes; error bars represent 95% confidence intervals

N100 and P200

For the Flanker and Oddball task, analyses revealed no significant effects of general sports participation on N100 or P200 amplitudes or latencies (tables 35 & 36, figures 16 - 19). For the Switching task, there was a marginal main effect of general sports participation on P200 amplitudes (F [1, 85] = 3.63, p = .06, η^2 = .041; males: F [1, 62] = 4.86, p = .03, η^2 = .073; females: F [1, 21] = .72, p = .41, η^2 = .033) with larger amplitudes for athletes (4.30 µV, 95 % CI 3.93 - 4.68) than nonathletes (3.59 µV, 95 % CI 2.95 - 4.23) and for male athletes (4.17 µV, 95 % CI 3.78 - 4.57) than male nonathletes (3.20 µV, 95 % CI 2.42 - 3.99). There was no effect on P200 latencies.

Table 35:   Mean N100 and P200 amplitudes for athletes and nonathletes in the Flanker, Switching and Oddball task

			Athletes		Nonathletes			
			n	μV (95 % CI)	n	μV (95 % CI)	p	η^2
Flanker	N100	♂	78	-1.73 (-1.93 - -1.53)	15	-1.37 (-1.82 - -.92)	.15	.022
		♀	23	-1.83 (-2.22 - -1.44)	11	-1.98 (-2.54 - -1.42)	.66	.006
		Total	101	-1.75 (-1.93 - -1.58)	26	-1.63 (-1.98 - -1.28)	.53	.003
	P200	♂	78	2.97 (2.64 - 3.30)	15	3.08 (2.33 - 3.84)	.78	.001
		♀	23	3.60 (2.83 - 4.36)	11	4.19 (3.09 - 5.29)	.37	.025
		Total	101	3.11 (2.80 - 3.42)	26	3.55 (2.94 - 4.17)	.21	.013
Switching	P200	♂	51	4.17 (3.78 - 4.57)	13	3.20 (2.42 - 3.99)	.03	.073
		♀	14	4.77 (3.81 - 5.74)	9	4.15 (2.95 - 5.35)	.41	.033
		Total	65	4.30 (3.93 - 4.68)	22	3.59 (2.95 - 4.23)	.06	.041
Oddball	N100	♂	61	-4.49 (-5.05 - -3.93)	13	-5.17 (-6.39 - -3.96)	.31	.014
		♀	22	-5.69 (-6.56 - -4.83)	8	-4.91 (-6.34 - -3.47)	.34	.032
		Total	83	-4.81 (-5.28 - -4.34)	21	-5.07 (-6.01 - -4.13)	.62	.002
	P200	♂	61	3.11 (2.61 - 3.61)	13	2.88 (1.79 - 3.96)	.70	.002
		♀	22	2.59 (1.53 - 3.65)	8	3.10 (1.35 - 4.86)	.61	.009
		Total	83	2.97 (2.51 - 3.42)	21	2.96 (2.06 - 3.87)	.99	.000

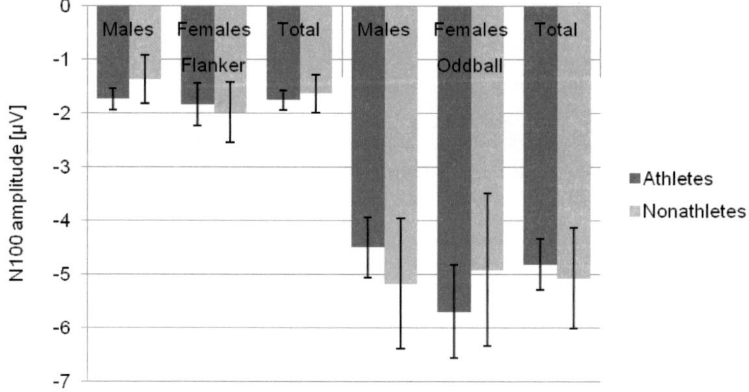

Figure 16:   Mean N100 amplitudes during Flanker and Oddball task for athletes and nonathletes; error bars represent 95% confidence intervals

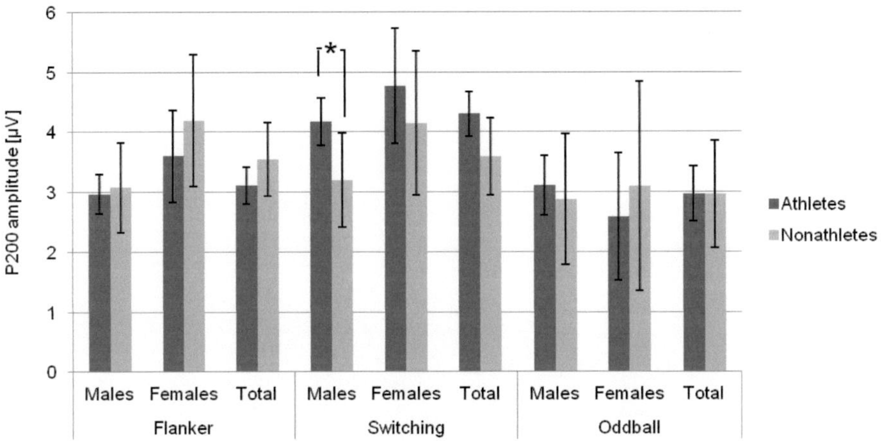

Figure 17:   Mean P200 amplitudes during Flanker, Switching and Oddball task for athletes and nonathletes; error bars represent 95% confidence intervals

Table 36:   Mean N100 and P200 latencies for athletes and nonathletes in the Flanker, Switching and Oddball task

			Athletes		Nonathletes			
			n	ms (95 % CI/ ± SD)	n	ms (95 % CI/ ± SD)	p	η^2
Flanker	N100	♂	27	140.17 (134.39 - 145.94)	5	139.26 (125.84 - 152.68)	.90	.001
		♀	9	138.33 (130.39 - 146.26)	5	145.31 (134.67 - 155.96)	.27	.099
		Total	36	139.71 (135.17 - 144.24)	10	142.29 (133.68 - 150.89)	.60	.006
	P200	♂	58	220.38 (213.31 - 227.44)	13	231.64 (216.72 - 246.56)	.18	.026
		♀	16	225.54 (210.83 - 240.25)	9	224.05 (204.43 - 243.66)	.90	.001
		Total	74	221.49 (215.22 - 227.77)	22	228.53 (217.03 - 240.04)	.29	.012
Switching	P200	♂	46	187.91 (182.38 - 193.43)	11	192.67 (181.37 - 203.97)	.45	.010
		♀	14	192.79 (180.36 - 205.21)	9	187.26 (171.77 - 202.76)	.57	.016
		Total	60	189.05 (183.99 - 194.10)	20	190.23 (181.49 - 198.98)	.82	.001
Oddball	N100	♂	53	169.03 (164.34 - 173.72)	13	172.00 (162.52 - 181.47)	.58	.005
		♀	21	177.60 (172.41 - 182.80)	7	172.77 (163.77 - 181.76)	.35	.034
		Total	74	171.46 (167.79 - 175.14)	20	172.27 (165.20 - 179.33)	.84	.000
	P200	♂	46	257.97	11	265.20	.29	.021

		(252.05 - 263.89)		(253.09 - 277.31)		
♀	11	259.52	5	263.28	.74	.008
		(246.09 - 272.95)		(243.36 - 283.20)		
Total	57	258.26	16	264.60	.27	.017
		(253.01 - 263.52)		(254.67 - 274.53)		

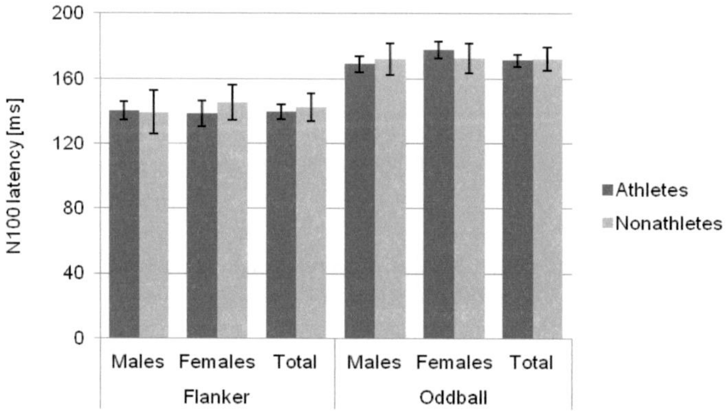

Figure 18:   Mean N100 latencies during Flanker and Oddball task for athletes and nonathletes; error bars represent 95% confidence intervals

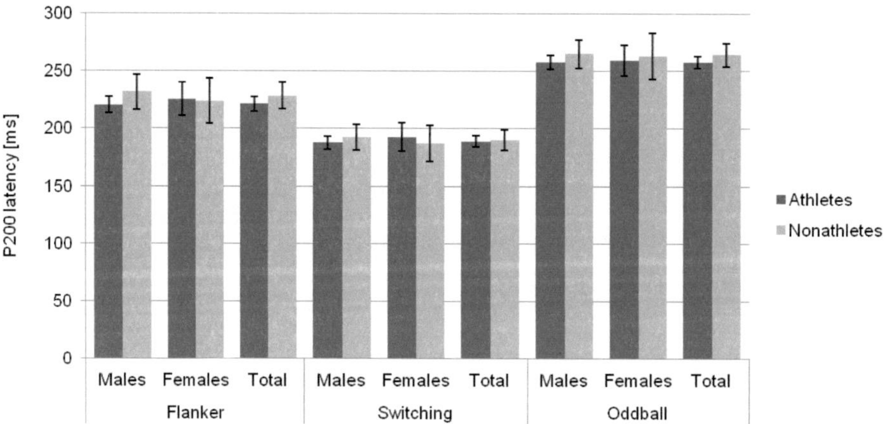

Figure 19:   Mean P200 latencies during Flanker, Switching and Oddball task for athletes and nonathletes; error bars represent 95% confidence intervals

4.3.1.2 Discussion

During the Oddball task, athletes had faster response times and larger P300 amplitudes than nonathletes but there was no effect for P300 latencies. This finding is consistent with the study from Polich and Lardon (1997) in which young adults with regular physical activity also exhibited greater P300 amplitudes in an Oddball paradigm compared with less active adults, whereas P300 latencies were not significantly affected by exercise. However, Polich and Lardon (1997) observed a tendency for faster latencies in athletes compared to nonathletes which was not confirmed in this study. McDowell et al. (2003) also report of larger Oddball P300 amplitudes for young active compared to inactive participants but found no significant effects on P300 latencies. In contrast to these findings, the study from Magnié et al. (2000) revealed no effect of regular physical activity on cognitive performance during an auditory Oddball task. However, it must be considered that they compared only one specific group of athletes (cyclists) with sedentary adults and thus their results have only a limited power. For the Flanker task, P300 amplitudes were significantly larger over central and parietal scalp regions and P300 latencies were faster during congruent trials for athletes compared to nonathletes. For females and the total sample, P300 latencies were significantly faster for athletes compared to nonathletes in the Flanker task. With regard to behavioral measures, no significant effects but a tendency for faster response times for athletes was found for the Flanker task. This is to some extent in line with a study by Hillman, Motl et al. (2006) who found physical activity only related to better response accuracy during the Flanker task for older compared to younger adults. Taken together the findings for the Oddball and Flanker task and given that P300 amplitude is assumed to detect attentional processes and working memory resource allocation devoted to a stimulus and faster P300 latencies are supposed to reflect a faster stimulus evaluation time and cognitive processing speed, it may be concluded that athletes in this sample of young adults have better executive functions and cognitive control. The fact that athletes revealed faster response times during the Oddball task might be linked to a better response-related processing speed but could also be explained by a better efficiency of peripheral motor processes. To clarify this problem it needs to be investigated whether there is still an effect on response speed if a task does not require a motor response.

Interestingly, there was no main effect on P300 component or response times during the Switching task. However, a tendency for faster response times for athletes compared to nonathletes was observed. The lack of significant differences between athletes and nonathletes is in contrast to a study from Hillman, Kramer et al. (2006) who found effects of regular physical activity on Switching response times, P300 amplitudes and latencies. However, in their study they tested both young and older adults and thus, these effects may be influenced by the age of the participants although the authors report no age x fitness interactions. In addition, a study by Kamijo and Takeda (2010) also showed that regular physical activity in young adults is related to faster response times in the heterogeneous condition, smaller switch costs and also affects P300 amplitude. It is not clear why the current study failed to detect any relationships between physical (in-) activity and performance in the Switching task. The only significant finding is that athletes had a better response accuracy under the heterogeneous condition but not under the homogenous condition. This partly supports the results by Kamijo and Takeda (2010) who observed faster response times only for the heterogeneous condition of the Switching task. These selective results might indicate that physical activity in young adulthood is primarily related to cognitive trials and conditions requiring greater amounts of executive top-down control and thus supports the finding. On the other hand, faster P300 latencies for athletes during the Flanker task were detected only for congruent but not incongruent trials. Given this finding, one might suggest that regular physical activity specifically increases cognitive performance during tasks or conditions which require smaller amounts of selective attention.

No effects were found in the current sample for N100 and P200 components except for larger P200 amplitudes for athletes compared to nonathletes in the Switching task. This effect was significant for males. However, there was a tendency for larger P200 amplitudes for nonathletes in the Flanker task.

This study involved a relatively large number of participants compared to other research that has been published. This allowed for the investigation of gender dependent differences regarding the effects of physical activity and cognition. It was found that the results were often not consistent and that there were differences between males and females. For example, the main effect of general physical activity on P300 amplitude in the Oddball task was significant for the total sample and for males but not females. This may be caused by large individual variations in the cognitive measures and the different sizes for the male and female subsamples. However, it is important to mention that there was a coinciding tendency for males and females for the significant results observed for the total sample (i.e. both male and female athletes had larger amplitudes than nonathletes, but the difference became only significant for males).

4.3.2 Participation in Competitive Sports

In this chapter, the differences between athletes participating in competitive sports, athletes not participating in competitive sports and nonathletes are described concerning their cognitive performance. The independent variable is competitive sports participation (athletes in competitive sports vs. athletes in non-competitive sports vs. nonathletes). The differences in physical activity and fitness variables between the groups are given in tables 16-19.

> *RQ 2*: Is the participation in competitive sports related to cognitive performance in young adulthood?

4.3.2.1 Results

Behavioral Measures
For the Flanker and Switching tasks, a 2 (condition) x 2 (congruency) repeated measures ANOVA revealed no significant effect of competitive sports participation on response times (table 37 & figure 20). However, numerical trends detect faster response times for athletes not participating in competitive sports compared to those who participate and to nonathletes. There was a main effect in the Oddball task (F [2, 130] = 3.39, p = .04, η^2 = .050; males: F [2, 90] = 1.24, p = .30, η^2 = .027; females: F [2, 37] = 2.42, p = .10, η^2 = .116). Post-hoc analyses revealed a significant difference in Oddball response times between non-competitive athletes and nonathletes (p = .03) with non-competitive athletes responding faster. No effects were observed for errors in the Switching task (table 38 & figure 21).

Table 37:   Mean response times for competitive sports groups and nonathletes in the Flanker, Switching and Oddball task

		\multicolumn{2}{l}{Competitive sports}		\multicolumn{2}{l}{No competitive sports}		\multicolumn{2}{l}{Nonathletes}		
		n	ms (95 % CI)	n	ms (95 % CI)	n	ms (95 % CI)	p	η^2
Flan-ker	♂	38	499.90 (486.53 - 513.27)	42	482.28 (469.56 - 495.00)	15	499.22 (477.94 - 520.51)	.13	.043
	♀	8	529.27 (495.39 - 563.15)	17	502.71 (479.46 - 525.95)	12	519.41 (491.75 - 547.08)	.38	.055
	Tot.	46	505.01 (492.23 - 517.79)	59	488.16 (476.88 - 499.45)	27	508.20 (491.52 - 524.88)	.06	.042
Swit-ching	♂	35	704.85 (677.51 - 732.19)	39	692.81 (666.90 - 718.71)	16	721.86 (681.42 - 762.30)	.48	.017
	♀	9	694.03 (624.87 - 763.18)	16	700.76 (648.89 - 752.62)	13	712.45 (654.91 - 769.99)	.91	.005
	Tot.	44	702.63 (676.71 - 728.56)	55	695.12 (671.93 - 718.31)	29	717.64 (685.71 - 749.58)	.53	.010
Odd-ball	♂	37	512.31 (482.79 - 541.83)	40	485.09 (456.70 - 513.48)	16	519.43 (474.54 - 564.33)	.30	.027
	♀	9	512.83 (448.84 - 576.82)	18	506.45 (461.20 - 551.70)	13	578.74 (525.49 - 631.98)	.10	.116
	Tot.	46	512.41 (485.60 - 539.23)	58	491.72 (467.84 - 515.60)	29	546.02 (512.25 - 579.79)	.04	.050

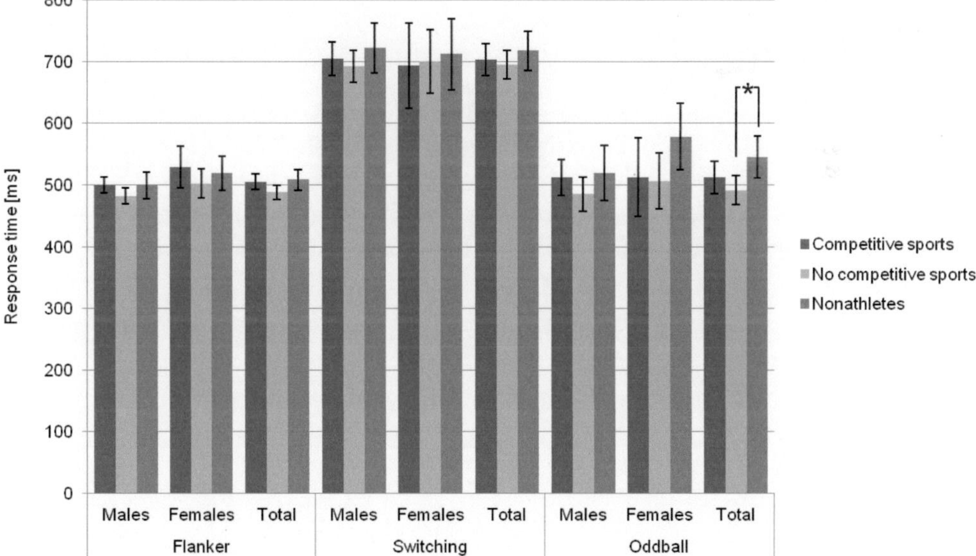

Figure 20:   Mean response times during Flanker, Switching and Oddball task for competitive sports groups and nonathletes; error bars represent 95% confidence intervals

Table 38:   Errors in the Switching task for competitive sports groups and nonathletes

		Competitive sports		No competitive sports		Nonathletes			
		n	No. (95 % CI)	n	No. (95 % CI)	n	No. (95 % CI)	p	η^2
Errors	♂	36	3.99 (3.28 - 4.71)	41	3.76 (3.09 - 4.23)	16	3.81 (2.74 - 4.89)	.89	.003
	♀	9	3.83 (1.99 - 5.68)	18	4.14 (2.83 - 5.45)	12	4.13 (2.52 - 5.73)	.96	.002
	Total	45	3.96 (3.28 - 4.65)	59	3.87 (3.28 - 4.47)	28	3.95 (3.08 - 4.81)	.98	.000

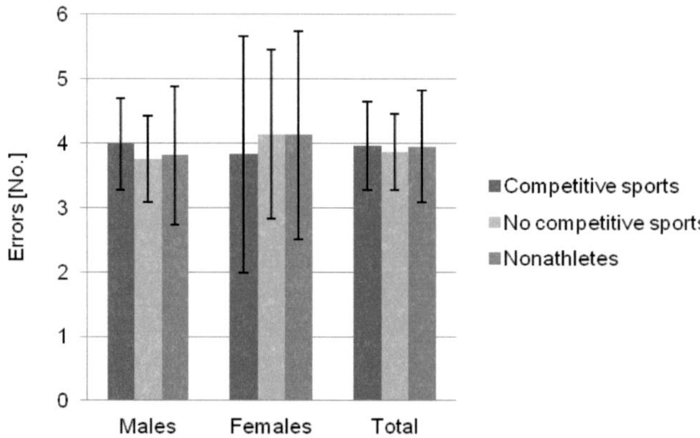

Figure 21:   Errors during the Switching task for competitive sports groups and nonathletes; error bars represent 95% confidence intervals

P300

For the Flanker P300 amplitude, there was a significant electrode x competitive sports interaction (F [4, 242] = 2.90, p = .02, η^2 = .046). For both athletes groups, amplitudes at Fz electrode were significantly lower than for Cz and Pz electrode whereas for nonathletes, there was only a slight difference between frontal and medial-lateral electrodes. There were no further effects for P300 amplitudes. The mean P300 amplitudes for the cognitive tasks are presented in table 39 and figure 22.

Table 39:   Mean P300 amplitudes for competitive sports groups and nonathletes in the Flanker, Switching and Oddball task

		Competitive sports		No competitive sports		Nonathletes			
		n	µV (95 % CI)	n	µV (95 % CI)	n	µV (95 % CI)	p	η^2
Flanker	♂	36	4.91 (4.46 - 5.36)	40	4.69 (4.27 - 5.12)	15	4.26 (3.56 - 4.95)	.30	.027
	♀	7	4.55 (3.41 - 5.69)	16	4.52 (3.76 - 5.27)	10	4.10 (3.14 - 5.05)	.75	.019
	Total	43	4.85 (4.44 - 5.27)	56	4.64 (4.28 - 5.01)	25	4.19 (3.65 - 4.74)	.17	.029
Switching	♂	24	3.94 (3.56 - 4.33)	24	3.69 (3.31 - 4.08)	13	3.72 (3.21 - 4.24)	.62	.016
	♀	5	3.90 (2.73 - 5.07)	9	4.44 (3.57 - 5.31)	9	4.26 (3.39 - 5.13)	.75	.029

		n		n		n		p	η^2
	Total	29	3.94 (3.55 - 4.32)	33	3.90 (3.54 - 4.26)	22	3.94 (3.50 - 4.38)	.98	.000
Oddball	♂	30	6.18 (5.30 - 7.06)	31	5.78 (4.91 - 6.64)	13	4.50 (3.17 - 5.83)	.12	.059
	♀	7	6.25 (4.51 - 7.98)	15	6.90 (5.71 - 8.08)	7	5.54 (3.81 - 7.28)	.42	.065
	Total	37	6.19 (5.42 - 6.96)	46	6.14 (5.45 - 6.83)	20	4.86 (3.81 - 5.91)	.09	.047

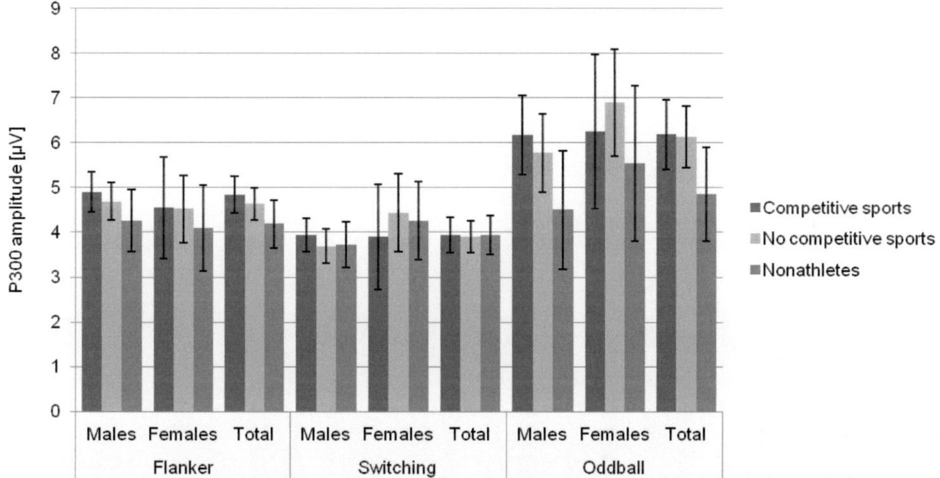

Figure 22: Mean P300 amplitudes during Flanker, Switching and Oddball task for competitive sports groups and nonathletes; error bars represent 95% confidence intervals

For P300 latencies during the Flanker task, there was a significant main effect (F [2, 99] = 3.76, p = .03, η^2 = .071; males: F [2, 71] = .03, p = .97, η^2 = .001; females: F [2, 25] = 5.86, p = <.01, η^2 = .319) with non-competitive athletes having faster latencies than nonathletes (p = .02). A condition x competitive sports (F [2, 99] = 7.68, p = .01, η^2 = .134) and congruency x competitive sports interaction (F [2, 99] = 3.91, p = .02, η^2 = .073) was also observed. The first interaction might be explained by the fact that competitive athletes had faster latencies during the speed condition (386.93 ms, 95 % CI 374.69 - 399.18; accuracy: 397.59 ms, 95 % CI 384.20 - 410.97), whereas non-competitive athletes had faster latencies during the accuracy condition (374.89 ms, 95 % CI 362.00 - 387.77; speed: 388.37 ms, 95 % CI 376.58 - 400.16). There was no difference in latencies for nonathletes. No effects were observed for P300 latencies during the Switching and Oddball tasks. The mean P300 latencies for the cognitive tasks are presented in table 40 and figure 23.

Table 40: Mean P300 latencies for competitive sports groups and nonathletes in the Flanker, Switching and Oddball task

		Competitive sports		No competitive sports		Nonathletes			
		n	ms (95 % CI)	n	ms (95 % CI)	n	ms (95 % CI)	p	η^2
Flanker	♂	32	392.01 (381.15 - 402.88)	28	391.04 (379.43 - 402.66)	14	393.68 (377.25 - 410.11)	.97	.001

		n	Competitive sports	n	No competitive sports	n	Nonathletes	p	η^2
	♀	6	393.59 (354.12 - 433.05)	13	361.35 (334.54 - 388.16)	9	430.93 (398.71 - 463.16)	<.01	.319
	Total	38	392.25 (380.24 - 404.28)	41	381.63 (370.06 - 393.20)	23	408.26 (392.81 - 423.70)	.03	.071
Swit-ching	♂	14	385.31 (359.18 - 411.44)	12	397.30 (369.08 - 425.53)	9	366.05 (333.46 - 398.63)	.35	.064
	♀	4	348.67 (300.20 - 397.13)	5	401.04 (357.70 - 444.39)	7	396.49 (359.86 - 433.13)	.20	.222
	Total	18	377.17 (354.71 - 399.62)	17	398.40 (375.30 - 421.51)	16	379.37 (355.55 - 403.18)	.36	.042
Odd-ball	♂	29	431.11 (401.80 - 460.41)	29	443.99 (414.69 - 473.30)	11	418.61 (371.02 - 466.19)	.63	.014
	♀	7	412.91 (372.46 - 453.36)	14	375.24 (346.64 - 403.84)	5	401.82 (353.97 - 449.68)	.27	.108
	Total	36	427.57 (402.65 - 452.49)	43	421.61 (398.81 - 444.41)	16	413.36 (375.98 - 450.74)	.82	.004

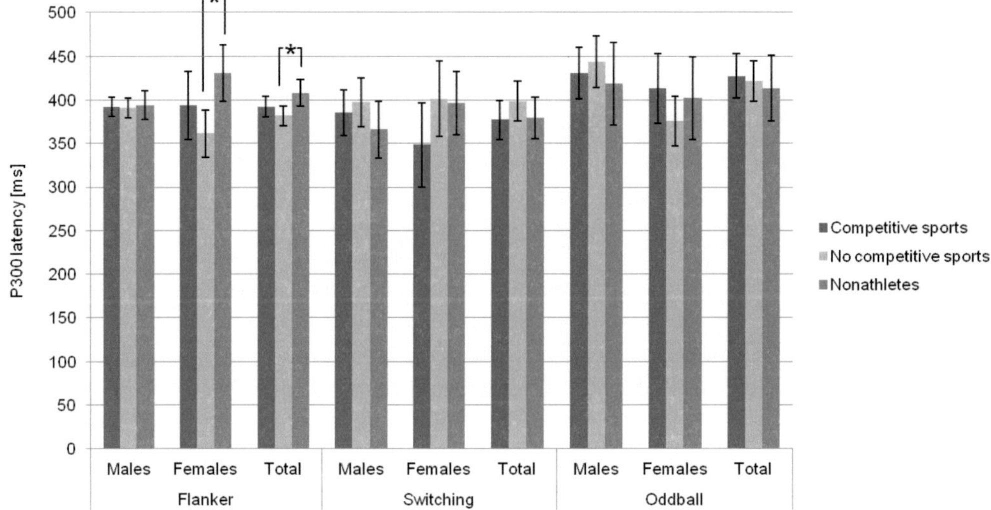

Figure 23:   Mean P300 latencies during Flanker, Switching and Oddball task for competitive sports groups and nonathletes; error bars represent 95% confidence intervals

N100 and P200

Participation in competitive sports did not have an effect on N100 and P200 amplitudes or latencies in this study sample except for P200 latencies in the Flanker task (F [2, 93] = 3.11, p = .05, η^2 = .063; males: F [2, 68] = 3.87, p = .03, η^2 = .102, females: F [2, 20] = 2.97, p = .07, η^2 = .22). Post-hoc analyses revealed a marginally significant (p = .06) difference in P200 latencies for males between athletes participating in competitive sports and those who do not participate. The mean amplitudes and latencies are presented in tables 41 and 42 and in figures 24 - 27.

Table 41: Mean N100 and P200 amplitudes for competitive sports groups and nonathletes in the Flanker, Switching and Oddball task

			Competitive sports		No competitive sports		Nonathletes			
			n	μV (95 % CI)	n	μV (95 % CI)	n	μV (95 % CI)	p	η^2
Flan-ker	N 100	♂	37	-1.81 (-2.10 - -1.52)	41	-1.66 (-1.93 - -1.38)	15	-1.37 (-1.83 - -.92)	.27	.029
		♀	7	-1.66 (-2.37 - -.95)	16	-1.91 (-2.38 - -1.44)	11	-1.98 (-2.55 - -1.41)	.76	.018
		Tot.	44	-1.79 (-2.06 - -1.52)	57	-1.73 (-1.96 - -1.49)	26	-1.63 (-1.98 - -1.28)	.78	.004
	P 200	♂	37	2.79 (2.31 - 3.27)	41	3.13 (2.68 - 3.59)	15	3.08 (2.33 - 3.84)	.56	.013
		♀	7	4.48 (3.13 - 5.84)	16	3.21 (2.32 - 4.11)	11	4.19 (3.11 - 5.27)	.20	.099
		Tot.	44	3.06 (2.58 - 3.53)	57	3.15 (2.74 - 3.57)	26	3.55 (2.93 - 4.17)	.43	.013
Swit-ching	P 200	♂	25	4.09 (3.52 - 4.66)	26	4.25 (3.69 - 4.81)	13	3.20 (2.41 - 3.99)	.09	.075
		♀	5	3.78 (2.23 - 5.33)	9	5.33 (4.17 - 6.48)	9	4.15 (2.99 - 5.30)	.19	.151
		Tot.	30	4.04 (3.49 - 4.59)	35	4.53 (4.02 - 5.03)	22	3.59 (2.95 - 4.23)	.08	.060
Odd-ball	N 100	♂	30	-4.10 (-4.90 - -3.31)	31	-4.86 (-5.65 - -4.08)	13	-5.17 (-6.38 - -3.97)	.24	.039
		♀	7	-6.16 (-7.01 - -4.61)	15	-5.48 (-6.53 - -4.42)	8	-4.91 (-6.35 - -3.46)	.49	.052
		Tot.	37	-4.49 (-5.20 - -3.78)	46	-5.06 (-5.70 - -4.43)	21	-5.07 (-6.01 - -4.13)	.44	.016
	P 200	♂	30	2.70 (2.00 - 3.41)	31	3.50 (2.81 - 4.19)	13	2.88 (1.81 - 3.95)	.26	.037
		♀	7	1.53 (-.32 - 3.37)	15	3.09 (1.83 - 4.34)	8	3.10 (1.38 - 4.83)	.33	.079
		Tot.	37	2.48 (1.81 - 3.15)	46	3.36 (2.76 - 3.97)	21	2.96 (2.07 - 3.86)	.16	.036

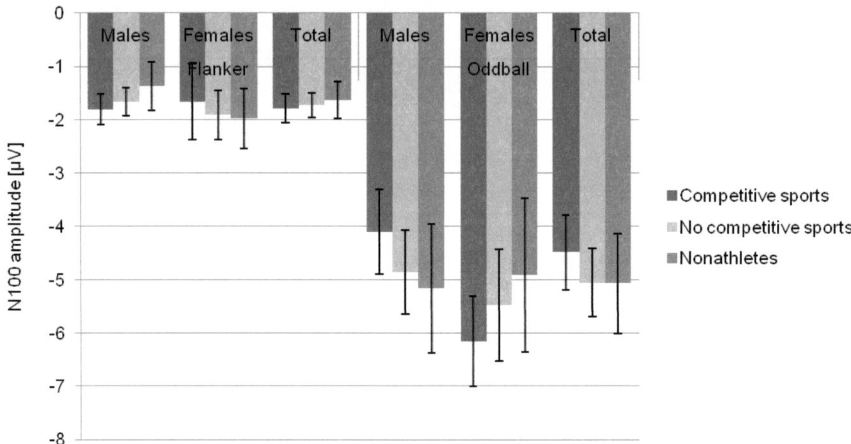

Figure 24:   Mean N100 amplitudes during Flanker and Oddball task for competitive sports groups and nonathletes; error bars represent 95% confidence intervals

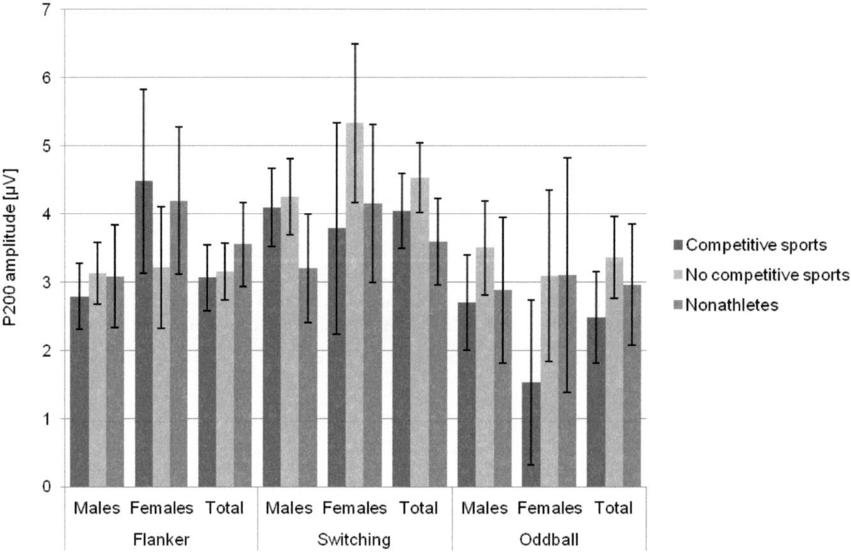

Figure 25:   Mean P200 amplitudes during Flanker, Switching and Oddball task for competitive sports groups and nonathletes; error bars represent 95% confidence intervals

Table 42:   Mean N100 and P200 latencies for competitive sports groups and nonathletes in the Flanker, Switching and Oddball task

			Competitive sports		No competitive sports		Nonathletes			
			n	ms (95 % CI)	n	ms (95 % CI)	n	ms (95 % CI)	p	η^2
Flan-ker	N 100	♂	15	143.03 (135.31 - 150.76)	12	136.59 (127.96 - 145.22)	5	139.26 (125.88 - 152.63)	.53	.043
		♀	2	152.05 (137.61 - 166.49)	7	134.40 (126.68 - 142.12)	5	145.31 (136.18 - 154.45)	.06	.404

		Tot.	17	144.10 (137.68 - 150.51)	19	135.78 (129.71 - 141.85)	10	142.29 (133.92 - 150.65)	.16	.083
	P 200	♂	26	229.50 (219.29 - 239.71)	32	212.96 (203.76 - 222.17)	13	231.64 (217.20 - 246.08)	.03	.102
		♀	6	229.20 (204.66 - 253.74)	10	223.34 (204.33 - 242.35)	9	224.05 (204.01 - 244.08)	.92	.008
		Tot.	32	229.44 (220.10 - 238.79)	42	215.43 (207.28 - 223.59)	22	228.53 (217.26 - 239.80)	.05	063
Swit-ching	P 200	♂	22	192.71 (184.85 - 200.57)	24	183.50 (175.98 - 191.03)	11	192.67 (181.55 - 203.78)	.19	.060
		♀	5	203.91 (183.54 - 224.27)	9	186.61 (171.43 - 201.79)	9	187.26 (172.08 - 202.44)	.33	.106
		Tot.	27	194.78 (187.41 - 202.16)	33	184.35 (177.68 - 191.02)	20	190.23 (181.67 - 198.80)	.12	.054
Odd-ball	N 100	♂	25	168.44 (161.56 - 175.32)	28	169.56 (163.06 - 176.06)	13	172.00 (162.45 - 181.54)	.83	.006
		♀	7	177.79 (168.60 - 186.98)	14	177.51 (171.01 - 184.01)	7	172.77 (163.58 - 181.96)	.65	.034
		Tot.	32	170.48 (164.87 - 176.10)	42	172.21 (167.31 - 177.11)	20	172.27 (165.17 - 179.36)	.88	.003
	P 200	♂	21	255.36 (246.56 - 264.15)	25	260.16 (252.10 - 268.22)	11	265.20 (253.05 - 277.35)	.41	.032
		♀	3	268.75 (242.66 - 294.84)	8	256.06 (240.08 - 272.03)	5	263.28 (243.07 - 283.49)	.64	.066
		Tot.	24	257.03 (248.88 - 265.19)	33	259.16 (252.21 - 266.12)	16	264.60 (254.61 - 274.59)	.50	.020

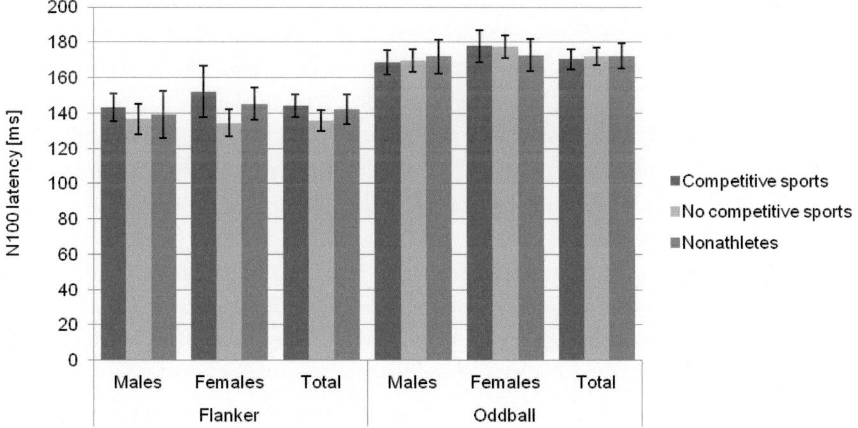

Figure 26: Mean N100 latencies during Flanker and Oddball task for competitive sports groups and nonathletes; error bars represent 95% confidence intervals

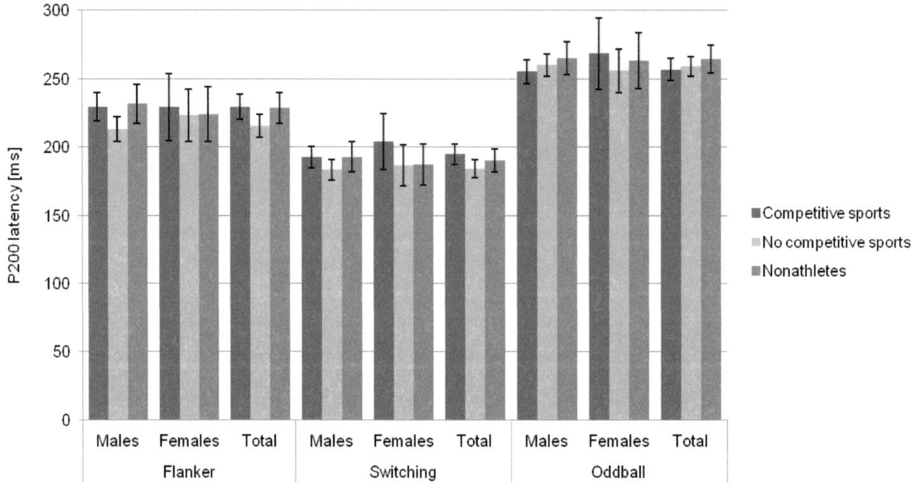

Figure 27:   Mean P200 latencies during Flanker, Switching and Oddball task for competitive sports groups and nonathletes; error bars represent 95% confidence intervals

4.3.2.2 Discussion

Response times during all cognitive tasks did not differ significantly between groups except for the Oddball task. Athletes not participating in competitive sports had the lowest response times followed by athletes participating in competitive sports. Thus, regular sports activity without high efforts for practicing and without pressure and expectations coming along with challenging situations and competitions might be beneficial for faster response preparations. By trend, nonathletes had the longest response times during the Switching and Oddball task compared to the two athletes groups.

With regard to event-related brain potentials, there were no significant effects for amplitudes but a tendency for larger amplitudes for competitors compared to non-competitors and nonathletes for males and the total sample in the Flanker and Oddball task. The results on P300 latencies are inconsistent. In the Flanker task, non-competitive athletes had significantly faster latencies than nonathletes, whereas there is no consistent tendency observed in the Switching and Oddball task. Interestingly, competitors had faster latencies in the speed condition but slower latencies in the accuracy condition of the Flanker task compared to non-competitive athletes. This might indicate that they have better adaptation skills since participants were asked to respond as quickly as possible during the speed condition.

In the Oddball task, female competitors tended to have larger N100 amplitudes whereas in the male subsample, nonathletes revealed the largest amplitudes. Since the number of participants per group was very low for this analysis, no significant effects were found. For N100 latencies, a consistent tendency was only found for the Flanker task with non-competitive athletes revealing the fastest latencies. For P200 amplitude, a large but also nonsignificant effect was found for females in the Switching task with the largest amplitudes for non-competitive athletes. There was further a general tendency for larger P200 amplitudes for non-competitive athletes compared to competitors and nonathetes in the Switching and Oddball task. However, due to small sample sizes, confidence intervals are very large.

4.3.3 Amount of Physical Activity

In this chapter, the relationship between the amount of physical activity and cognitive performance is described. The independent variables are amount of sports activity (kcal/week) and amount of habitual physical activity (kcal/week). For sports and habitual physical activity, three groups were built individually for the study sample.

In the low sports activity group, males and females have an energy expenditure of $\leq$ 1870 kcal/week ($\leq \sim$ 4.5 hours), in the medium group of 1871 - 3225 kcal/week ($\sim$ 4.5 - $\sim$ 8 hours) and in the high group, they have an energy expenditure of $\geq$ 3226 kcal/week ($\geq\sim$ 8 hours). Post-hoc analyses revealed significant differences between all groups (table 43).

Table 43:   Classification of sports activity groups

	Low		Medium		High		Total	
	n (%)	M (SD)	n (%)	M (SD)	n (%)	M (SD)	n	p
♂	26 (30)	1339 ± 427	29 (33)	2435 ± 391	33 (37)	4797 ± 1670	88	.00
♀	12 (41)	1361 ± 383	10 (35)	2487 ± 486	7 (24)	4177 ± 970	29	.00
Total	38 (33)	1346 ± 408	39 (33)	2448 ± 412	40 (34)	4688 ± 1578	117	.00

In the low habitual physical activity group, males and females have an energy expenditure of $\leq$ 440 kcal/week ($\leq \sim$ 1 hour), in the medium group of 440 - 800 kcal/week ($\sim$ 1 - $\sim$ 2 hours) and in the high group, they have an energy expenditure of $\geq$ 800 kcal/week ($\geq\sim$ 2 hours). Post-hoc analyses revealed significant differences between all groups (table 44).

Table 44:   Classification of habitual physical activity groups

	Low		Medium		High		Total	
	n (%)	M (SD)	n (%)	M (SD)	n (%)	M (SD)	n	p
♂	36 (35)	260 ± 145	47 (45)	621 ± 106	21 (20)	1266 ± 575	104	.00
♀	13 (29)	316 ± 74	21 (48)	607 ± 102	10 (23)	1139 ± 472	44	.00
Total	49 (33)	268 ± 134	68 (46)	617 ± 105	31 (21)	1225 ± 539	148	.00

> *RQ 3*: Is the amount of sports and physical activity related to cognitive performance in young adulthood?

4.3.3.1 Amount of Sports Activity (kcal/week) - Results

Behavioral Measures
Sports activity (kcal/week) did not have any effect on behavioral measures in the Flanker, Switching or Oddball task (table 45 & 46, figure 28 & 29).

Table 45:   Mean response times for sports activity groups in the Flanker, Switching and Oddball task

		Low		Medium		High			
		n	ms (95 % CI)	n	ms (95 % CI)	n	ms (95 % CI)	p	η^2
Flan-ker	♂	23	479.29 (463.42 - 495.17)	24	496.91 (481.37 - 512.45)	31	496.64 (482.97 - 510.32)	.19	.043
	♀	11	500.42 (468.33 - 532.52)	9	521.98 (486.50 - 557.46)	5	515.54 (467.93 - 563.14)	.64	.040
	Tot.	34	486.13 (471.84 - 500.42)	33	503.75 (489.24 - 518.26)	36	499.27 (485.38 - 513.16)	.21	.031
Swit-ching	♂	23	694.38 (660.55 - 728.21)	21	706.54 (671.14 - 741.94)	28	697.67 (667.01 - 728.33)	.88	.004
	♀	9	664.11 (586.51 - 741.72)	10	721.74 (648.11 - 795.36)	6	710.67 (615.62 - 805.72)	.52	.058
	Tot.	32	685.86 (654.68 - 717.05)	31	711.44 (679.76 - 743.12)	34	699.96 (669.71 - 730.22)	.52	.014
Odd-ball	♂	23	479.20 (439.31 - 519.10)	22	486.76 (445.97 - 527.55)	30	519.22 (484.29 - 554.16)	.27	.035
	♀	12	511.62 (458.50 - 564.74)	9	498.21 (436.88 - 559.55)	6	518.04 (442.92 - 593.15)	.90	.008
	Tot.	35	490.32 (459.00 - 521.63)	31	490.08 (456.81 - 523.36)	36	519.03 (488.15 - 549.90)	.33	.022

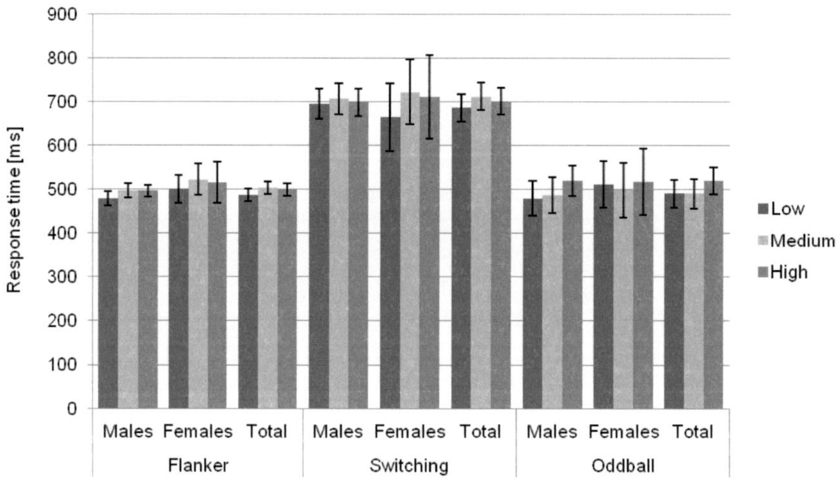

Figure 28:   Mean response times during Flanker, Switching and Oddball task for sports activity groups; error bars represent 95% confidence intervals

Table 46:   Errors in the Switching task for sports activity groups

		Low		Medium		High			
		n	No. (95 % CI)	n	No. (95 % CI)	n	No. (95 % CI)	p	η^2
Errors	♂	23	3.41 (2.50 - 4.32)	22	3.94 (3.01 - 4.87)	30	3.97 (3.17 - 4.76)	.61	.013
	♀	11	4.98 (3.38 - 6.57)	10	3.38 (1.70 - 5.05)	6	3.42 (1.26 - 5.58)	.30	.094

| | | Total | 34 | 3.92 (3.14 - 4.70) | 32 | 3.77 (2.96 - 4.57) | 36 | 3.88 (3.11 - 4.64) | .96 | .001 |

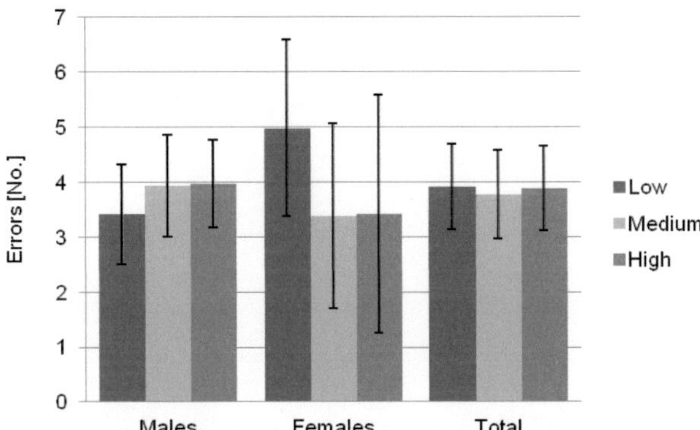

Figure 29: Errors during the Switching task for sports activity groups; error bars represent 95% confidence intervals

P300

Sports activity (kcal/week) did not have any effects on P300 amplitudes and latencies in either the Flanker, Switching or Oddball task except for a condition x sports activity interaction for amplitudes during Switching task (F [2, 59] = 3.89, p = .03, η^2 = .117). This interaction indicates that there was a large difference between homogenous (4.55 µV, 95 % CI 4.02 - 5.09) and heterogeneous trials (3.70 µV, 95 % CI 3.25 - 4.16) for the high activity group whereas there was no difference between conditions for the other groups. Also, there was a significant main effect for amplitudes during the Switching task for females (F [2, 59] = .85, p = .43, η^2 = .028; males: F [2, 45] = .35, p = .70, η^2 = .015, females: F [2, 11] = 4.89, p = .03, η^2 = .471). Post-hoc analyses showed significant differences between the medium and high sports activity group (p = .03). However, it must be considered that the group sizes for females are very small. The mean P300 amplitudes for the cognitive tasks are presented in table 47 and figure 30, the latencies are given in table 48 and figure 31.

Table 47: Mean P300 amplitudes for sports activity groups in the Flanker, Switching and Oddball task

		Low		Medium		High			
		n	µV (95 % CI)	n	µV (95 % CI)	n	µV (95 % CI)	p	η^2
Flanker	♂	22	4.83 (4.25 - 5.41)	22	4.26 (3.68 - 4.84)	31	5.13 (4.64 - 5.61)	.08	.068
	♀	10	4.56 (3.51 - 5.62)	8	4.37 (3.19 - 5.54)	5	4.71 (3.23 - 6.20)	.93	.008
	Total	32	4.75 (4.26 - 5.24)	30	4.29 (3.78 - 4.79)	36	5.07 (4.61 - 5.53)	.08	.051
Switching	♂	12	3.62 (3.05 - 4.19)	14	3.85 (3.32 - 4.37)	22	3.91 (3.49 - 4.33)	.70	.015
	♀	4	3.90 (2.68 - 5.13)	8	3.85 (2.98 - 4.72)	2	6.53 (4.79 - 8.27)	.03	.471
	Total	16	3.69 (3.15 - 4.23)	22	3.85 (3.38 - 4.31)	24	4.13 (3.69 - 4.57)	.43	.028

		n	Low	n	Medium	n	High	p	η²
Oddball	♂	15	5.81 (4.52 - 7.10)	19	5.97 (4.82 - 7.12)	25	6.28 (5.28 - 7.28)	.83	.007
	♀	10	6.94 (5.42 - 8.45)	8	5.69 (4.00 - 7.39)	4	8.07 (5.67- 10.47)	.24	.139
	Total	25	6.26 (5.28 - 7.24)	27	5.89 (4.95 - 6.83)	29	6.53 (5.62 - 7.43)	.62	.012

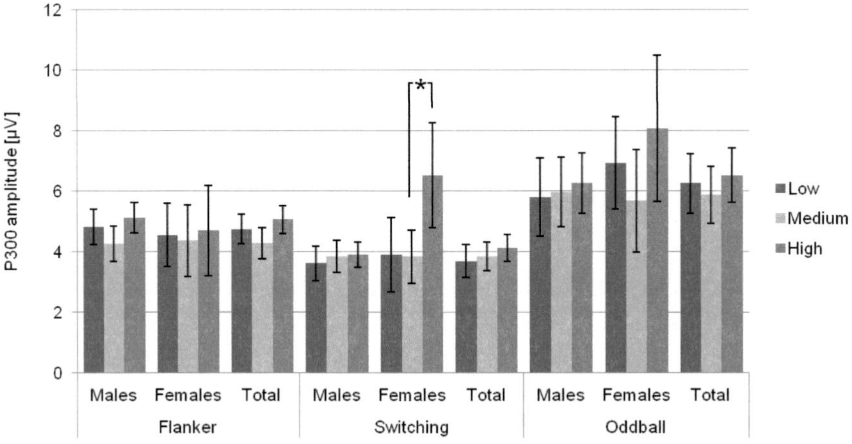

Figure 30:   Mean P300 amplitudes during Flanker, Switching and Oddball task for sports activity groups; error bars represent 95% confidence intervals

Table 48:   Mean P300 latencies for sports activity groups in the Flanker, Switching and Oddball task

			Low		Medium		High		
		n	ms (95 % CI)	n	ms (95 % CI)	n	ms (95 % CI)	p	η²
Flanker	♂	18	390.66 (377.46 - 403.86)	15	385.81 (371.35 - 400.27)	26	395.12 (384.13 - 406.10)	.59	.019
	♀	8	368.48 (337.76 - 399.20)	8	371.53 (340.81 - 402.25)	3	379.64 (329.48 - 429.81)	.92	.010
	Tot.	26	383.83 (371.47 - 396.20)	23	380.84 (367.70 - 393.99)	29	393.51 (381.81 - 405.22)	.32	.030
Switching	♂	6	408.67 (365.18 - 452.16)	7	378.87 (338.61 - 419.13)	13	389.07 (359.52 - 418.61)	.58	.046
	♀	3	345.68 (297.15 - 394.21)	4	389.36 (347.33 - 431.39)	2	402.70 (343.26 - 462.14)	.21	.407
	Tot.	9	387.67 (354.45 - 420.90)	11	382.68 (352.63 - 412.73)	15	390.89 (365.15 - 416.62)	.92	.006
Oddball	♂	15	460.76 (417.58 - 503.95)	19	426.58 (388.20 - 464.95)	23	432.47 (397.60 - 467.35)	.46	.028
	♀	9	389.12 (349.32 - 428.92)	8	389.68 (347.47 - 431.89)	4	381.06 (321.36 - 440.75)	.97	.004
	Tot.	24	433.90 (401.38 - 466.41)	27	415.64 (384.99 - 446.30)	27	424.86 (394.20 - 455.51)	.72	.009

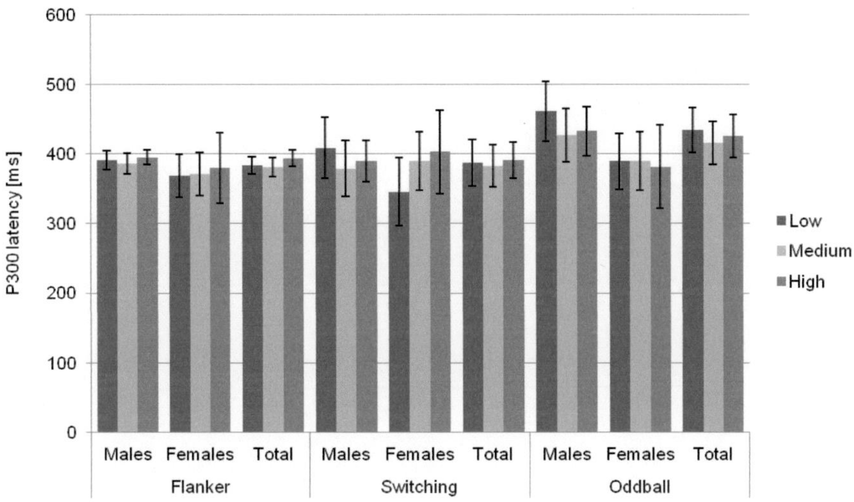

Figure 31:   Mean P300 latencies during Flanker, Switching and Oddball task for sports activity groups; error bars represent 95% confidence intervals

N100 and P200

Sports activity (kcal/week) did not have any effect on N100 and P200 amplitudes or latencies in this study sample except for P200 amplitudes for males in the Flanker task (F [2, 96] = .65, p = .59, η^2 = .013; males: F [2, 73] = 3.37, p = .04, η^2 = .084, females: F [2, 20] = 2.97, p = .07, η^2 = .22). Post-hoc analyses revealed a marginally significant (p = .06) difference in P200 amplitudes between the low and the high sports activity group in males. The mean amplitudes and latencies are given in table 49 and 50 and figure 32 - 35.

Table 49:   Mean N100 and P200 amplitudes for sports activity groups in the Flanker, Switching and Oddball task

			Low		Medium		High			
			n	μV (95 % CI)	n	μV (95 % CI)	n	μV (95 % CI)	p	η^2
Flanker	N 100	♂	22	-1.65 (-2.5 - -1.26)	23	-1.71 (-2.10 - -1.33)	31	-1.83 (-2.16 - -1.49)	.78	.007
		♀	10	-1.57 (-2.17 - -.96)	8	-1.94 (-2.61 - -1.26)	5	-2.19 (-3.04 - -1.33)	.45	.077
		Tot.	32	-1.63 (-1.95 - -1.30)	31	-1.77 (-2.10 - -1.44)	36	-1.88 (-2.18 - -1.57)	.53	.013
	P 200	♂	22	3.67 (3.03 - 4.30)	23	2.70 (2.09 - 3.32)	31	2.68 (2.15 - 3.21)	.04	.084
		♀	10	2.69 (1.65 - 3.73)	8	4.23 (3.07 - 5.38)	5	4.41 (2.94 - 5.87)	.07	.229
		Tot.	32	3.36 (2.80 - 3.92)	31	3.10 (2.53 - 3.66)	36	2.92 (2.39 - 3.45)	.53	.013
Switching	P 200	♂	14	4.20 (3.38 - 5.02)	14	4.17 (3.35 - 4.99)	23	4.16 (3.52 - 4.80)	.99	.000
		♀	4	4.02 (2.03 - 6.01)	8	5.04 (3.63 - 6.45)	2	5.23 (2.41 - 8.05)	.62	.082

			n			n			n		
		Tot.	18	4.16 (3.42 - 4.90)		22	4.48 (3.81 - 5.16)		25	4.24 (3.61 - 4.87)	.79	.008
Oddball	N 100	♂	15	-4.13 (-5.25 - -3.01)		19	-4.53 (-5.53 - -3.54)		25	-4.72 (-5.59 - -3.86)	.70	.012
		♀	10	-6.08 (-7.18 - -4.98)		8	-5.31 (-6.55 - -4.08)		4	-5.48 (-7.23 - -3.74)	.61	051
		Total	25	-4.91 (-5.75 - -4.08)		27	-4.76 (-5.57 - -3.96)		29	-4.83 (-5.60 - -4.05)	.97	.001
	P 200	♂	15	2.68 (1.62 - 3.74)		19	3.23 (2.28 - 4.17)		25	3.12 (2.30 - 3.94)	.72	.012
		♀	10	2.12 (.75 - 3.49)		8	3.71 (2.18 - 5.24)		4	1.52 (-.64 - 3.69)	.17	.171
		Total	25	2.46 (1.64 - 3.28)		27	3.37 (2.58 - 4.16)		29	2.90 (2.14 - 3.66)	.28	.032

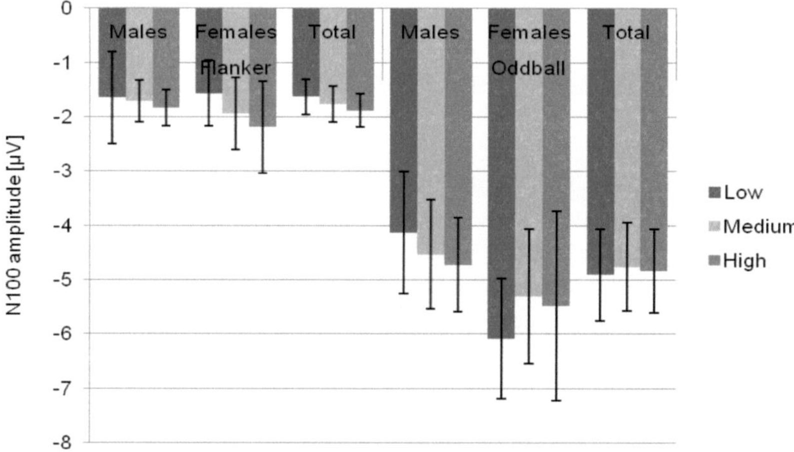

Figure 32: Mean N100 amplitudes during Flanker and Oddball task for sports activity groups; error bars represent 95% confidence intervals

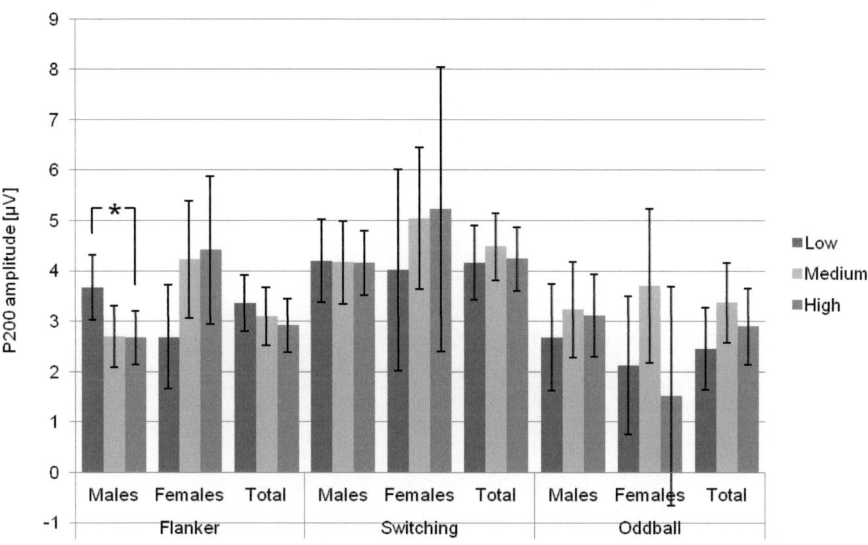

Figure 33: Mean P200 amplitudes during Flanker, Switching and Oddball task for sports activity groups; error bars represent 95% confidence intervals

Table 50: Mean N100 and P200 latencies for sports activity groups in the Flanker, Switching and Oddball task

			Low		Medium		High			
			n	ms (95 % CI)	n	ms (95 % CI)	n	ms (95 % CI)	p	η^2
Flan-ker	N 100	♂	4	133.98 (117.73 - 156.24)	6	145.05 (131.78 - 158.32)	17	139.90 (132.02 - 147.79)	.56	.048
		♀	4	139.60 (122.46 - 156.74)	3	134.31 (114.51 - 154.11)	2	141.80 (117.55 - 166.04)	.82	.063
		Tot.	8	136.79 (126.00 - 147.59)	9	141.47 (131.29 - 151.65)	19	140.10 (133.10 - 147.11)	.80	.013
	P 200	♂	19	219.61 (206.71 - 232.51)	15	219.60 (205.08 - 234.12)	23	221.11 (209.39 - 232.84)	.98	.001
		♀	6	225.78 (202.69 - 248.88)	6	221.71 (198.62 - 244.81)	4	230.91 (202.62 - 359.19)	.86	.022
		Tot.	25	221.09 (210.23 - 231.96)	21	220.20 (208.35 - 232.05)	27	222.56 (212.11 - 233.02)	.96	.001
Swit-ching	P 200	♂	12	184.12 (172.80 - 195.43)	13	188.75 (177.88 - 199.62)	21	189.56 (181.00 - 198.11)	.73	.014
		♀	4	194.78 (166.77 - 222.78)	8	187.09 (167.28 - 206.89)	2	211.62 (172.01 - 251.23)	.49	.121
		Tot.	16	186.78 (176.38 - 197.18)	21	188.11 (179.03 - 197.19)	23	191.47 (182.80 - 200.15)	.76	.009
Oddball	N 100	♂	12	160.03 (149.62 - 170.43)	16	171.83 (162.82 - 180.84)	24	171.58 (164.23 - 178.94)	.15	.074
		♀	10	171.88 (164.60 - 179.15)	7	182.81 (174.11 - 191.51)	4	182.81 (171.30 - 194.32)	.10	.225
		Tot.	22	165.41 (158.28 - 172.55)	23	175.17 (168.19 - 182.15)	28	173.19 (166.86 - 179.51)	.13	.058

P 200	♂	11	253.48 (240.45 - 266.51)	13	259.74 (247.75 - 271.72)	20	258.79 (249.13 - 268.45)	.74	.014
	♀	4	248.24 (232.04 - 264.45)	6	261.59 (248.36 - 274.82)	1	292.19 (259.77 - 324.60)	.06	.503
	Tot.	15	252.08 (241.54 - 262.62)	19	260.32 (250.96 - 269.69)	21	260.38 (251.47 - 269.29)	.41	.032

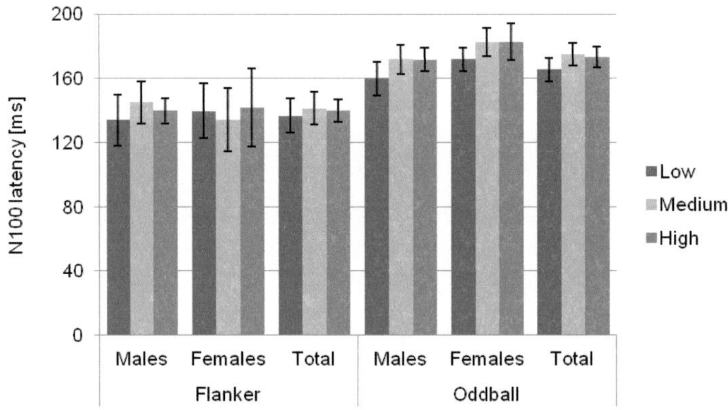

Figure 34:   Mean N100 latencies during Flanker and Oddball task for sports activity groups; error bars represent 95% confidence intervals

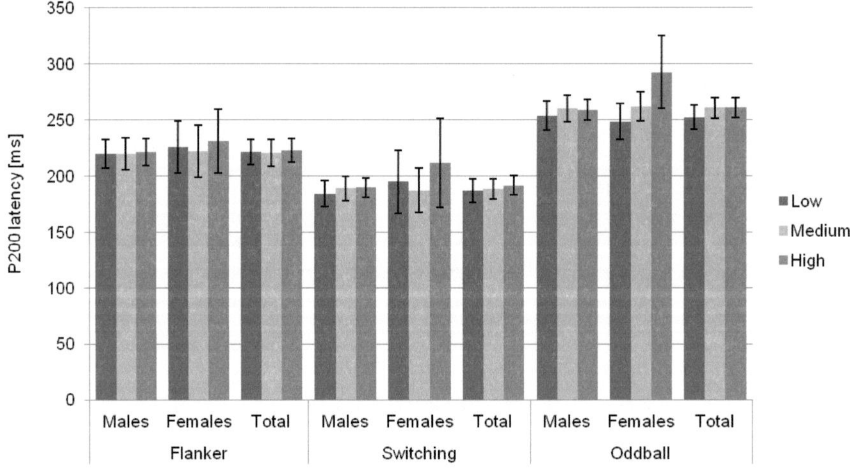

Figure 35:   Mean P200 latencies during Flanker, Switching and Oddball task for sports activity groups; error bars represent 95% confidence intervals

4.3.3.2 Amount of Habitual Physical Activity (kcal/week) - Results

Behavioral Measures

Habitual physical activity (kcal/week) had no effect on behavioral measures in the Flanker, Switching or Oddball task (table 51 & 52, figure 36 & 37).

Table 51:   Mean response times for habitual physical activity groups in the Flanker, Switching and Oddball task

		Low		Medium		High			
		n	ms (95 % CI)	n	ms (95 % CI)	n	ms (95 % CI)	p	η^2
Flan-ker	♂	30	491.86 (476.45 - 507.28)	43	492.26 (479.39 - 505.14)	19	494.19 (474.82 - 513.56)	.98	.000
	♀	11	506.95 (478.39 - 535.51)	17	515.87 (492.90 - 538.85)	8	508.23 (474.74 - 541.73)	.86	.009
	Tot.	41	495.91 (482.31 - 509.51)	60	498.95 (487.71 - 510.19)	27	498.35 (481.59 - 515.11)	.94	.001
Swit-ching	♂	28	679.20 (649.22 - 709.17)	41	709.61 (684.84 - 734.38)	18	721.10 (683.72 - 758.49)	.16	.042
	♀	12	675.21 (615.59 - 734.82)	15	715.53 (662.21 - 768.85)	10	711.89 (646.58 - 777.19)	.56	.034
	Tot.	40	678.00 (651.21 - 704.79)	56	711.19 (688.55 - 733.84)	28	717.81 (685.79 - 749.84)	.10	.038
Odd-ball	♂	29	488.87 (454.89 - 522.85)	43	503.53 (475.62 - 531.43)	18	514.73 (471.60 - 557.86)	.63	.011
	♀	12	511.64 (452.70 - 570.59)	17	539.08 (489.55 - 588.60)	10	535.00 (470.43 - 599.57)	.76	.015
	Tot.	41	495.53 (466.33 - 524.74)	60	513.60 (489.46 - 537.74)	28	521.97 (486.64 - 557.30)	.48	.012

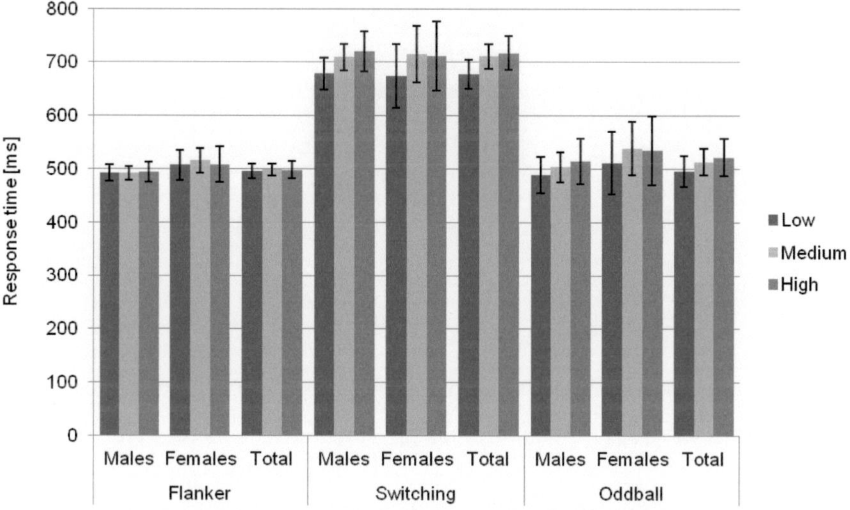

Figure 36:   Mean response times during Flanker, Switching and Oddball task for habitual physical activity groups; error bars represent 95% confidence intervals

Table 52:   Errors in the Switching task for habitual physical activity groups

		Low		Medium		High			
		n	No. (95 % CI)	n	No. (95 % CI)	n	No. (95 % CI)	p	η^2
Errors	♂	29	4.34 (3.58 - 5.10)	42	3.35 (2.72 - 3.98)	19	3.72 (2.78 - 4.66)	.15	.043
	♀	11	4.25 (2.59 - 5.91)	17	4.31 (2.98 - 5.64)	10	3.78 (2.04 - 5.51)	.88	.008
	Total	40	4.31 (3.61 - 5.02)	59	3.63 (3.05 - 4.21)	29	3.74 (2.91 - 4.57)	.32	.018

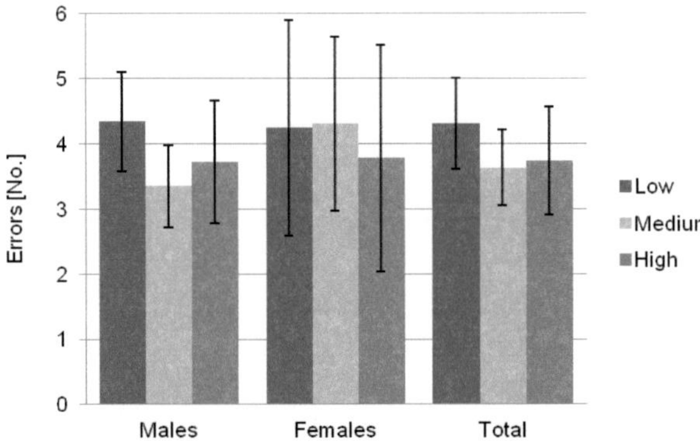

Figure 37:   Errors during the Switching task for habitual physical activity groups; error bars represent 95% confidence intervals

P300

Habitual physical activity (kcal/week) did not have any effect on P300 in the Flanker and Oddball task (table 53 & figure 38). There was a marginally significant main effect of physical activity on P300 amplitudes in the Switching task (F [2, 80] = 2.96, p = .06, η^2 = .069; males: F [2, 58] = 1.40, p = .26, η^2 = .046; females: F [2, 19] = 1.05, p = .37, η^2 = .100). Post-hoc analyses revealed a significant difference between the low and high activity group (p = .05). No effects were observed for P300 latencies in the Flanker, Switching or Oddball task (table 54 & figure 39).

Table 53:   Mean P300 amplitudes for habitual physical activity groups in the Flanker, Switching and Oddball task

		Low		Medium		High			
		n	µV (95 % CI)	n	µV (95 % CI)	n	µV (95 % CI)	p	η^2
Flanker	♂	28	4.54 (4.02 - 5.05)	42	4.75 (4.33 - 5.17)	19	4.91 (4.29 - 5.54)	.64	.010
	♀	10	4.48 (3.56 - 5.40)	15	3.97 (3.21 - 4.72)	7	5.21 (4.10 - 6.31)	.18	.111
	Tot.	38	4.52 (4.08 - 4.97)	57	4.55 (4.18 - 4.91)	26	4.99 (4.45 - 5.53)	.34	.018
Switching	♂	20	3.59 (3.18 - 4.01)	31	3.81 (3.48 - 4.14)	10	4.19 (3.61 - 4.77)	.26	.046
	♀	6	3.67 (2.62 - 4.72)	8	4.45 (3.54 - 5.36)	8	4.58 (3.67 - 5.49)	.37	.100

		n	Low	n	Medium	n	High	p	η^2
	Tot.	26	3.61 (3.22 - 4.00)	39	3.94 (3.62 - 4.26)	18	4.36 (3.89 - 4.83)	.06	.069
Oddball	♂	25	5.59 (4.61 - 6.56)	36	6.06 (5.25 - 6.88)	11	5.28 (3.81 - 6.75)	.58	.016
	♀	8	6.81 (5.19 - 8.42)	13	5.75 (4.49 - 7.02)	8	7.09 (5.48 - 8.71)	.35	.077
	Tot.	33	5.88 (5.05 - 6.71)	49	5.98 (5.30 - 6.66)	19	6.05 (4.95 - 7.14)	.97	.001

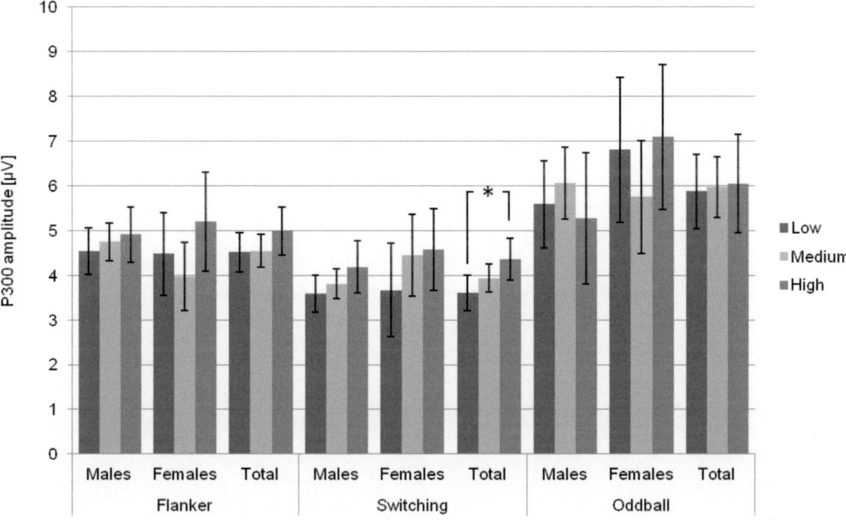

Figure 38: Mean P300 amplitudes during Flanker, Switching and Oddball task for habitual physical activity groups; error bars represent 95% confidence intervals

Table 54: Mean P300 latencies for habitual physical activity groups in the Flanker, Switching and Oddball task

		Low n	Low ms (95 % CI)	Medium n	Medium ms (95 % CI)	High n	High ms (95 % CI)	p	η^2
Flan-ker	♂	25	391.12 (378.95 - 403.29)	31	387.18 (376.24 - 398.11)	17	401.35 (386.59 - 416.12)	.31	.033
	♀	10	371.68 (335.38 - 407.98)	10	398.63 (362.33 - 434.94)	7	408.35 (364.96 - 451.74)	.37	.080
	Tot.	35	385.57 (372.67 - 398.46)	41	389.97 (378.05 - 401.89)	24	403.39 (387.82 - 418.97)	.21	.032
Swit-ching	♂	12	379.10 (350.32 - 407.89)	17	392.12 (367.94 - 416.31)	6	373.51 (332.80 - 414.22)	.65	.026
	♀	3	371.18 (311.57 - 430.79)	6	410.46 (368.31 - 452.61)	6	376.01 (333.86 - 418.16)	.38	.149
	Tot.	15	377.52 (352.93 - 402.10)	23	396.91 (377.05 - 416.76)	12	374.76 (347.27 - 402.25)	.31	.048
Odd-	♂	25	438.02	33	427.29	10	451.69	.68	.012

ball			(406.19 - 469.85)		(399.59 - 454.99)		(401.37 - 502.02)		
	♀	6	415.89 (373.42 - 458.35)	13	395.35 (366.50 - 424.20)	7	359.71 (320.39 - 399.03)	.14	.157
	Tot.	31	433.74 (406.83 - 460.64)	46	418.26 (396.18 - 440.35)	17	413.82 (377.48 - 450.15)	.59	.011

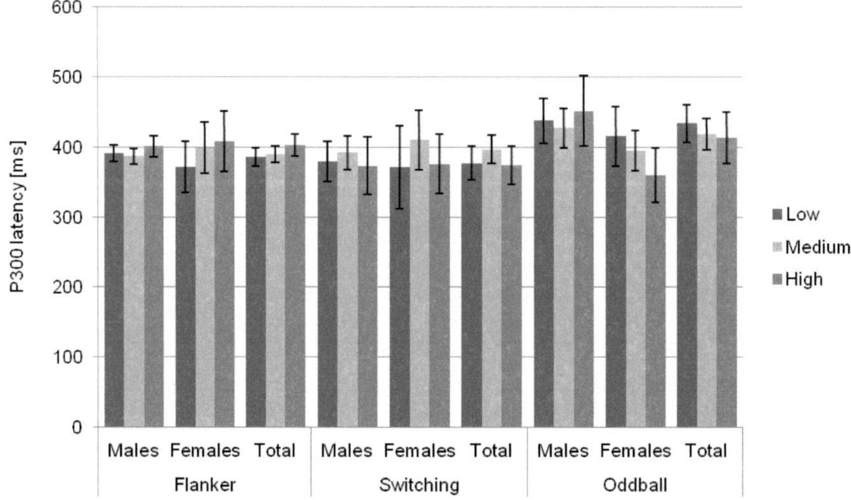

Figure 39:   Mean P300 latencies during Flanker, Switching and Oddball task for habitual physical activity groups; error bars represent 95% confidence intervals

N100 and P200

For P200 amplitude during the Switching task, no main effect but a significant type of trials x physical activity interaction (F [2, 83] = 3.64, p = .03, η^2 = .081) was observed. Further analyses revealed a difference between odd/even (3.96 µV, 95 % CI 3.23 - 4.68) and </>5 trials (4.41 µV, 95 % CI 3.67 - 5.16) for the high active group but no differences between the other two groups. No further effects were found (table 55 & 56, figure 40 - 43).

Table 55:   Mean N100 and P200 amplitudes for habitual physical activity groups in the Flanker, Switching and Oddball task

			Low		Medium		High			
			n	µV (95 % CI)	n	µV (95 % CI)	n	µV (95 % CI)	p	η^2
Flan-ker	N 100	♂	29	-1.73 (-2.07 - -1.40)	42	-1.65 (-1.93 - -1.38)	19	-1.70 (-2.12 - -1.29)	.94	.002
		♀	10	-1.86 (-2.46 - -1.26)	15	-1.79 (-2.28 - -1.31)	8	-2.15 (-2.82 - -1.49)	.67	.027
		Tot.	39	-1.76 (-2.05 - -1.48)	57	-1.69 (-1.93 - -1.45)	27	-1.84 (-2.18 - -1.49)	.78	.004
	P 200	♂	29	2.63 (2.08 - 3.17)	42	3.17 (2.71 - 3.62)	19	3.12 (2.45 - 3.80)	.29	.028
		♀	10	3.52 (2.36 - 4.67)	15	3.57 (2.62 - 4.51)	8	4.71 (3.41 - 6.00)	.30	.078
		Tot.	39	2.85	57	3.27	27	3.59	.17	.029

			n		n		n			
				(2.35 - 3.36)		(2.85 - 3.69)		(2.98 - 4.20)		
Swit-ching	P 200	♂	20	3.77 (3.11 - 4.42)	33	4.16 (3.65 - 4.68)	11	3.79 (2.90 - 4.67)	.57	.018
		♀	6	4.31 (2.73 - 5.88)	8	4.49 (3.12 - 5.85)	8	4.73 (3.37 - 6.09)	.91	.010
		Tot.	26	3.89 (3.28 - 4.50)	41	4.23 (3.74 - 4.71)	19	4.18 (3.47 - 4.90)	.68	.009
Odd-ball	N 100	♂	25	-5.28 (-6.14 - -4.43)	36	-4.46 (-5.17 - -3.75)	11	-3.72 (-5.01 - -2.44)	.11	.061
		♀	8	-6.01 (-7.46 - -4.55)	13	-5.48 (-6.63 - -4.34)	9	-5.02 (-6.39 - -3.64)	.61	.036
		Tot.	33	-5.46 (-6.19 - -4.72)	49	-4.73 (-5.33 - -4.13)	20	-4.31 (-5.25 - -3.36)	.13	.040
	P 200	♂	25	2.79 (2.01 - 3.56)	36	3.27 (2.62 - 3.91)	11	2.69 (1.52 - 3.86)	.54	.018
		♀	8	1.81 (.11 - 3.52)	13	3.54 (2.21 - 4.88)	9	2.36 (.75 - 3.97)	.24	.100
		Tot.	33	2.55 (1.84 - 3.26)	49	3.34 (2.76 - 3.92)	20	2.54 (1.63 - 3.45)	.15	.037

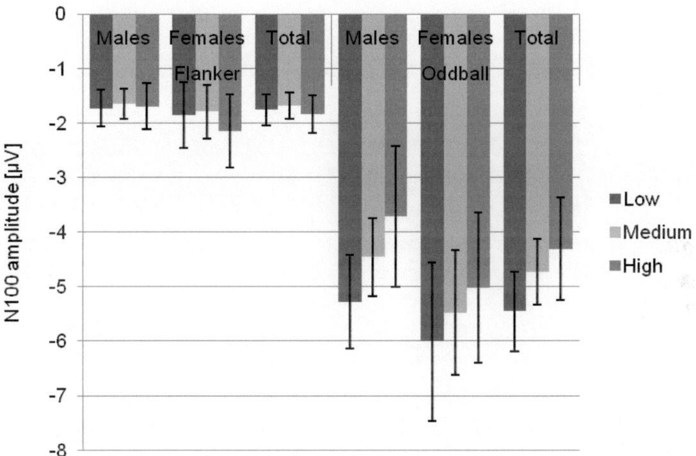

Figure 40:   Mean N100 amplitudes during Flanker and Oddball task for habitual physical activity groups; error bars represent 95% confidence intervals

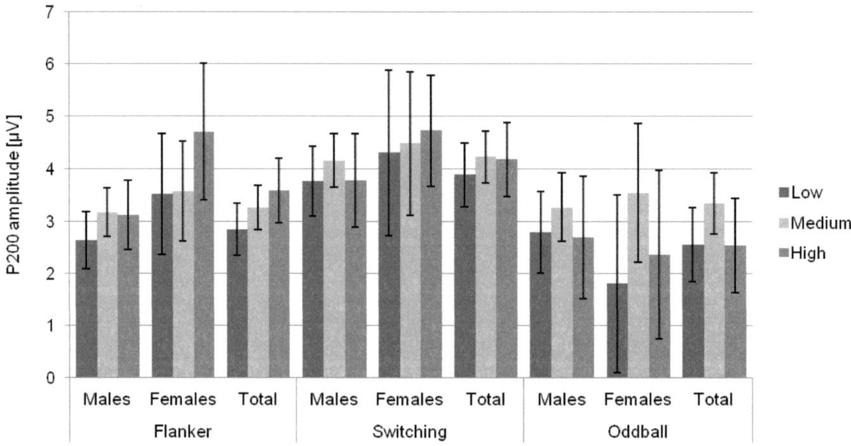

Figure 41:   Mean P200 amplitudes during Flanker, Switching and Oddball task for habitual physical activity groups; error bars represent 95% confidence intervals

Table 56:   Mean N100 and P200 latencies for habitual physical activity groups in the Flanker, Switching and Oddball task

			Low		Medium		High			
			n	ms (95 % CI)	n	ms (95 % CI)	n	ms (95 % CI)	p	η^2
Flanker	N 100	♂	12	144.56 (136.29 - 152.84)	15	134.77 (127.36 - 142.17)	5	144.92 (132.10 - 157.74)	.16	.121
		♀	4	136.43 (123.72 - 149.13)	5	143.75 (132.38 - 155.12)	5	141.41 (130.04 - 152.77)	.65	.077
		Tot.	16	142.53 (135.78 - 149.28)	20	137.01 (130.98 - 143.05)	10	143.16 (134.63 - 151.70)	.36	.047
	P 200	♂	19	226.71 (214.21 - 239.20)	36	217.37 (208.29 - 226.45)	14	230.25 (215.69 - 244.81)	.25	.041
		♀	7	219.36 (197.51 - 241.22)	9	229.36 (210.09 - 248.63)	8	230.30 (209.86 - 250.74)	.71	.032
		Tot.	26	224.73 (214.13 - 235.33)	45	219.77 (211.71 - 227.83)	22	230.27 (218.74 - 241.79)	.33	.024
Switching	P 200	♂	18	191.93 (183.05 - 200.80)	30	186.36 (179.49 - 193.24)	9	190.84 (178.29 - 203.39)	.58	.020
		♀	6	181.35 (162.02 - 200.67)	8	196.61 (179.87 - 213.34)	8	190.14 (173.40 - 206.87)	.47	.076
		Tot.	24	189.28 (181.21 - 197.35)	38	188.52 (182.11 - 194.93)	17	190.51 (180.92 - 200.10)	.94	.002
Oddball	N 100	♂	24	167.35 (160.29 - 174.41)	31	170.46 (164.25 - 176.68)	10	172.27 (161.33 - 183.21)	.70	.011
		♀	7	172.77 (163.72 - 181.82)	12	175.65 (168.74 - 182.56)	9	180.21 (172.23 - 188.19)	.44	.064
		Tot.	31	168.57 (162.91 - 174.23)	43	171.91 (167.11 - 176.72)	19	176.03 (168.80 - 183.26)	.27	.028
	P 200	♂	19	263.82 (254.42 - 273.21)	28	256.19 (248.45 - 263.94)	8	258.98 (244.50 - 273.47)	.46	.029
		♀	3	250.52 (224.91	8	266.31 (250.62	5	257.81 (237.97	.51	.099

		- 276.14)		- 282.00)		- 277.65)		
Tot.	22	262.00 (253.33	36	258.44 (251.66	13	258.53 (247.25	.80	.007
		- 270.68)		- 265.22)		- 269.82)		

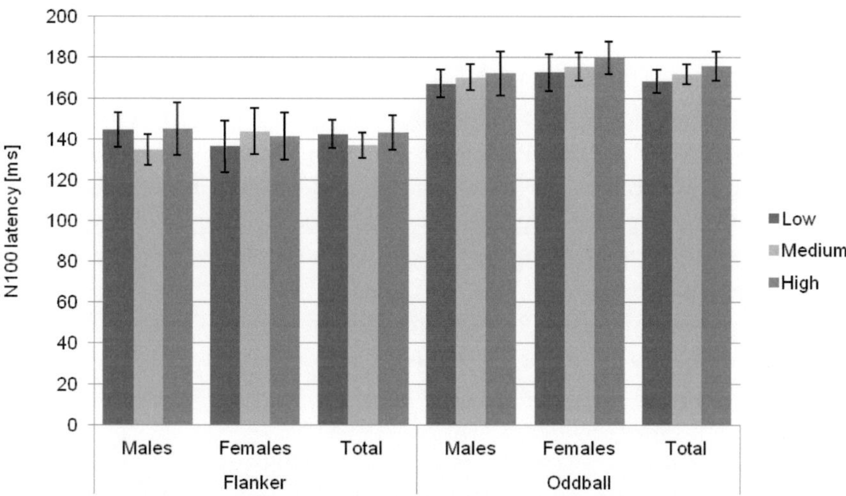

Figure 42: Mean N100 latencies during Flanker and Oddball task for habitual physical activity groups; error bars represent 95% confidence intervals

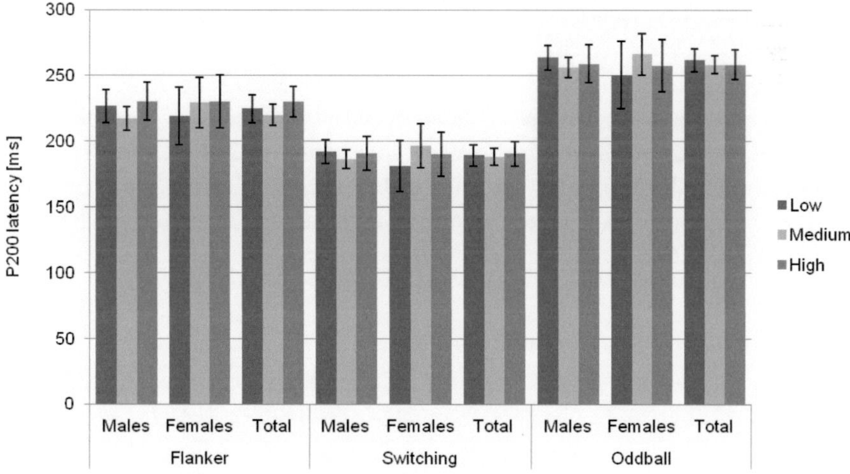

Figure 43: Mean P200 latencies during Flanker, Switching and Oddball task for habitual physical activity groups; error bars represent 95% confidence intervals

4.3.3.3 Discussion

The amount of sports and habitual physical activity was not related to response times during all cognitive tasks. However, there was a tendency for faster response times for the low activity groups. This indicates that the amount of sports and physical activity might even impair response-related processing speed during cognitive tasks, although further correlation analyses revealed low coefficients ($r = <.30$).

With regard to event-related brain potentials, there was no consistent relationship between sports activity and P300 component. For habitual physical activity, the only effect was found for P300 amplitudes during the Switching task. Participants with a high amount of habitual physical activity had larger amplitudes than those with a low amount which could be an indicator for improved attention resources induced by regular habitual activity. In addition, a tendency for larger amplitudes for the high habitual physical activity during the Flanker task and for the high sports activity group during all cognitive tasks was found.

An interesting effect was investigated for P200 amplitudes during the Flanker task. A high amount of sports activity was related to smaller P200 amplitude for males, but to higher amplitudes for females. This might be an indicator that attention modulation and stimulus classification is improved by the amount of sports activity for females but impaired for males.

When analyzing the data, there was no indication for the assumption that the relationship between the amount of activity and cognitive performance might be following an inverted U-shaped curve similar to the correlation between physical activity and various health parameters. A major limitation of this study is the calculation of the amount of sports and habitual physical activity based on questionnaire data. Therefore, it should be investigated in further studies whether objectively measured activity amounts are related to executive functions in young adults.

4.3.4 Type of Sport Activity

In this chapter, the differences between athletes from different types of sport activity with regard to their cognitive performance are described. The independent variable is the type of preferred sport activity which is categorized to four subgroups: nonathletes, endurance athletes, field athletes, strength athletes. An analysis separated for males and females was not possible since the group sizes are imbalanced and there would have been too less females in the strength athletes group for some calculations (table 57).

Table 57:   Number of athletes per sport activity group

	Nonathletes n (%)	Endurance athletes n (%)	Field athletes n (%)	Strength athletes n (%)	Total n
♂	17 (17)	45 (44)	25 (24)	15 (15)	102
♀	16 (43)	11 (30)	6 (16)	4 (11)	37
Total	33 (24)	56 (40)	31 (22)	19 (14)	139

> *RQ 4*: Is there a difference in cognitive performance between athletes engaging in different types of sport?

4.3.4.1 Results

Behavioral Measures

2 (condition) x 2 (congruency) repeated measures ANOVA with between-subject factor type of sport activity revealed no significant effects of type of activity on response times during Flanker and Switching task. Though, there was a main effect on response times during the Oddball task (F [3, 117] = 4.01, p = <.01, η^2 = .093). Post-hoc tests revealed a significant difference between nonathletes and strength athletes regarding their Oddball response times (p = <.01). Furthermore, there was a significant condition x type of sport interaction (F [3, 116] = 3.18, p = .03, η^2 = .076) in the Flanker task indicating the longest response times for field athletes during the accuracy condition but the fastest response times for this group during the speed condition. Mean response times are presented in table 58 and figure 44. There was no effect on errors in the Switching task observed (table 59 & figure 45).

Table 58:   Mean response times for nonathletes and different types of athletes in the Flanker, Switching and Oddball task

	NA		EA		FA		SA		p	η^2
	n	ms (95 % CI)	n	ms (95 % CI)	n	ms (95 % CI)	n	ms (95 % CI)		
Flanker	27	508.20 (491.65 - 524.74)	49	491.05 (478.76 - 503.33)	26	503.90 (487.04 - 520.76)	18	484.58 (464.32 - 504.84)	.19	.040
Switching	29	717.64 (687.59 - 747.69)	45	697.08 (672.95 - 721.21)	25	717.72 (685.35 - 750.09)	18	664.47 (626.32 - 702.62)	.12	.050
Oddball	29	546.02 (513.85 - 578.19)	50	498.96 (474.46 - 523.46)	25	516.48 (481.84 - 551.13)	17	457.05 (415.03 - 499.07)	<.01	.093

Note: NA = nonathletes, EA = endurance athletes, FA = field athletes, SA = strength athletes

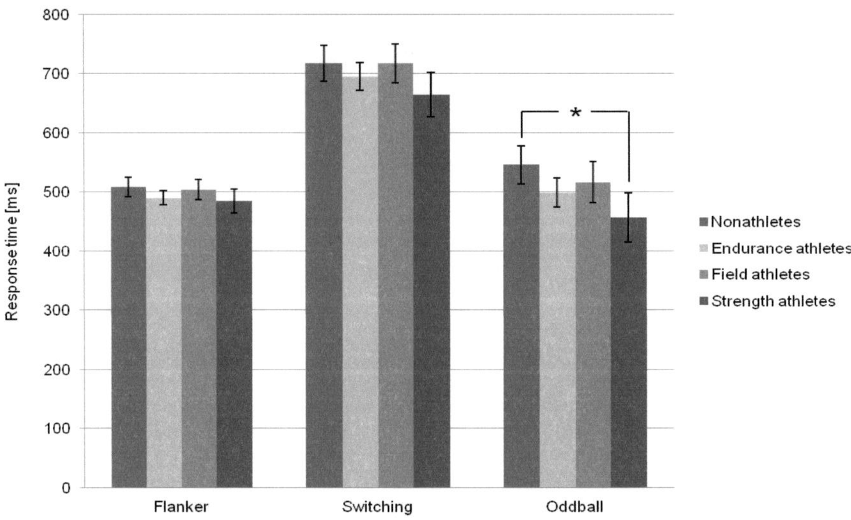

Figure 44:   Mean response times during Flanker, Switching and Oddball task for nonathletes and different types of athletes; error bars represent 95% confidence intervals

Table 59:   Errors in the Switching task for nonathletes and different types of athletes

	NA		EA		FA		SA			
	n	No. (95 % CI)	n	No. (95 % CI)	n	No. (95 % CI)	n	No. (95 % CI)	p	η^2
Errors	28	3.95 (3.10 - 4.80)	48	3.84 (3.19 - 4.49)	25	3.84 (2.94 - 4.74)	18	3.79 (2.73 - 4.85)	.99	.001

Note: NA = nonathletes, EA = endurance athletes, FA = field athletes, SA = strength athletes

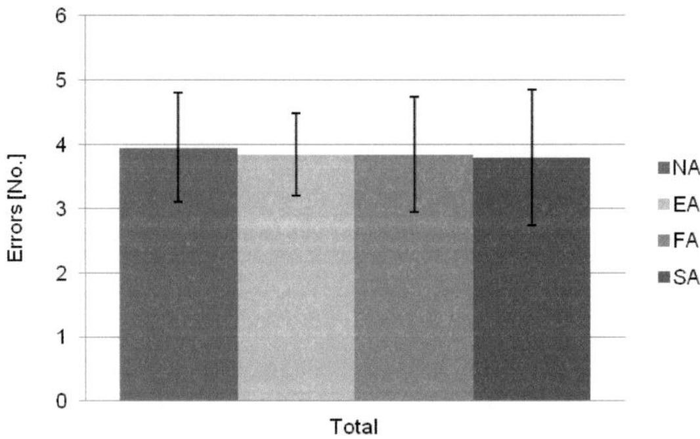

Figure 45:   Errors during the Switching task for nonathletes and different types of athletes; error bars represent 95% confidence intervals

P300

There were no effects of type of sport on P300 amplitudes in the Flanker, Switching or Oddball task (table 60 & figure 46) except for a significant electrode x type of sport interaction for the Flanker task (F [6, 218] = 2.30, p = .04, η^2 = .060). However, post-hoc analyses revealed no significant differences.

Table 60:   Mean P300 amplitudes for nonathletes and different types of athletes in the Flanker, Switching and Oddball task

	NA		EA		FA		SA			
	n	µV (95% CI)	n	µV (95% CI)	n	µV (95% CI)	n	µV (95% CI)	p	η^2
Flanker	25	4.19 (3.64 - 4.75)	46	4.85 (4.44 - 5.27)	24	4.32 (3.75 - 4.89)	17	4.88 (4.20 - 5.55)	.16	.046
Switching	22	3.94 (3.49 - 4.40)	27	3.77 (3.36 - 4.18)	16	3.91 (3.38 - 4.45)	11	4.01 (3.37 - 4.65)	.91	.008
Oddball	20	4.86 (3.79 - 5.93)	38	6.32 (5.54 - 7.10)	22	5.87 (4.85 - 6.89)	13	6.01 (4.68 - 7.34)	.19	.052

Note: NA = nonathletes, EA = endurance athletes, FA = field athletes, SA = strength athletes

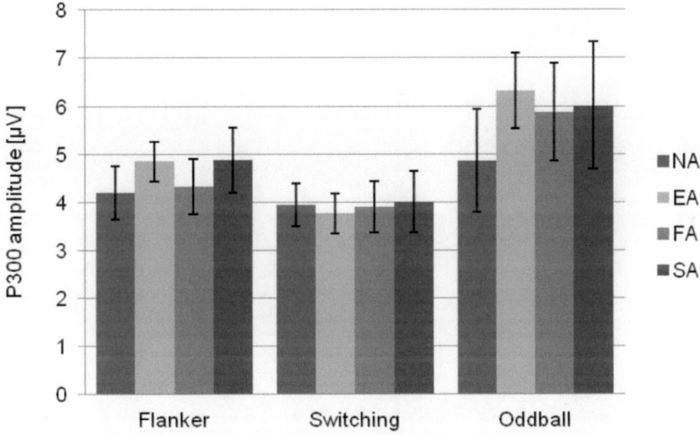

Figure 46:   Mean P300 amplitudes during Flanker, Switching and Oddball task for nonathletes and different types of athletes; error bars represent 95% confidence intervals

There were no effects of type of sport on P300 latencies in the Flanker, Switching or Oddball task (table 61 & figure 47). Post-hoc analyses however revealed a significant difference between nonathletes and endurance athletes for P300 latencies during the Flanker task (p = .05).

Table 61:   Mean P300 latencies for nonathletes and different types of athletes in the Flanker, Switching and Oddball task

| | NA | | EA | | FA | | SA | | | |
	n	ms (95 % CI)	n	ms (95 % CI)	n	ms (95 % CI)	n	ms (95 % CI)	p	η^2
Flanker	23	408.26 (392.53 - 423.98)	37	381.99 (369.59 - 394.38)	19	393.13 (375.84 - 410.43)	13	395.60 (374.69 - 416.52)	.08	.073
Switching	16	379.37 (355.87 - 402.86)	15	383.38 (359.11 - 407.64)	9	401.82 (370.49 - 433.14)	5	396.55 (354.52 - 438.58)	.65	.039
Oddball	16	413.36 (375.51 - 451.22)	36	431.22 (405.98 - 456.46)	20	442.19 (408.33 - 476.05)	13	399.96 (357.96 - 441.96)	.39	.036

Note: NA = nonathletes, EA = endurance athletes, FA = field athletes, SA = strength athletes

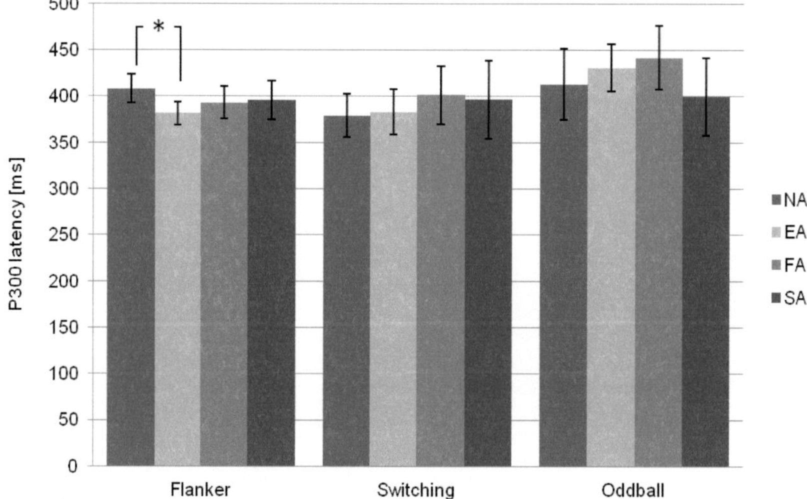

Figure 47:   Mean P300 latencies during Flanker, Switching and Oddball task for nonathletes and different types of athletes; error bars represent 95% confidence intervals

N100 and P200

There was no effect of type of sport on N100 and P200 amplitudes and latencies in the Flanker, Switching and Oddball task except for a significant main effect for N100 amplitude in the Oddball task (F [3, 90] = 4.17, p = <.01, η^2 = .122). Post-hoc analyses revealed a significant difference between field and strength athletes (p <.01) and a marginally significant difference between endurance and field athletes (p = .07). For P200 amplitude during the Switching task, there was a condition x type of sport interaction (F [3, 75] = 2.98, p = .04, η^2 = .107). Larger amplitudes for the homogenous compared to the heterogeneous condition were found for endurance (homogenous: 4.57 µV; 95% CI 3.95 - 5.20; heterogeneous: 4.48 µV; 95% CI 3.89 - 5.07) and strength athletes (homogenous: 4.51 µV; 95% CI 3.53 - 5.48; heterogeneous: 3.95 µV; 95% CI 3.04 - 4.87) whereas nonathletes (homogenous: 3.55 µV; 95% CI 2.83 - 4.26; heterogeneous: 3.63 µV; 95% CI 2.96 -

4.31) and field athletes (homogenous: 3.55 µV; 95% CI 2.71 - 4.39; heterogeneous: 4.19 µV; 95% CI 3.40 - 4.99) had larger amplitudes during the heterogeneous conditions. Mean N100 and P200 amplitudes are given in table 62 and 63 and figure 48 - 51.

Table 62: Mean N100 and P200 amplitudes for nonathletes and different types of athletes in the Flanker, Switching and Oddball task

		NA		EA		FA		SA		p	η^2
		n	µV (95 % CI)	n	µV (95 % CI)	n	µV (95 % CI)	n	µV (95 % CI)		
Flan-ker	N 100	26	-1.63 (-1.97 - -1.29)	47	-1.82 (-2.08 - -1.57)	25	-1.72 (-2.07 - -1.37)	17	-1.71 (-2.13 - -1.28)	.84	.008
	P 200	26	3.55 (2.94 - 4.16)	47	3.12 (2.67 - 3.58)	25	2.61 (1.99 - 3.24)	17	3.13 (2.38 - 3.88)	.21	.040
Swit-ching	P 200	22	3.59 (2.93 - 4.25)	29	4.52 (3.95 - 5.10)	16	3.87 (3.10 - 4.64)	12	4.23 (3.34 - 5.12)	.18	.062
Odd-ball	N 100	21	-5.07 (-5.99 - -4.16)	38	-5.01 (-5.69 - -4.33)	22	-3.56 (-4.45 - -2.66)	13	-5.99 (-7.16 - -4.83)	<.01	.122
	P 200	21	2.96 (2.07 - 3.86)	38	3.05 (2.39 - 3.72)	22	2.63 (1.75 - 3.50)	13	3.86 (2.73 - 5.00)	.40	.032

Note: NA = nonathletes, EA = endurance athletes, FA = field athletes, SA = strength athletes

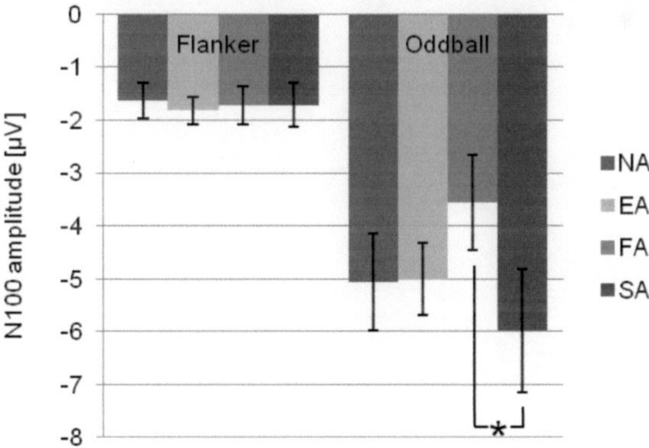

Figure 48: Mean N100 amplitudes during Flanker and Oddball task for nonathletes and different types of athletes; error bars represent 95% confidence intervals

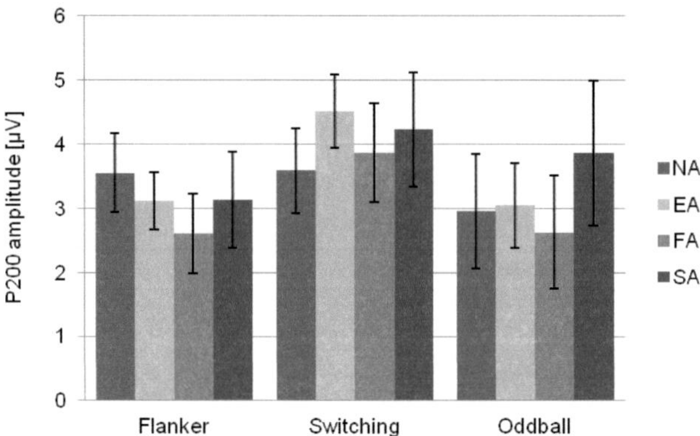

Figure 49: Mean P200 amplitudes during Flanker, Switching and Oddball task for nonathletes and different types of athletes; error bars represent 95% confidence intervals

Table 63: Mean N100 and P200 latencies for nonathletes and different types of athletes in the Flanker, Switching and Oddball task

| | | NA | | EA | | FA | | SA | | p | η^2 |
		n	ms (95 % CI)	n	ms (95 % CI)	n	ms (95 % CI)	n	ms (95 % CI)		
Flanker	N 100	10	142.29 (133.79 - 150.78)	17	137.55 (131.03 - 144.06)	7	144.73 (134.57 - 154.88)	8	135.94 (126.44 - 145.44)	.49	.061
	P 200	22	228.53 (217.26 - 239.81)	39	215.89 (207.42 - 224.36)	14	229.69 (215.56 - 243.82)	11	217.26 (201.32 - 233.20)	.19	.056
Swit-ching	P 200	20	190.23 (181.46 - 199.01)	29	186.52 (179.23 - 193.81)	12	192.66 (181.33 - 203.99)	11	186.81 (174.97 - 198.64)	.79	.015
Oddball	N 100	20	172.27 (164.92 - 179.61)	36	170.55 (165.08 - 176.02)	15	172.14 (163.66 - 180.61)	13	171.09 (161.99 - 180.20)	.98	.002
	P 200	16	264.60 (254.51 - 274.69)	30	260.81 (253.44 - 268.18)	13	253.13 (241.93 - 264.32)	10	257.42 (244.66 - 270.19)	.48	.037

Note: NA = nonathletes, EA = endurance athletes, FA = field athletes, SA = strength athletes

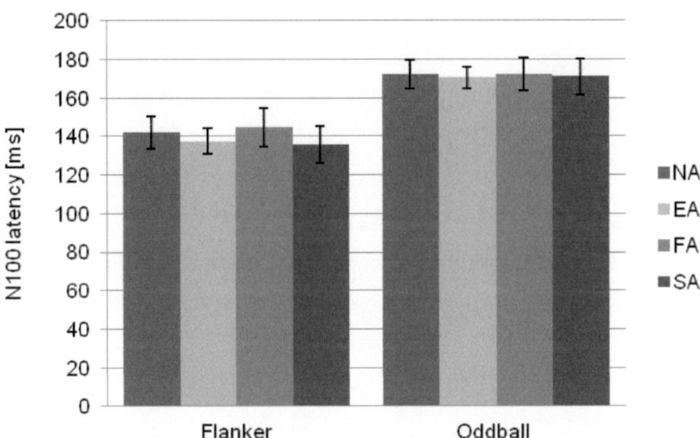

Figure 50:   Mean N100 latencies during Flanker and Oddball task for nonathletes and different types of athletes; error bars represent 95% confidence intervals

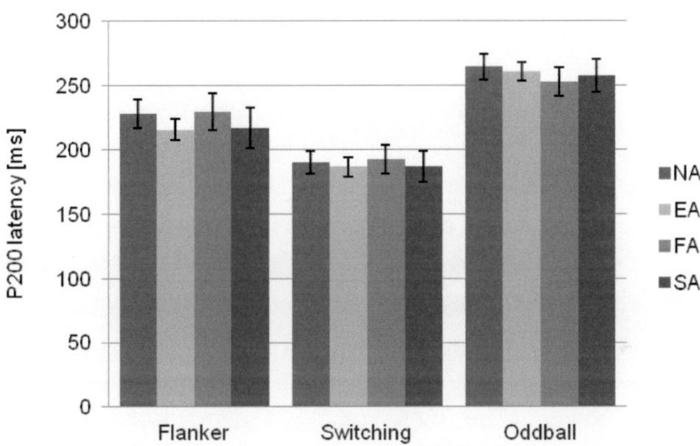

Figure 51:   Mean P200 latencies during Flanker, Switching and Oddball task for nonathletes and different types of athletes; error bars represent 95% confidence intervals

4.3.4.2 Discussion

Strength athletes had the lowest response times compared to athletes from other types of sport and nonathletes during all cognitive tasks but the effect was only significant for the Oddball task. Further, there was a tendency for larger P300 amplitudes for strength athletes compared to other types in the Flanker task and for faster P300 latencies during the Oddball task. Strength athletes also had the largest N100 and P200 amplitudes by trend in the Oddball task. These are interesting findings since only few studies have evaluated the effects of resistance training on cognitive performance. In young adults, a study by Goekint et al. (2010) tested the effects of resistance training on working memory and found no significant influences. However, the intervention phase could have been too short to elicit any exercise-related cognitive improvements. Furthermore, there is some evidence in older adults for a positive relationship between resistance training and cognitive performance (Cassilhas et al., 2007; Liu-Ambrose & Donaldson, 2009).

Interestingly, field athletes had the longest response times during the accuracy condition of the Flanker task which required answering as precisely as possible, but the fastest during the speed condition which required responding as quickly as possible. This could be explained by a good adaptation to task-specific demands that might be especially trained in field sports such as soccer, handball or volleyball. On the other hand, field athletes had the smallest N100 amplitudes during the Oddball task which might indicate a worse selective attention compared to other athletes. Thus, the findings are contradictory and further research is needed to clearly detect any possible effects.

Given that field sport is generally performed together with other athletes and requires much tactic skills, anticipation, interaction and interplay with team members, one could assume that this type of sport requires a continuous activation of executive control functions, which might play a dominant role in triggering selected neurophysiologic adaptations (Memmert & Weickgenannt, 2006, p. 82). These findings can be interpreted in light of an enriched environment in studies with rodents. This has shown to be related to structural changes in the brain and an enhanced survival of neurons (Van Praag et al., 1999) that might further promote enhanced cognitive performance. Another important aspect related to these questions could be the social interaction during an exercise session. Cacioppo and Hawkley (2009) examined that the degree of social interaction is highly associated with cognitive performance and that social isolation may contribute to poorer executive functioning. Therefore it may be suggested, that physical exercise performed together with others facilitates cognitive performance. However, these hypotheses are not supported by the findings in the current study.

Since many studies among young adults have found correlations between cardiovascular fitness and cognitive performance in the recent past (see also chapter 4.4.1), it is surprising that there are no significant effects observed in this study that might confirm this hypothesis. Endurance athletes tended to elicit the largest P300 amplitudes during the Oddball task, had significantly faster P300 latencies than nonathletes during the Flanker task and tended to reveal the largest P200 amplitudes during the Switching task. Nevertheless, these weak and mainly non significant effects might indicate that the potential benefits of endurance sport and cardiovascular fitness on cognitive performance in young adults have been overvalued in previous studies. One may therefore suggest that type of sport activity is not or only slightly related to executive functions in young adulthood. Based on the current data, strength athletes seem to reveal the largest cognitive benefits. It must be noted though that only the preferred type of sport was analyzed in this study. Many athletes (65%) though reported to practice other types of sport, too. This could be a reason for the few findings regarding this factor.

4.3.5 Training Characteristics

In this chapter, the relationship between training characteristics and cognitive performance is described. The independent variables are intensity of preferred sport (two groups: moderate intensity, vigorous intensity; light intensity group was not included due to too less participants), frequency of sport (three groups: less than two sessions per week, two to three sessions per week, four or more sessions per week; averaged for winter and summer season), and duration of exercise sessions (three groups: one hour per session or less, one and a half hour per session, two hours and more per session; averaged for summer and winter season).

4.3.5.1 Intensity of Preferred Sport – Results

The number of participants per intensity group is given in table 64.

Table 64:   Number of participants per intensity group

	Light intensity n (%)	Moderate intensity n (%)	Vigorous intensity n (%)	Total n
♂	1 (1)	46 (51)	43 (48)	90
♀	2 (7)	19 (65)	8 (28)	29
Total	3 (2)	65 (55)	51 (43)	119

RQ 5: Is exercise intensity related to cognitive performance in young adulthood?

Behavioral Measures
Repeated measures ANOVAs with between-subject factor intensity of exercise revealed no significant effects on response times during the Flanker, Switching or Oddball task (table 65). There was also no effect on errors in the Switching task.

Table 65:   Mean response times for exercise intensity groups in the Flanker, Switching and Oddball task

		Moderate intensity		Vigorous intensity			
		n	ms (95 % CI)	n	ms (95 % CI)	p	η^2
Flanker	♂	41	489.08 (476.90 - 501.24)	38	491.92 (479.27 - 504.56)	.75	.001
	♀	15	511.73 (483.38 - 540.08)	8	517.06 (478.24 - 555.87)	.82	.003
	Total	56	495.14 (483.70 - 506.58)	46	496.29 (483.66 - 508.91)	.89	.000
Switching	♂	38	693.44 (667.09 - 719.77)	35	702.22 (674.78 - 729.67)	.65	.003
	♀	15	719.87 (658.23 - 781.51)	8	664.78 (580.38 - 749.19)	.29	.054
	Total	53	700.92 (676.33 - 725.51)	43	695.26 (667.96 - 722.56)	.76	.001
Oddball	♂	40	493.32 (462.78 - 523.87)	36	503.78 (471.58 - 535.98)	.64	.003
	♀	17	514.69 (469.30 - 560.07)	8	501.22 (435.06 - 567.38)	.73	.005
	Total	57	499.70 (474.79 - 524.60)	44	503.31 (474.96 - 531.66)	.85	.000

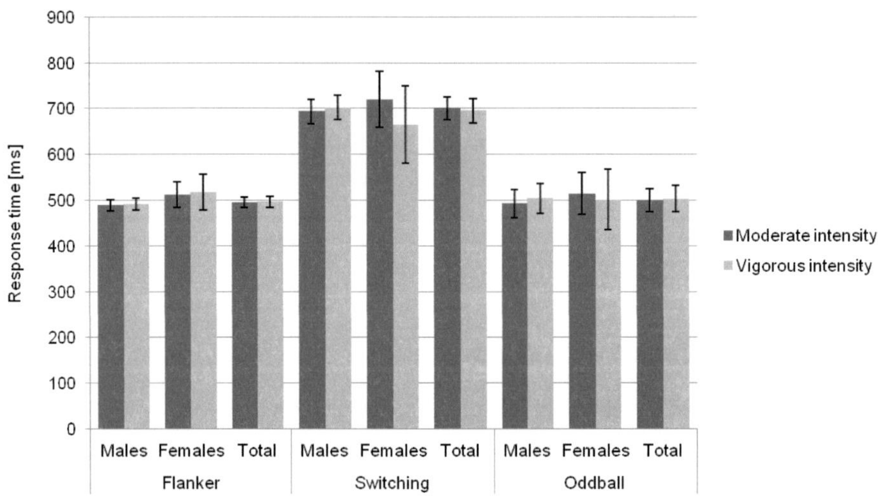

Figure 52:   Mean response times during Flanker, Switching and Oddball task for exercise intensity groups; error bars represent 95% confidence intervals

Table 66:   Errors in the Switching task for exercise intensity groups

		\multicolumn{2}{Moderate intensity}		\multicolumn{2}{Vigorous intensity}		p	η^2
		n	No. (95 % CI)	n	No. (95 % CI)		
Errors	♂	40	4.11 (3.42 - 4.81)	36	3.63 (2.90 - 4.37)	.35	.012
	♀	17	4.02 (2.67 - 5.36)	8	4.31 (2.35 - 6.28)	.80	.003
	Total	57	4.08 (3.47 - 4.69)	44	3.76 (3.06 - 4.45)	.48	.005

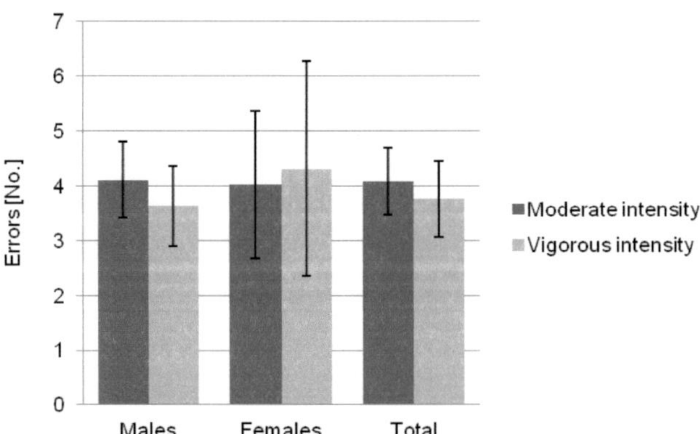

Figure 53:   Errors during the Switching task for exercise intensity groups; error bars represent 95% confidence intervals

P300

There was no significant main effect on P300 amplitudes (table 67 & figure 54) or latencies (table 68 & figure 55) in the Flanker, Switching or Oddball task. There was only a significant condition x intensity interaction for P300 latencies during the Flanker task observed (F [1, 75] = 4.53, p = .04,

$\eta^2 = .057$). Further analysis revealed a longer latency during the accuracy condition (385.95 ms, 95% CI 374.50 - 397.40) compared to the speed condition (381.13 ms, 95% CI 370.98 - 391.27) for participants with moderate intensity, whereas participants with vigorous intensity had longer latencies during the speed condition (394.18 ms, 95% CI 383.36 - 405.01) compared to the accuracy condition (384.80 ms, 95% CI 372.58 - 397.02).

Table 67:   Mean P300 amplitudes for exercise intensity groups in the Flanker, Switching and Oddball task

		Moderate intensity		Vigorous intensity			
		n	µV (95 % CI)	n	µV (95 % CI)	p	η^2
Flanker	♂	39	4.72 (4.28 - 5.17)	36	4.90 (4.43 - 5.36)	.59	.004
	♀	15	4.65 (3.80 - 5.51)	7	4.21 (2.96 - 5.46)	.55	.019
	Total	54	4.70 (4.32 - 5.09)	43	4.78 (4.35 - 5.22)	.79	.001
Switching	♂	21	3.78 (3.36 - 4.21)	26	3.89 (3.51 - 4.28)	.70	.003
	♀	8	4.63 (3.53 - 5.74)	5	3.57 (2.17 - 4.97)	.22	.134
	Total	29	4.02 (3.61 - 4.42)	31	3.84 (3.45 - 4.23)	.54	.007
Oddball	♂	29	6.07 (5.13 - 7.02)	31	5.88 (4.97 - 6.80)	.77	.001
	♀	14	6.97 (5.59 - 8.35)	6	5.89 (3.78 - 7.99)	.38	.043
	Total	43	6.37 (5.60 - 7.13)	37	5.88 (5.06 - 6.70)	.39	.009

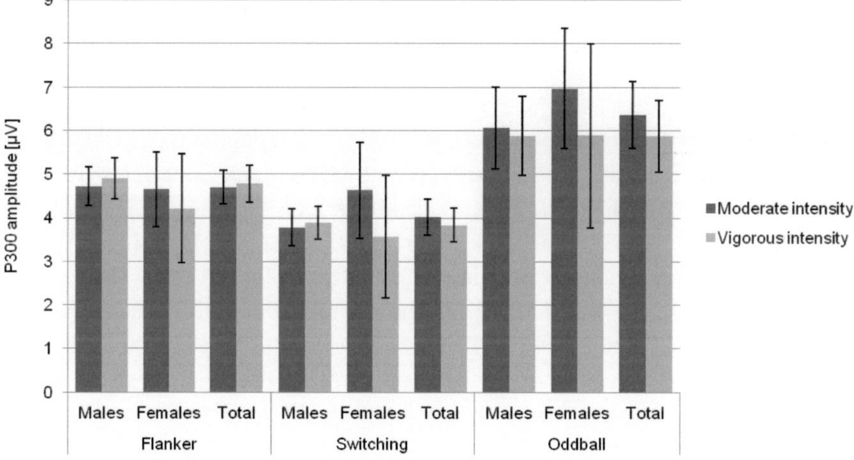

Figure 54:   Mean P300 amplitudes during Flanker, Switching and Oddball task for exercise intensity groups; error bars represent 95% confidence intervals

Table 68:   Mean P300 latencies for exercise intensity groups in the Flanker, Switching and Oddball task

		Moderate intensity		Vigorous intensity			
		n	ms (95 % CI)	n	ms (95 % CI)	p	η^2
Flanker	♂	30	389.36 (379.33 - 399.38)	29	392.41 (382.22 - 402.61)	.67	.003
	♀	11	367.68 (341.55 - 393.81)	7	377.38 (344.62 - 410.14)	.63	.015
	Total	41	383.54 (373.72 - 393.36)	36	389.49 (379.02 - 399.97)	.41	.009
Switching	♂	15	404.82 (378.84 - 430.79)	11	371.79 (341.46 - 402.12)	.10	.108
	♀	6	369.66 (327.58 - 411.74)	2	398.47 (325.58 - 471.36)	.43	.105

	Total	21	394.77 (373.37 - 416.17)	13	375.90 (348.70 - 403.09)	.28	.037
Oddball	♂	28	420.63 (389.58 - 451.68)	29	454.90 (424.39 - 485.42)	.12	.043
	♀	13	389.24 (355.53 - 422.95)	6	394.18 (344.57 - 443.80)	.86	.002
	Total	41	410.68 (386.32 - 435.03)	35	444.49 (418.14 - 470.85)	.06	.045

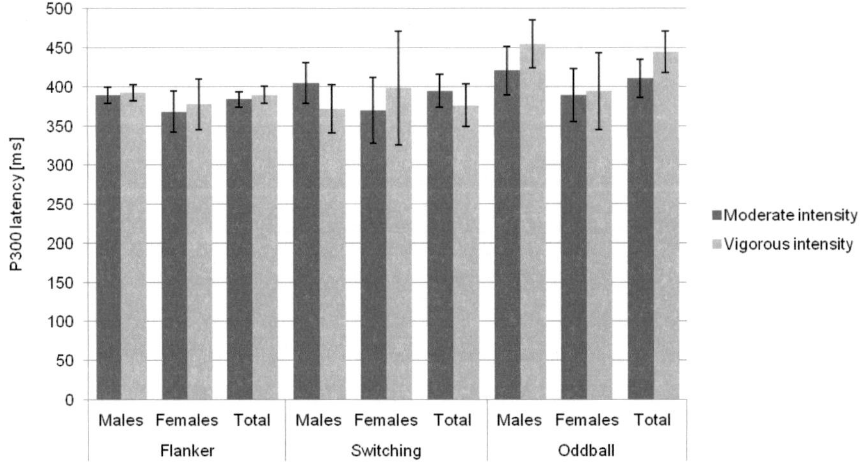

Figure 55:   Mean P300 latencies during Flanker, Switching and Oddball task for exercise intensity groups; error bars represent 95% confidence intervals

N100 and P200

There was no effect of intensity of sport on N100 and P200 amplitudes (table 69, figure 56 & 57) and latencies (table 70, figure 58 & 59) in the Flanker, Switching and Oddball task except for a significant main effect for P200 amplitude in the Oddball task for males and females (F [1, 78] = .35, p = .56, η^2 = .004; males: F [1, 58] = 4.01, p = .05, η^2 = .065, females: F [1, 18] = 4.30, p = .05, η^2 = .193). Larger amplitudes for males performing with moderate (3.65 µV, 95% CI 2.90 - 4.40) than vigorous intensity (2.61 µV, 95% CI 1.89 - 3.33) but larger amplitudes for females performing with vigorous (3.92 µV, 95% CI 2.26 - 5.59) than moderate intensity (1.96 µV, 95% CI .87 - 3.05) were found.

Table 69:   Mean N100 and P200 amplitudes for exercise intensity groups in the Flanker, Switching and Oddball task

			Moderate intensity		Vigorous intensity			
			n	µV (95 % CI)	n	µV (95 % CI)	p	η^2
Flanker	N100	♂	40	-1.79 (-2.08 - -1.51)	37	-1.63 (-1.93 - -1.34)	.44	.008
		♀	15	-1.70 (-2.21 - -1.20)	7	-2.12 (-2.86 - -1.39)	.34	.047
		Total	55	-1.77 (-2.01 - -1.53)	44	-1.71 (-1.98 - -1.44)	.75	.001
	P200	♂	40	2.98 (2.50 - 3.47)	37	2.96 (2.46 - 3.46)	.95	.000
		♀	15	3.49 (2.54 - 4.44)	7	3.94 (2.55 - 5.33)	.58	.015
		Total	55	3.12 (2.70 - 3.55)	44	3.12 (2.64 - 3.59)	.99	.000
Switching	P200	♂	22	4.00 (3.37 - 4.64)	28	4.37 (3.81 - 4.94)	.39	.016
		♀	8	4.99 (3.54 - 6.45)	5	4.51 (2.67 - 6.35)	.66	.018

		Total	30	4.27 (3.69 - 4.84)	33	4.39 (3.85 - 4.94)	.75	.002
Oddball	N100	♂	29	-4.18 (-4.98 - -3.37)	31	-4.71 (-5.49 - -3.93)	.34	.016
		♀	14	-5.62 (-6.57 - -4.67)	6	-5.59 (-7.04 - -4.14)	.98	.000
		Total	43	-4.64 (-5.29 - -4.00)	37	-4.85 (-5.54 - -4.16)	.66	.002
	P200	♂	29	3.65 (2.90 - 4.40)	31	2.61 (1.89 - 3.33)	.05	.065
		♀	14	1.96 (.87 - 3.05)	6	3.92 (2.26 - 5.59)	.05	.193
		Total	43	3.10 (2.47 - 3.73)	37	2.82 (2.14 - 3.51)	.56	.004

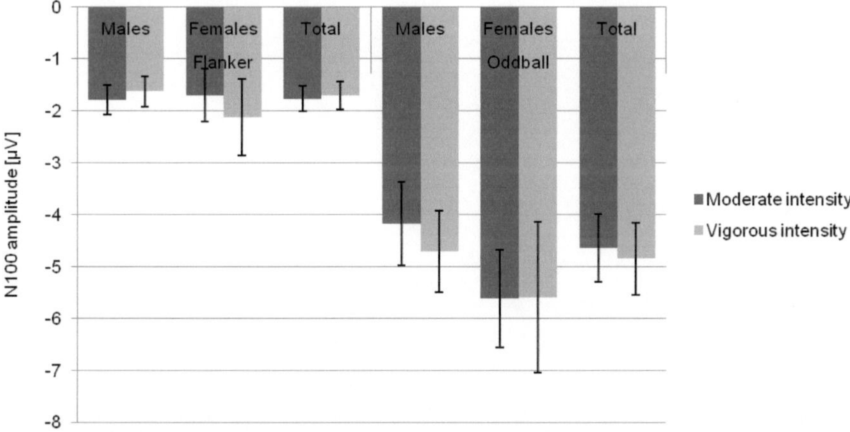

Figure 56: Mean N100 amplitudes during Flanker and Oddball task for exercise intensity groups; error bars represent 95% confidence intervals

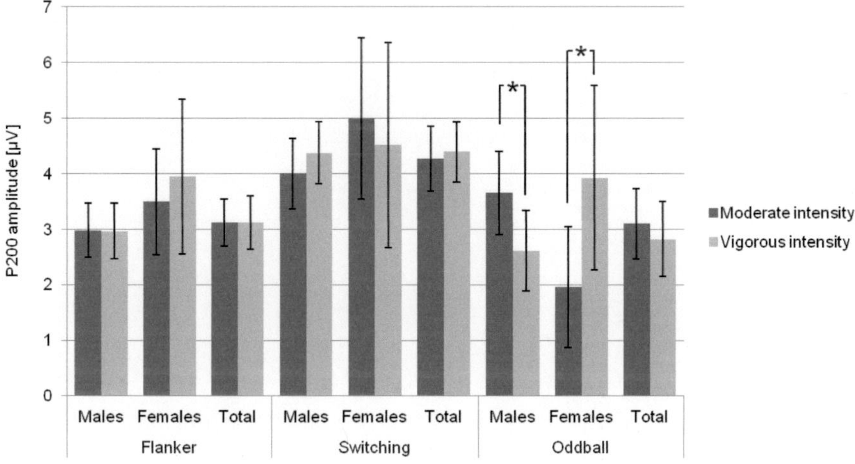

Figure 57: Mean P200 amplitudes during Flanker, Switching and Oddball task for exercise intensity groups; error bars represent 95% confidence intervals

Table 70:   Mean N100 and P200 latencies for exercise intensity groups in the Flanker, Switching- and Oddball task

			Moderate intensity		Vigorous intensity			
			n	ms (95 % CI)	n	ms (95 % CI)	p	η^2
Flanker	N100	♂	16	137.52 (129.47 - 145.58)	10	142.97 (132.78 - 153.16)	.40	.030
		♀	6	138.38 (124.71 - 152.05)	2	133.50 (109.82 - 157.17)	.68	.031
		Total	22	137.76 (131.28 - 144.24)	12	141.39 (132.61 - 150.17)	.50	.014
	P200	♂	32	221.63 (211.86 - 231.40)	25	217.85 (206.79 - 228.91)	.61	.005
		♀	11	223.30 (207.52 - 239.08)	4	222.36 (196.20 - 248.53)	.95	.000
		Total	43	222.06 (213.96 - 230.15)	29	218.47 (208.62 - 228.33)	.58	.004
Switching	P200	♂	22	189.96 (181.62 - 198.30)	23	185.53 (177.38 - 193.68)	.45	.013
		♀	8	190.02 (176.00 - 204.03)	5	184.38 (166.65 - 202.10)	.59	.027
		Total	30	189.97 (183.11 - 196.84)	28	185.32 (178.22 - 192.43)	.35	.016
Oddball	N100	♂	23	171.43 (164.00 - 178.87)	29	166.11 (159.49 - 172.73)	.29	.023
		♀	13	175.60 (168.89 - 182.31)	6	182.16 (172.28 - 192.04)	.26	.073
		Total	36	172.94 (167.36 - 178.52)	35	168.86 (163.20 - 174.52)	.31	.015
	P200	♂	23	258.05 (249.23 - 266.87)	22	257.03 (248.01 - 266.05)	.87	.001
		♀	5	263.28 (243.18 - 283.38)	5	257.03 (236.93 - 277.13)	.63	.031
		Total	28	258.98 (251.25 - 266.72)	27	257.03 (249.15 - 264.91)	.72	.002

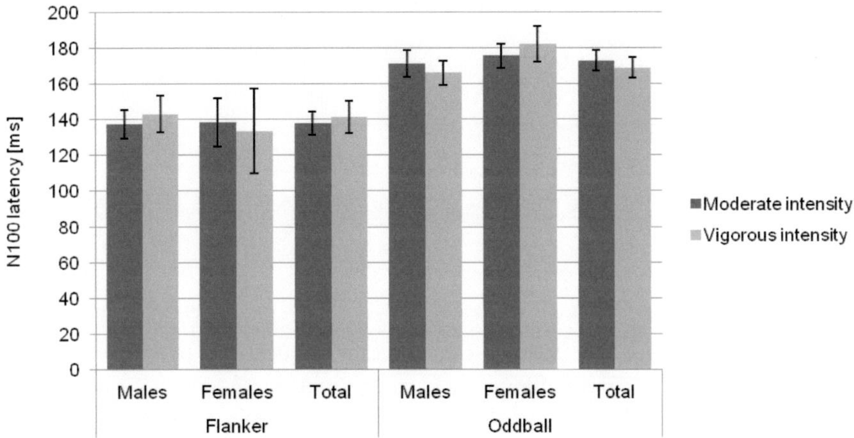

Figure 58:   Mean N100 latencies during Flanker and Oddball task for exercise intensity groups; error bars represent 95% confidence intervals

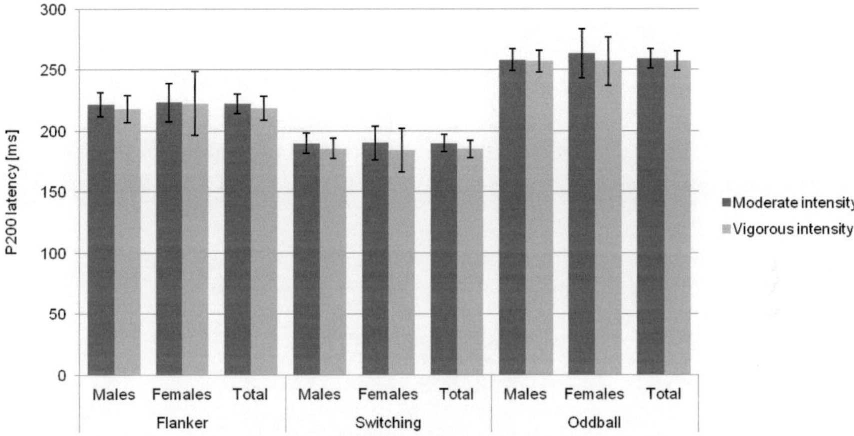

Figure 59:   Mean P200 latencies during Flanker, Switching and Oddball task for exercise intensity groups; error bars represent 95% confidence intervals

4.3.5.2 Frequency of Sport - Results

An analysis separated for males and females was not possible since the group sizes are imbalanced and there would have been too less females in the < 2 sessions and > 4 sessions groups. The number of participants per frequency group is given in table 71.

Table 71:   Number of participants per frequency group

	< 2 sessions n (%)	2 - 3 sessions n (%)	≥ 4 sessions n (%)	Total n
Total	16 (14)	86 (72)	17 (14)	119

> **RQ 6:**   Is exercise frequency per week related to cognitive performance in young adulthood?

Behavioral Measures

Repeated measures ANOVAs with between-subject factor frequency of preferred sport revealed no significant effects on response times during the Flanker, Switching or Oddball task (table 72 & figure 60). There was also no effect on errors in the Switching task (table 73 & figure 61).

Table 72:   Mean response times for different frequency groups during Flanker, Switching and Oddball task

	n	< 2 sessions ms (95 % CI)	n	2 - 3 sessions ms (95 % CI)	n	≥ 4 sessions ms (95 % CI)	p	η^2
Flanker	14	499.54 (476.89 - 522.20)	74	494.19 (484.33 - 504.04)	17	498.14 (477.58 - 518.70)	.88	.003
Switching	13	723.17 (674.22 - 772.11)	69	692.02 (670.78 - 713.26)	17	705.70 (662.90 - 748.50)	.48	.015
Oddball	14	506.37 (456.68 - 556.06)	74	503.30 (481.69 - 524.92)	16	484.81 (438.33 - 531.29)	.75	.006

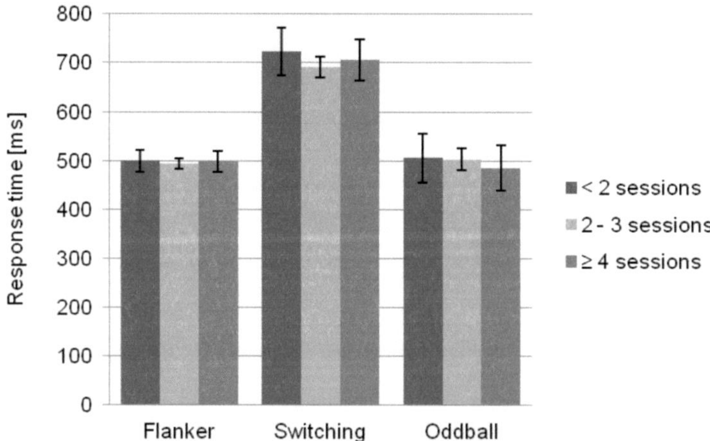

Figure 60:   Mean response times during Flanker, Switching and Oddball task for different frequency groups; error bars represent 95% confidence intervals

Table 73:   Errors in the Switching task for different frequency groups

	< 2 sessions		2 - 3 sessions		≥ 4 sessions			
	n	No. (95 % CI)	n	No. (95 % CI)	n	No. (95 % CI)	p	η^2
Errors	13	3.10 (1.84 - 4.35)	74	4.17 (3.64 - 4.70)	17	3.41 (2.31 - 4.51)	.19	.033

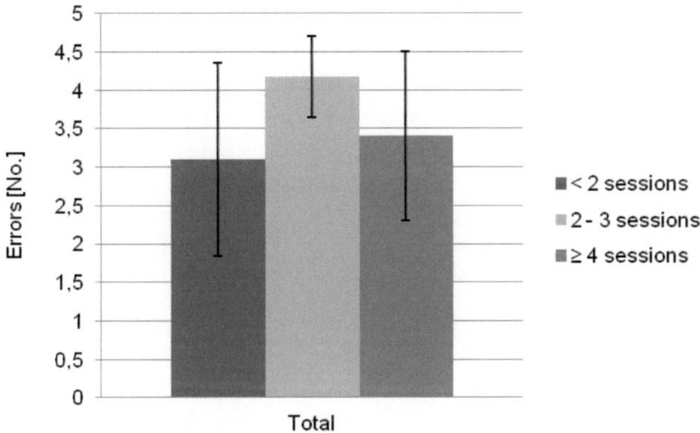

Figure 61:   Errors during the Switching task for different frequency groups; error bars represent 95% confidence intervals

P300

There were significant effects of frequency of sport and P300 amplitude in the Flanker (F [2, 96] = 5.41, p = <.01, η^2 = .101) and Oddball task (F [2, 80] = 7.34, p = .00, η^2 = .155). For the Switching task, no significant effect could be observed (table 74 & figure 62). Post-hoc analyses revealed significant differences in P300 amplitude for the Flanker task between athletes with less than two and athletes with four or more sessions per week (p = .03) as well as between athletes with two to three and athletes with four or more sessions per week (p = <.01). For the Oddball task, post-hoc analyses also showed differences between athletes with less than two and athletes with four or more sessions per week (p = .01) as well as between athletes with two to three and athletes with four or more sessions per week (p = .00).

Table 74:   Mean P300 amplitudes for different frequency groups during Flanker, Switching and Oddball task

	< 2 sessions		2 - 3 sessions		≥ 4 sessions			
	n	µV (95 % CI)	n	µV (95 % CI)	n	µV (95 % CI)	p	η^2
Flanker	14	4.39 (3.68 - 5.11)	68	4.56 (4.23 - 4.89)	17	5.71 (5.06 - 6.36)	<.01	.101
Switching	8	3.54 (2.79 - 4.28)	43	3.82 (3.50 - 4.14)	11	4.57 (3.94 - 5.21)	.07	.088
Oddball	11	5.83 (4.45 - 7.21)	62	5.80 (5.22 - 6.38)	10	8.77 (7.33 - 10.21)	.00	.155

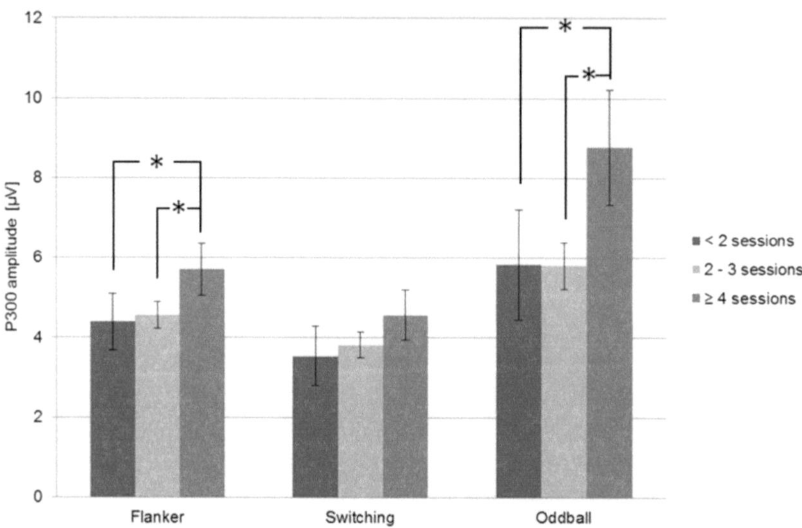

Figure 62:   Mean P300 amplitudes during Flanker, Switching and Oddball task for different frequency groups; error bars represent 95% confidence intervals

There were no significant effects of frequency of sport on P300 latency in the Flanker and Switching task. Yet, there was a significant main effect for the Oddball task (F [2, 76] = 4.71, p = .01, η^2 = .110). Post-hoc analyses revealed significant differences in P300 latency between athletes with less than two and athletes with two to three sessions per week (p = .04) as well as between athletes with less than two and athletes with four or more sessions per week (p = .01). Mean P300 latencies are given in table 75 and figure 63.

Table 75:   Mean P300 latencies for different frequency groups during Flanker, Switching and Oddball task

	< 2 sessions		2 - 3 sessions		≥ 4 sessions		p	η^2
	n	µV (95 % CI)	n	µV (95 % CI)	n	µV (95 % CI)		
Flanker	11	387.24 (368.05 - 406.44)	54	386.79 (378.13 - 395.46)	14	386.15 (369.13 - 403.16)	.99	.000
Switching	4	401.16 (352.96 - 449.35)	27	381.07 (362.52 - 399.62)	4	417.11 (368.91 - 465.30)	.32	.070
Oddball	11	483.00 (437.80 - 528.20)	59	419.42 (399.90 - 438.93)	9	384.78 (334.81 - 434.75)	.01	.110

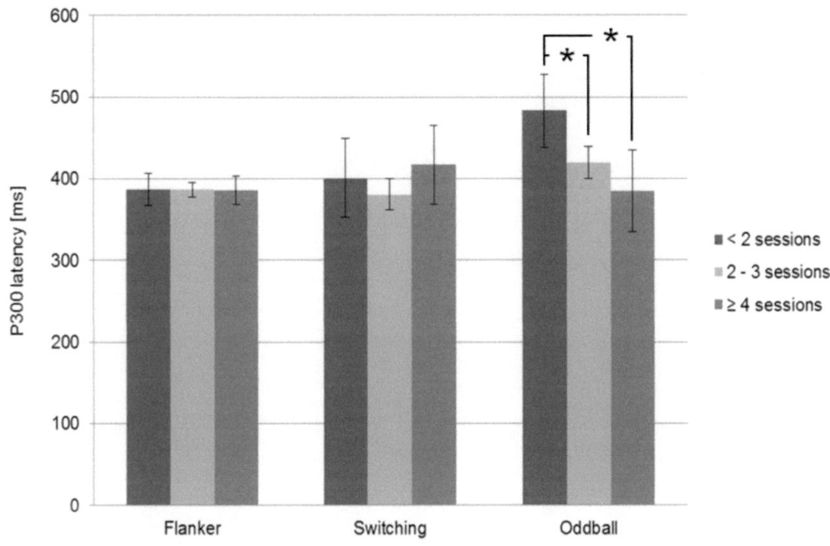

Figure 63:   Mean P300 latencies during Flanker, Switching and Oddball task for different frequency groups; error bars represent 95% confidence intervals

N100 and P200

There was a significant effect of frequency of sport on P200 latencies in the Switching task (F [2, 57] = 3.21, p = .05, η^2 = .101) with significantly (p = .05) shorter latencies for athletes performing less than 2 sessions per week than athletes performing 2-3 sessions. However, no other effects were observed (table 76 & 77, figure 64 - 67).

Table 76:   Mean N100 and P200 amplitudes for different frequency groups in the Flanker, Switching and Oddball task

		< 2 sessions		2-3 sessions		≥ 4 sessions			
		n	µV (95 % CI)	n	µV (95 % CI)	n	µV (95 % CI)	p	η^2
Flanker	N100	14	-1.32 (-1.79 - -.84)	70	-1.81 (-2.02 - -1.60)	17	-1.89 (-2.32 - -1.46)	.14	.039
	P200	14	3.76 (2.94 - 4.58)	70	2.89 (2.52 - 3.26)	17	3.49 (2.75 - 4.24)	.09	.048
Switching	P200	10	4.63 (3.66 - 5.60)	43	4.05 (3.59 - 4.52)	12	4.91 (4.03 - 5.80)	.18	.053
Oddball	N100	11	-4.22 (-5.48 - -2.95)	62	-4.89 (-6.28 - -4.36)	10	-4.95 (-6.28 - -3.63)	.61	.012
	P200	11	3.03 (1.78 - 4.28)	62	3.07 (2.54 - 3.59)	10	2.30 (.99 - 3.62)	.56	.014

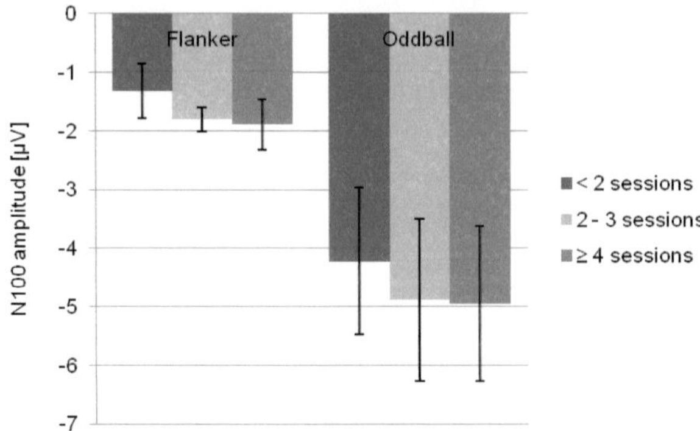

Figure 64:   Mean N100 amplitudes during Flanker and Oddball task for different frequency groups; error bars represent 95% confidence intervals

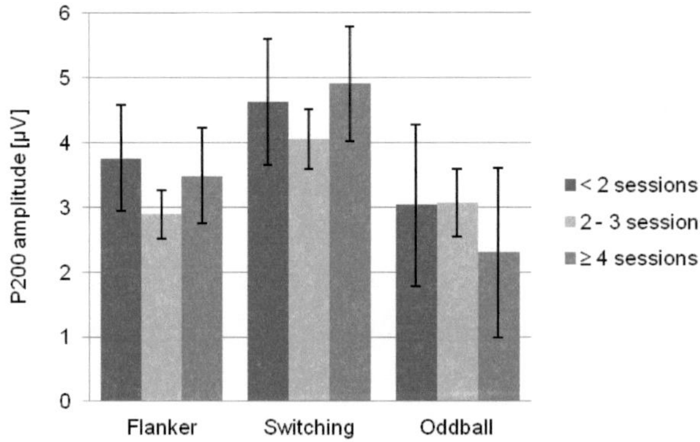

Figure 65:   Mean P200 amplitudes during Flanker, Switching and Oddball task for different frequency groups; error bars represent 95% confidence intervals

Table 77:   Mean N100 and P200 latencies for different frequency groups in the Flanker, Switching and Oddball task

			< 2 sessions		2-3 sessions		≥ 4 sessions	p	η^2
		n	ms (95 % CI)	n	ms (95 % CI)	n	ms (95 % CI)		
Flanker	N100	1	138.87 (108.44 - 169.30)	27	138.61 (132.76 - 144.47)	6	143.51 (132.75 - 154.26)	.72	.020
	P200	12	210.32 (195.02 - 225.62)	48	224.38 (216.73 - 232.03)	14	221.18 (207.02 - 235.34)	.27	.036
Switching	P200	9	174.89 (161.68 - 188.11)	39	193.08 (186.73 - 199.43)	12	186.56 (175.11 - 198.00)	.05	.101
Oddball	N100	9	165.89 (154.57 - 177.20)	56	172.21 (167.67 - 176.75)	9	172.40 (161.08 - 183.72)	.58	.015
	P200	8	258.50 (244.44 - 272.56)	42	260.29 (254.15 - 266.42)	7	245.87 (230.84 - 260.90)	.21	.055

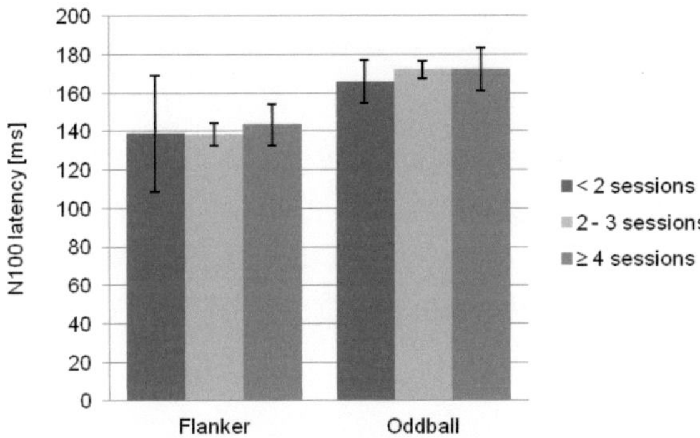

Figure 66:   Mean N100 latencies during Flanker and Oddball task for different frequency groups; error bars represent 95% confidence intervals

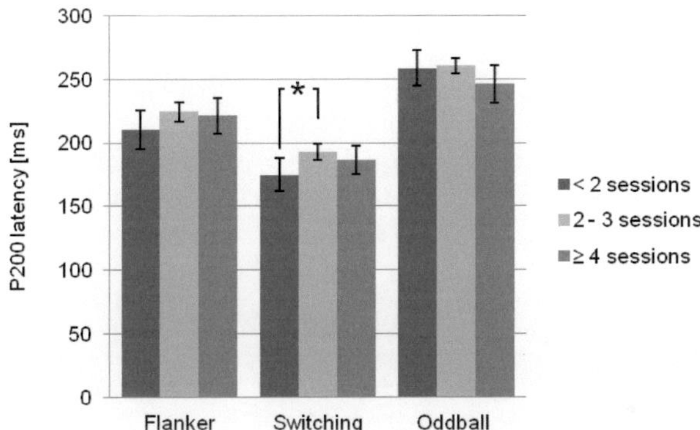

Figure 67:   Mean P200 latencies during Flanker, Switching and Oddball task for different frequency groups; error bars represent 95% confidence intervals

4.3.5.3 Duration of Exercise Sessions - Results

The number of participants per duration group is given in table 78.

Table 78:   Number of participants per duration group

	≤ 1 hour/session n (%)	1.5 hours/session n (%)	≥ 2 hours/session n (%)	Total n
♂	35 (39)	37 (41)	18 (20)	90
♀	9 (31)	10 (34.5)	10 (34.5)	29
Total	44 (37)	47 (39)	28 (24)	119

> *RQ 7:*   Is exercise duration per session related to cognitive performance in young adulthood?

Behavioral Measures

Repeated measures ANOVA with between-subject factor exercise duration revealed no significant effects of duration of exercise sessions on response times during Flanker and Switching task (table 79 & figure 68). But there was a main effect on response times during the Oddball task (F [2, 101] = 3.83, p = .03, η^2 = .070; males: F [2, 74] = 2.76, p = .07, η^2 = .069; females: F [2, 24] = .94, p = .41, η^2 = .072). Post-hoc tests indicated a marginally significant difference between athletes with a duration of 1 hours or less per session and those with a duration of two hours or more per session regarding their Oddball response times (p = .06). There was no effect on errors in the Switching task (table 80 & figure 69).

Table 79:   Mean response times for athletes with durations of ≤ 1 hour/session, 1.5 hours/session and ≥ 2 hours/session in the Flanker, Switching and Oddball task

		≤ 1 hour/session		1.5 hours/session		≥ 2 hours/session			
		n	ms (95 % CI)	n	ms (95 % CI)	n	ms (95 % CI)	p	η^2
Flan-ker	♂	31	487.06 (473.08 - 501.05)	35	492.81 (479.65 - 505.97)	14	493.19 (472.38 - 514.00)	.81	.006
	♀	9	497.09 (461.75 - 532.43)	7	516.85 (476.78 - 556.93)	9	520.93 (485.59 - 556.27)	.59	.048
	Tot.	40	489.32 (476.02 - 502.62)	42	496.81 (483.83 - 509.80)	23	504.05 (486.50 - 521.59)	.41	.018
Swit-ching	♂	31	699.32 (670.03 - 728.61)	31	701.33 (672.04 - 730.62)	12	689.08 (642.00 - 736.15)	.91	.003
	♀	7	645.85 (561.40 - 730.30)	8	747.69 (668.69 - 826.68)	10	695.60 (624.94 - 766.25)	.21	.132
	Tot.	38	689.47 (660.81 - 718.13)	39	710.84 (682.55 - 739.13)	22	692.04 (654.38 - 729.71)	.54	.013
Odd-ball	♂	31	468.23 (434.70 - 501.75)	32	514.12 (481.13 - 547.12)	14	528.01 (478.13 - 577.89)	.07	.069
	♀	9	476.56 (417.23 - 535.88)	8	526.75 (463.82 - 589.67)	10	522.86 (466.58 - 579.14)	.41	.072
	Tot.	40	470.10 (441.68 - 498.52)	40	516.65 (488.22 - 545.07)	24	525.87 (489.17 - 562.56)	.03	.070

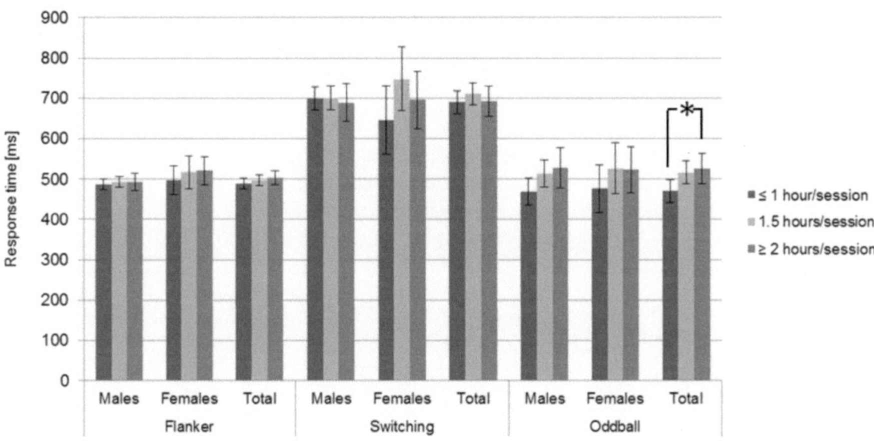

Figure 68:   Mean response times during Flanker, Switching and Oddball task for athletes with durations of
≤ 1 hour/session, 1.5 hours/session and ≥ 2 hours/session; error bars represent 95% confidence
intervals

Table 80:   Errors in the Switching task for athletes with durations of ≤ 1 hour/session, 1.5 hours/session
and ≥ 2 hours/session

		≤ 1 hour/session		1.5 hours/session		≥ 2 hours/session			
		n	No. (95 % CI)	n	No. (95 % CI)	n	No. (95 % CI)	p	η^2
Errors	♂	31	3.82 (3.03 - 4.62)	33	3.78 (3.01 - 4.55)	13	4.19 (2.96 - 5.42)	.84	.005
	♀	8	4.50 (2.55 - 6.45)	9	3.83 (1.99 - 5.67)	10	3.85 (2.11 - 5.60)	.84	.014
	Tot.	39	3.96 (3.23 - 4.70)	42	3.79 (3.08 - 4.50)	23	4.04 (3.09 - 5.00)	.90	.002

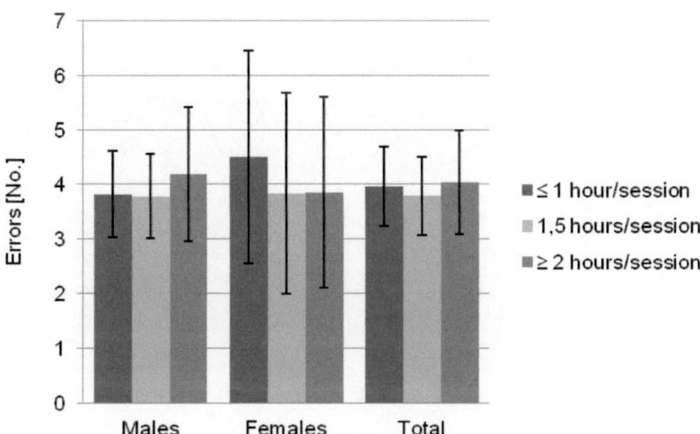

Figure 69:   Errors during the Switching task for athletes with durations of ≤ 1 hour/session, 1.5
hours/session and ≥ 2 hours/session; error bars represent 95% confidence intervals

P300

There was no significant main effect or interaction on P300 amplitudes (table 81 & figure 70) or latencies (table 82 & figure 71) in the Flanker, Switching or Oddball task.

Table 81:   Mean P300 amplitudes for athletes with durations of ≤ 1 hour/session, 1.5 hours/session and ≥ 2 hours/session in the Flanker, Switching and Oddball task

		≤ 1 hour/session		1.5 hours/session		≥ 2 hours/session			
		n	μV (95 % CI)	n	μV (95 % CI)	n	μV (95 % CI)	p	η^2
Flanker	♂	29	4.91 (4.39 - 5.42)	33	4.64 (4.16 - 5.13)	14	4.93 (4.18 - 5.67)	.71	.009
	♀	8	4.88 (3.77 - 5.98)	7	3.76 (2.57 - 4.94)	8	4.86 (3.75 - 5.96)	.29	.117
	Total	37	4.90 (4.44 - 5.36)	40	4.49 (4.04 - 4.93)	22	4.90 (4.30 - 5.49)	.37	.021
Switching	♂	20	3.65 (3.22 - 4.08)	18	3.75 (3.30 - 4.21)	10	4.28 (3.67 - 4.88)	.24	.062
	♀	4	4.70 (3.08 - 6.34)	4	3.87 (2.23 - 5.51)	6	4.20 (2.86 - 5.54)	.73	.055
	Total	24	3.83 (3.38 - 4.27)	22	3.77 (3.31 - 4.24)	16	4.25 (3.70 - 4.79)	.37	.033
Oddball	♂	25	6.09 (5.08 - 7.10)	26	6.05 (5.05 - 7.04)	10	5.50 (3.90 - 7.10)	.81	.007
	♀	8	6.95 (5.20 - 8.71)	7	7.33 (5.46 - 9.20)	7	5.75 (3.88 - 7.62)	.44	.083
	Total	33	6.30 (5.44 - 7.16)	33	6.32 (5.46 - 7.18)	17	5.60 (4.41 - 6.80)	.58	.014

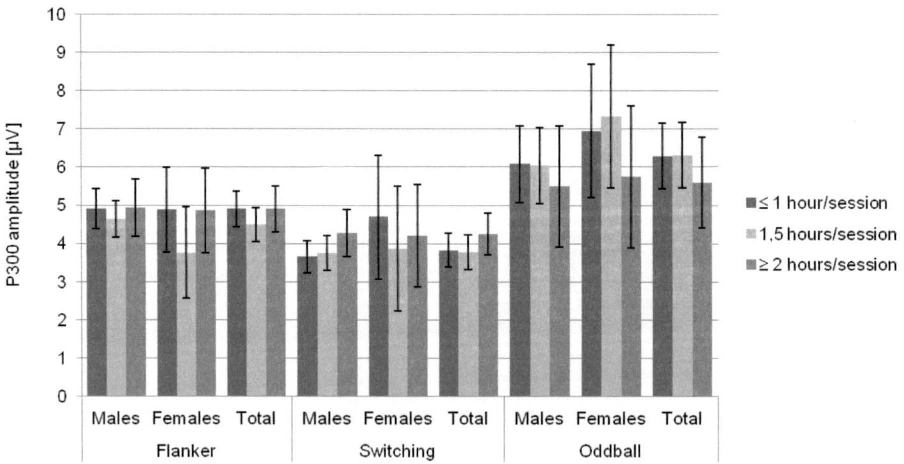

Figure 70:   Mean P300 amplitudes during Flanker, Switching and Oddball task for athletes with durations of ≤ 1 hour/session, 1.5 hours/session and ≥ 2 hours/session; error bars represent 95% confidence intervals

Table 82: Mean P300 latencies for athletes with durations of ≤ 1 hour/session, 1.5 hours/session and ≥ 2 hours/session in the Flanker, Switching and Oddball task

		≤ 1 hour/session		1.5 hours/session		≥ 2 hours/session			
		n	ms (95 % CI)	n	ms (95 % CI)	n	ms (95 % CI)	p	η^2
Flan-ker	♂	24	391.12 (379.67 - 402.56)	24	391.34 (379.89 - 402.78)	12	392.88 (376.70 - 409.07)	.98	.001
	♀	6	355.23 (321.59 - 388.87)	6	371.40 (337.76 - 405.03)	7	385.61 (354.47 - 416.75)	.39	.110
	Tot.	30	383.94 (372.35 - 395.53)	30	387.35 (375.76 - 398.94)	19	390.20 (375.64 - 404.77)	.79	.006
Swit-ching	♂	10	380.99 (347.81 - 414.17)	9	409.48 (374.51 - 444.46)	7	380.96 (341.31 - 420.62)	.41	.075
	♀	3	398.74 (348.19 - 449.29)	2	394.89 (332.98 - 456.80)	4	353.47 (309.69 - 397.25)	.27	.357
	Tot.	13	385.09 (358.69 - 411.48)	11	406.83 (378.14 - 435.52)	11	370.96 (342.27 - 399.66)	.21	.093
Odd-ball	♂	24	424.24 (390.38 - 458.11)	24	453.21 (419.35 - 487.08)	10	431.90 (379.44 - 484.36)	.47	.027
	♀	7	377.94 (333.29 - 422.59)	7	398.21 (353.56 - 442.87)	7	387.24 (342.59 - 431.89)	.80	.025
	Tot.	31	413.79 (385.65 - 441.92)	31	440.79 (412.66 - 468.93)	17	413.51 (375.52 - 451.50)	.33	.029

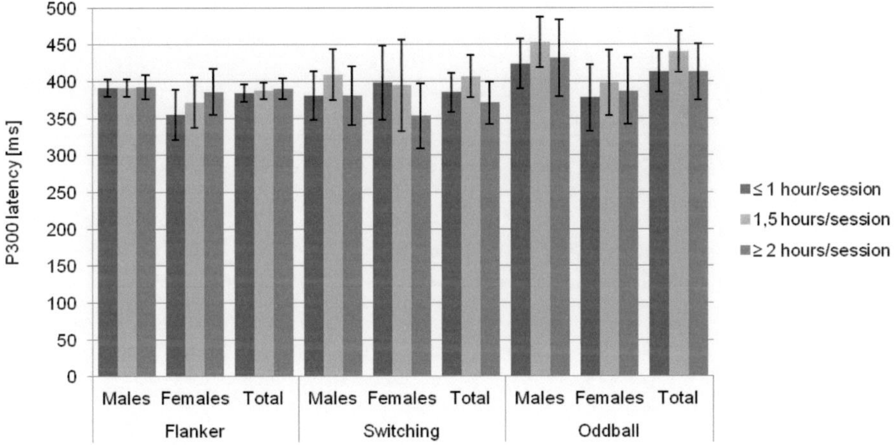

Figure 71: Mean P300 latencies during Flanker, Switching and Oddball task for athletes with durations of ≤ 1 hour/session, 1.5 hours/session and ≥ 2 hours/session; error bars represent 95% confidence intervals

N100 and P200

There was a significant main effect for N100 amplitude for females in the Flanker task (F [2, 98] = .24, p = .79, η^2 = .005; males: F [2, 75] = .50, p = .61, η^2 = .013; females: F [2, 20] = 5.01, p = .02, η^2 = .334) with larger amplitudes for athletes performing ≤ 1 hour per session compared to those performing 1.5 hours (p = .02). A second effect was found for males in the Oddball task (F [2, 80] =

2.66, p = .08, η^2 = .062; males: F [2, 58] = 3.07, p = .05, η^2 = .096; females: F [2, 19] = 1.29, p = .29, η^2 = .119) with significantly larger negative amplitudes for athletes performing ≤ 1 hour per session than ≥ 2 hours (p = .05). No further effects were observed for N100 and P200 components (table 83 & 84, figure 72 - 75).

Table 83: Mean N100 and P200 amplitudes for athletes with durations of ≤ 1 hour/session, 1.5 hours/session and ≥ 2 hours/session in the Flanker, Switching and Oddball task

| | | | ≤ 1 hour/session | | 1.5 hours/session | | ≥ 2 hours/session | | | |
			n	µV (95 % CI)	n	µV (95 % CI)	n	µV (95 % CI)	p	η^2
Flan-ker	N100	♂	30	-1.61 (-1.94 - -1.27)	34	-1.83 (-2.15 - -1.52)	14	-1.75 (-2.24 - -1.26)	.61	.013
		♀	8	-2.26 (-2.84 - -1.69)	7	-1.06 (-1.68 - -.45)	8	-2.07 (-2.65 - -1.50)	.02	.334
		Tot.	38	-1.74 (-2.04 - -1.45)	41	-1.70 (-1.99 - -1.42)	22	-1.87 (-2.25 - -1.48)	.79	.005
	P200	♂	30	3.13 (2.58 - 3.69)	34	2.98 (2.46 - 3.50)	14	2.59 (1.78 - 3.40)	.54	.016
		♀	8	3.37 (2.07 - 4.66)	7	3.37 (1.99 - 4.76)	8	4.02 (2.73 - 5.32)	.70	.035
		Tot.	38	3.18 (2.67 - 3.69)	41	3.05 (2.55 - 3.54)	22	3.11 (2.44 - 3.78)	.93	.001
Swit-ching	P200	♂	21	4.32 (3.65 - 4.98)	20	4.08 (3.39 - 4.76)	10	4.06 (3.09 - 5.02)	.85	.007
		♀	4	4.99 (3.17 - 6.82)	4	5.84 (4.01 - 7.66)	6	3.92 (2.43 - 5.41)	.24	.231
		Tot.	25	4.42 (3.80 - 5.05)	24	4.37 (3.73 - 5.01)	16	4.01 (3.22 - 4.79)	.68	.012
Odd-ball	N100	♂	25	-5.08 (-5.92 - -4.25)	26	-4.43 (-5.25 - -3.61)	10	-3.15 (-4.47 - -1.82)	.05	.096
		♀	8	-6.41 (-7.60 - -5.22)	7	-5.17 (-6.44 - -3.89)	7	-5.40 (-6.67 - -4.13)	.30	.119
		Tot.	33	-5.41 (-6.12 - -4.69)	33	-4.59 (-5.30 - -3.88)	17	-4.08 (-5.07 - -3.08)	.08	.062
	P200	♂	25	3.32 (2.49 - 4.14)	26	2.85 (2.03 - 3.66)	10	3.26 (1.95 - 4.57)	.70	.012
		♀	8	3.26 (1.63 - 4.89)	7	2.12 (.38 - 3.87)	7	2.29 (.54 - 4.03)	.56	.059
		Tot.	33	3.30 (2.58 - 4.02)	33	2.69 (1.97 - 3.41)	17	2.86 (1.86 - 3.87)	.48	.018

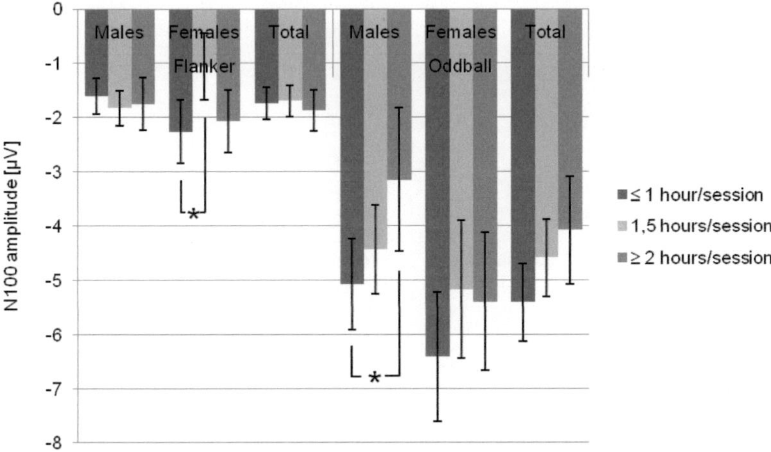

Figure 72: Mean N100 amplitudes during Flanker and Oddball task for athletes with durations of ≤ 1 hour/session, 1.5 hours/session and ≥ 2 hours/session; error bars represent 95% confidence intervals

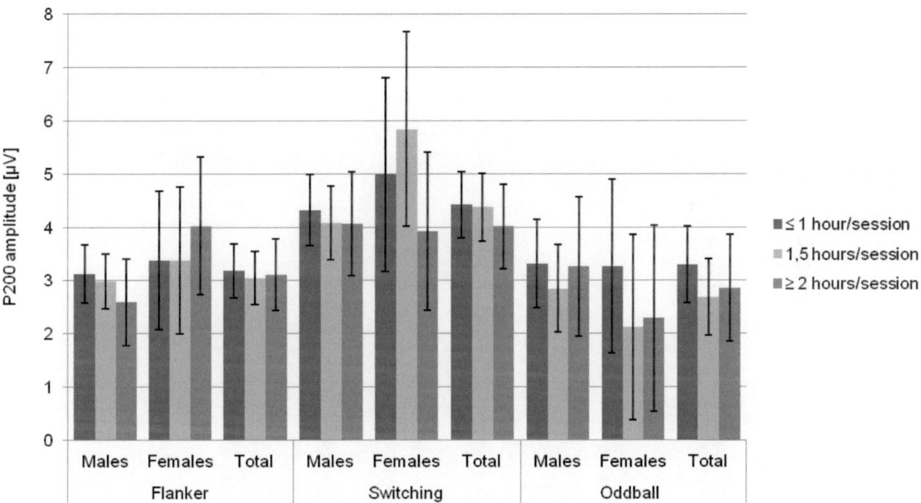

Figure 73: Mean P200 amplitudes during Flanker, Switching and Oddball task for athletes with durations of ≤ 1 hour/session, 1.5 hours/session and ≥ 2 hours/session; error bars represent 95% confidence intervals

Table 84: Mean N100 and P200 latencies for athletes with durations of ≤ 1 hour/session, 1.5 hours/session and ≥ 2 hours/session in the Flanker, Switching and Oddball task

			≤ 1 hour/session		1.5 hours/session		≥ 2 hours/session		p	η^2
			n	ms (95 % CI)	n	ms (95 % CI)	n	ms (95 % CI)		
Flanker	N100	♂	10	140.04 (129.51 - 150.57)	10	140.14 (129.61 - 150.67)	7	140.40 (127.81 - 152.99)	.99	.000
		♀	5	132.42 (120.67 - 144.18)	-	-	4	145.70 (132.56 - 158.85)	.12	.312

			n		n		n			
		Tot.	15	137.50 (129.64 - 145.36)	10	140.14 (130.52 - 149.76)	11	142.33 (133.16 - 151.50)	.72	.020
	P 200	♂	26	219.88 (209.18 - 230.59)	23	216.14 (204.76 - 227.53)	9	232.62 (214.42 - 250.82)	.31	.041
		♀	5	221.29 (196.64 - 245.94)	5	219.92 (195.27 - 244.58)	6	233.76 (211.25 - 256.26)	.62	.072
		Tot.	31	220.11 (210.67 - 229.55)	28	216.82 (206.88 - 226.75)	15	233.07 (219.50 - 246.65)	.15	.052
Swit-ching	P 200	♂	20	189.75 (181.03 - 198.47)	17	184.08 (174.62 - 193.53)	9	191.06 (178.06 - 204.06)	.59	.025
		♀	4	174.51 (151.15 - 197.88)	4	186.23 (162.87 - 209.59)	6	209.34 (190.27 - 228.42)	.07	.389
		Tot.	24	187.21 (178.99 - 195.42)	21	184.49 (175.70 - 193.27)	15	198.37 (187.98 - 208.76)	.11	.073
Odd-ball	N 100	♂	23	171.26 (163.58 - 178.94)	22	167.37 (159.51 - 175.22)	8	167.19 (154.17 - 180.21)	.74	.012
		♀	8	172.07 (163.74 - 180.40)	6	177.60 (167.99 - 187.22)	7	183.93 (175.03 - 192.83)	.15	.188
		Tot.	31	171.47 (165.37 - 177.58)	28	169.56 (163.14 - 175.98)	15	175.00 (166.23 - 183.77)	.61	.014
	P 200	♂	21	261.12 (252.00 - 270.24)	17	258.18 (248.04 - 268.32)	8	249.22 (234.44 - 264.00)	.39	.043
		♀	4	253.13 (234.51 - 271.74)	4	272.66 (254.04 - 291.27)	3	250.52 (229.03 - 272.02)	.18	.345
		Tot.	25	259.84 (251.85 - 267.84)	21	260.94 (252.22 - 269.66)	11	249.57 (237.53 - 261.62)	.28	.046

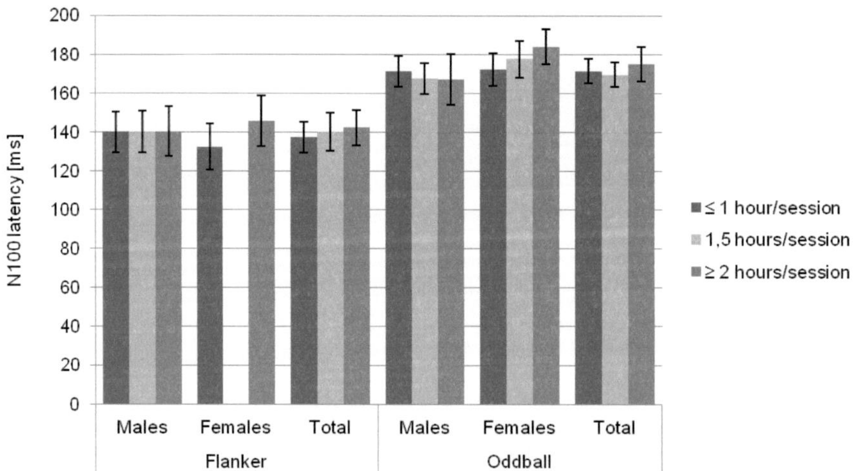

Figure 74: Mean N100 latencies during Flanker and Oddball task for athletes with durations of ≤ 1 hour/session, 1.5 hours/session and ≥ 2 hours/session; error bars represent 95% confidence intervals

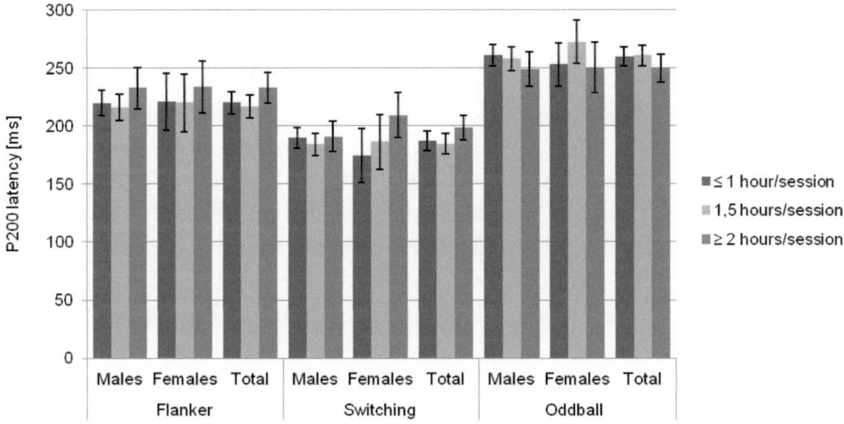

Figure 75:   Mean P200 latencies during Flanker, Switching and Oddball task for athletes with durations of ≤ 1 hour/session, 1.5 hours/session and ≥ 2 hours/session; error bars represent 95% confidence intervals

4.3.5.4 Discussion

Intensity of preferred sport was unrelated to cognitive performance except for P200 amplitudes in the Oddball task. Females with vigorous exercise intensity had larger amplitudes whereas males with vigorous intensity had smaller amplitudes compared to participants who exercise with moderate intensity. A consistent but not significant trend was observed for the Oddball task with larger P300 amplitudes and faster P300 latencies for the moderate intensity group. For the Flanker and Switching task, no consistent tendencies for P300 component were observed. In addition, no differences between both groups were found with regard to N100 and P20 component. This was surprising since exercise intensity is an important aspect in studies on the acute effects on cognitive performance and levels of neurotrophins. Both, high-intensity aerobic exercise (Ferris et al., 2007) and moderate aerobic exercise (Tang, Chu, Hui, Helmeste & Law, 2008) yielded enhanced BDNF levels after exercise cessation. Moreover, Ferris et al. (2007) could detect a beneficial influence of high-intensity exercise on activation and attentional performance. According to Hillman et al. (2003), high-intensity exercise seems to facilitate higher executive control and attentional processes after its cessation. In addition, inhibitory control and attention are positively influenced by medium-intensity exercise but have shown to be impaired by high-intensity exercise (Kamijo, Nishihira, Hatta, Kaneda, Wasaka et al., 2004). This finding might be explained by an inverted U-shaped model indicating an increased cognitive performance with increasing physical arousal until reaching its peak at moderate-intensity. An increase of exercise intensity above this point leads to fatigue and a decline in cognitive performance. However, exercise intensity seems to play a role for cognitive performance only in study assessing the acute but not regular effects of physical activity.

There was a significant relationship between frequency of sport and P300 amplitudes in the Flanker and Oddball task and a non-significant relationship for the Switching task having the same tendency. Athletes with a frequency of ≥ 4 sessions per week had significantly larger amplitudes than athletes performing < 2 and 2-3 sessions. Further, athletes with ≥ 4 sessions per week revealed significantly lower Oddball P300 latencies compared to the other two groups. Given that the amount of physical and sports activity as expressed in kcal/week was only slightly related to cognitive performance in young adults, this is an interesting finding that supports the hypothesis of better executive functions being related to a higher amount of sports activity. The results on P300 amplitude are very consistent across all cognitive tasks however no effects on a behavioral level could have been detected. This finding is in contrast to a study by Knab, Nieman, Sha, Boman-Fulks and Canu (2012) who tested adults between 18 and 85 years and found that exercise frequency was unrelated to neurocognitive function but related to psychopathologic measures such as depression or anxiety. They further report an inverse relationship between exercise frequency and perceived stress. However, neurocognitive function was solely assessed by behavioral measures in the study by Knab et al. (2012), and in the current study also no frequency-related effects could be observed for behavioral measures. This might indicate that exercise frequency selectively influences underlying brain structures as observable by event-related brain potentials, but that this positive effect cannot be confirmed by behavioral measures. For the N100 component in the Flanker and Oddball task, there was a non significant tendency with larger amplitudes but also longer latencies for athletes with a frequency of ≥ 4 sessions per week. Athletes performing less than two sessions per week had the shortest P200 latencies during the Switching task which might indicate that a lower number of exercise sessions per week might be beneficial for cognitive functions related to P200 component.

When looking at the effects of the duration of exercise sessions, significantly lower response times during the Oddball task were observed for athletes exercising up to 1 hour than those exercising 1.5 or ≥ 2 hours per session. In addition, there was also a tendency for faster response times for athletes exercising up to 1 hour per session during the Flanker and Switching task. Although there were no effects observed for P300 event-related brain potentials, female athletes with exercise durations of up to 1 hour per session also had significantly larger N100 amplitudes in the Flanker task and male athletes with exercise durations of up to 1 hour per session had significantly larger N100 amplitudes in the Oddball task. One could conclude from these findings on both, frequency of sport and duration of exercise session, that the most positive effects on cognitive performance might be elicited when exercise is performed more frequently (≥ 4 sessions per week), but with a relatively short duration of 1 hour per session. This is also in accordance with current physical activity recommendations.

4.3.6 Summary

The results of this chapter on the relationship between physical activity and cognition variables are summarized in the following table 85. In addition, tables 86 to 88 provide an overview of effect sizes for all analyses.

Table 85:   Summary of results on physical activity and cognition

Hypothesis	Summary of Results
RQ 1: Do young adults who regularly engage in physical activity have a higher cognitive performance than inactive young adults?	Significant findings: Difference between athletes and nonathletes in response times during Oddball task (athletes respond faster). Larger P300 amplitudes for athletes compared to nonathletes during Flanker task at Cz and Pz electrode and during Oddball task. Faster latencies for athletes compared to nonathletes in the congruent but not incongruent trials of the Flanker task. Trends: Tendency for faster response times for athletes in all tasks. Tendency for larger P300 amplitudes in the Flanker task. Tendency for larger P200 amplitudes for athletes compared to nonathletes in the Flanker and Switching task. Gender-specific: Larger P300 amplitudes during the Oddball task and larger P200 amplitudes during the Switching task for male athletes compared to nonathletes. Faster response times during Oddball and faster P300 latencies during Flanker task for female athletes compared to nonathletes. Generally larger effect sizes for females (mean $\eta^2 = .039$) than for males (mean $\eta^2 = .017$). → Research question can be partly answered in the affirmative. Regular engagement in physical activity has positive effects on selected measures of cognitive performance in young adults. However, gender-specific effects play an important role. The overall mean effect size is small ($\eta^2 = .015$).
RQ 2: Is the participation in competitive sports related to cognitive performance in young adulthood?	Significant findings: Difference between competitive and non-competitive athletes and nonathletes in response times during Oddball task (non-competitive athletes respond faster than competitors and nonathletes). Faster P300 latencies during Flanker task for non-competitive compared to nonathletes. Faster P200 latencies during Flanker task for non-competitive compared to competitive athletes. Trends: Tendency for longest response times during the Switching and Oddball task for nonathletes. Tendency for larger amplitudes for competitors compared to non-competitors and nonathletes in the Flanker and Oddball task. Faster N100 latencies for non-competitive athletes in the Flanker task. Gender-specific: Faster P200 latencies in the Flanker task for male non-competitive athletes compared to nonathletes. Faster P300 latencies during Flanker task for female non-competitive athletes compared to nonathletes. Generally larger effect sizes for females (mean $\eta^2 = .085$) than for males (mean $\eta^2 = .032$). → Research question can be partly answered in the affirmative. Participation in non-competitive sports compared to competitive sports has positive effects on selected measures of cognitive performance in young adulthood. The overall mean effect size is small ($\eta^2 = .032$).

RQ 3: Is the amount of sports and physical activity related to cognitive performance in young adulthood?	<u>Significant findings:</u> Habitual activity is significantly related to P300 amplitudes of the Switching task (higher amplitudes for high activity group). <u>Trends:</u> Tendency for faster response times in low sports activity groups. Tendency for larger P300 amplitudes for participants with a high amount of habitual physical activity during the Flanker task and for the high sports activity group during all cognitive tasks. <u>Gender-specific:</u> P200 amplitudes during the Flanker task for males are larger for low sports activity group compared to high sports activity group. Larger P300 amplitudes during Switching task for females with high sports activity compared to low sports activity. Generally larger effect sizes for females (sports activity: mean η^2 = .139; habitual physical activity: mean η^2 = .067) than for males (sports activity: mean η^2 = .025; habitual physical activity: mean η^2 = .028). → Research question can be mostly answered in the negative. The amount of sports and habitual physical activity is not related to cognitive performance in young adulthood. The overall mean effect sizes are small (sports activity: η^2 = .019; habitual physical activity: η^2 = .024).
RQ 4: Is there a difference in cognitive performance between athletes engaging in different types of sport?	<u>Significant findings:</u> Difference between types of sport in Oddball response times (longer response times for nonathletes compared to strength athletes). Effect on N100 amplitudes in the Oddball task (larger negative amplitude for strength compared to field athletes). Significantly faster P300 latencies for endurance athletes compared to nonathletes during the Flanker task. <u>Trends:</u> Tendency for lowest response times for strength athletes compared to athletes from other types of sport and nonathletes. Tendency for larger P300 amplitudes for strength athletes compared to other types in the Flanker task. Tendency for faster P300 latencies and larger P200 amplitudes during the Oddball task for strength athletes. Tendency for larger P300 amplitudes during the Oddball task, and larger P200 amplitudes during the Switching task for endurance athletes. → Research question can be partly answered in the affirmative. Strength athletes and, to some extent endurance athletes showed better results in selected cognitive measures than field athletes and nonathletes. The overall mean effect size is moderate (η^2 = .044).
RQ 5: Is exercise intensity related to cognitive performance in young adulthood?	<u>Significant findings:</u> none <u>Trends:</u> Larger P300 amplitudes and faster P300 latencies in the Oddball task for moderate compared to vigorous intensity group. <u>Gender-specific:</u> Effect on Oddball P200 amplitude with larger amplitudes for males with moderate intensity but larger amplitudes for females with vigorous intensity. Generally larger effect sizes for females (mean η^2 = .041) than for males (mean η^2 = .013). → Research question can be mostly answered in the negative. Exercise intensity is not related to cognitive performance in young adulthood. The overall mean effect size is small (η^2 = .009).
RQ 6: Is exercise frequency per week related to cognitive	<u>Significant findings:</u> Effects on Flanker and Oddball P300 amplitudes (largest amplitudes for athletes with ≥ 4 sessions per week compared to < 2 and 2-3 sessions per week). Athletes with ≥ 4 sessions per week have significantly lower Oddball

performance in young adulthood?	P300 latencies compared to the other two groups. Athletes performing less than two sessions per week have shortest P200 latencies during the Switching task. Trends: Tendency for larger P300 amplitudes for athletes with ≥ 4 sessions per week compared to < 2 and 2-3 sessions per week in the Switching task. Tendency for larger Flanker and Oddball N100 amplitudes but also longer latencies for athletes with a frequency of ≥ 4 sessions per week. $\rightarrow$ Research question can be partly answered in the affirmative. Exercise frequency is related to selected measures of cognitive performance in young adulthood with a higher frequency being more beneficial. The overall mean effect size is moderate ($\eta^2 = .049$).
RQ 7: Is exercise duration per session related to cognitive performance in young adulthood?	Significant findings: Shorter response times for athletes with ≤ 1 hour/session compared to ≥ 2 hours per session in the Oddball task. Trends: Tendency for faster response times for athletes exercising ≤ 1 hour per session during the Flanker and Switching task. Gender-specific: Significantly larger N100 amplitudes in the Flanker task for female athletes with exercise durations of ≤ 1 hour per session. Significantly larger N100 amplitudes in the Oddball task for male athletes with exercise durations of ≤ 1 hour per session. Generally larger effect sizes for females (mean $\eta^2 = .155$) than for males (mean $\eta^2 = .023$). $\rightarrow$ Research question can be partly answered in the affirmative. Exercise duration is related to selected measures of cognitive performance in young adulthood with a shorter duration being more beneficial. The overall mean effect size is small ($\eta^2 = .030$).

The largest effect sizes for the independent variables on cognitive performance with regard to the total study sample were found for exercise frequency, type of sport and competitive sports participations (table 86). These factors explain the highest variance in cognitive performance (frequency: 4.9 %; type: 4.4 %; competitive sports: 3.2 %). When looking at the dependent variables, the largest effects of physical activity were found on P300 amplitudes (mean explained variance: 3.6 %) and P300 latencies (mean explained variance: 3.6 %).

For the male subsample (table 87), the largest effects on cognitive performance were found for competitive sports participation (explained variance: 3.2 %) and the amount of habitual physical (explained variance: 2.8 %) and sports activity (explained variance: 2.5 %). The largest effects of physical activity variables were observed for N100 latencies (explained variance: 3.2 %) and P300 latencies (explained variance: 3.1 %).

For the female subsample (table 88), the effect sizes were consistently larger than those for males. However, due to the small number of female participants in this study, most effects were not significant. The largest effects on cognitive performance were found for exercise duration (explained variance: 15.5 %), the amount of sports activity (explained variance: 13.9 %) and competitive sports participation (explained variance: 8.5 %). The largest effects of physical activity variables were observed for N100 latencies (explained variance: 13.4 %) and P300 latencies (explained variance: 11.9 %).

Table 86: Summary of effect sizes (η^2) for physical activity measures for the total sample (only main effects, interactions are not displayed)

		General sports participation [η²]	Competitive sports [η²]	Sports activity [η²]	Habitual physical activity [η²]	Type of sport [η²]	Intensity [η²]	Frequency [η²]	Duration [η²]	Total (Mean η²)
Resp. Times	Flanker	.013	.042	.031	.001	.040	.000	.003	.018	.019
	Switching	.009	.010	.014	.038	.050	.001	.015	.013	.019
	Oddball	**.040***	**.050***	.022	.012	**.093***	.000	.006	**.070***	.037
Errors	Switching	.000	.000	.001	.018	.001	.005	.033	.002	.008
P300 ampl.	Flanker	.025	.029	.051	.018	.046	.001	**.101***	.021	.037
	Switching	.000	.000	.028	.069	.008	.007	.088	.033	.029
	Oddball	**.047***	.047	.012	.001	.052	.009	**.155***	.014	.042
P300 lat.	Flanker	**.056***	**.071***	.030	.032	.073	.009	.000	.006	.035
	Switching	.006	.042	.006	.048	.039	.037	.070	.093	.043
	Oddball	.003	.004	.009	.011	.036	.045	**.110***	.029	.031
N100 ampl.	Flanker	.003	.004	.013	.004	.008	.001	.039	.005	.010
	Oddball	.002	.016	.001	.040	**.122***	.002	.012	.062	.032
N100 lat.	Flanker	.006	.083	.013	.047	.061	.014	.020	.020	.033
	Oddball	.000	.003	.058	.028	.002	.015	.015	.014	.017
P200 ampl.	Flanker	.013	.013	.013	.029	.040	.000	.048	.001	.020
	Switching	.041	.060	.008	.009	.062	.002	.053	.012	.031
	Oddball	.000	.036	.032	.037	.032	.004	.014	.018	.022
P200 lat.	Flanker	.012	**.063***	.001	.024	.056	.004	.036	.052	.031
	Switching	.001	.054	.009	.002	.015	.016	**.101***	.073	.034
	Oddball	.017	.020	.032	.007	.037	.002	.055	.046	.027
Total (Mean η²)		.015	.032	.019	.024	.044	.009	.049	.030	

Table 87: Summary of effect sizes (η^2) for all cognitive and physical activity measures for males (only main effects, interactions are not displayed)

		General sports participation [η^2]	Competitive sports [η^2]	Sports activity [η^2]	Habitual physical activity [η^2]	Type of sport [η^2]	Intensity [η^2]	Frequency [η^2]	Duration [η^2]	Total (Mean η^2)
Resp. Times	Flanker	.006	.043	.043	.000		.001		.006	.017
	Switching	.012	.017	.004	.042		.003		.003	.014
	Oddball	.008	.027	.035	.011		.003		.069	.026
Errors	Switching	.000	.003	.013	.043		.012		.005	.013
P300 ampl.	Flanker	.022	.027	.068	.010		.004		.009	.023
	Switching	.002	.016	.015	.046		.003		.062	.024
	Oddball	**.054***	.059	.007	.016		.001		.007	.024
P300 lat.	Flanker	.001	.001	.019	.033		.003		.001	.010
	Switching	.052	.064	.046	.026		.108		.075	.062
	Oddball	.008	.014	.028	.012		.043		.027	.022
N100 ampl.	Flanker	.022	.029	.007	.002		.008		.013	.014
	Oddball	.014	.039	.012	.061		.016		**.096***	.040
N100 lat.	Flanker	.001	.043	.048	.121		.030		.000	.041
	Oddball	.005	.006	.074	.011		.023		.012	.022
P200 ampl.	Flanker	.001	.013	**.084***	.028		.000		.016	.024
	Switching	**.073***	.075	.000	.018		.016		.007	.032
	Oddball	.002	.037	.012	.018		**.065***		.012	.024
P200 lat.	Flanker	.026	**.102***	.001	.041		.005		.041	.036
	Switching	.010	.060	.014	.020		.013		.025	.024
	Oddball	.021	.032	.014	.029		.001		.043	.023
Total (Mean η^2)		.017	.032	.025	.028		.013		.023	

Table 88: Summary of effect sizes (η^2) for all cognitive and physical activity measures for females (only main effects, interactions are not displayed)

		General sports participation [η^2]	Competitive sports [η^2]	Sports activity [η^2]	Habitual physical activity [η^2]	Type of sport [η^2]	Intensity [η^2]	Frequency [η^2]	Duration [η^2]	Total (Mean η^2)
Resp. Times	Flanker	.007	.055	.040	.009		.003		.048	.027
	Switching	.005	.005	.058	.034		.054		.132	.048
	Oddball	**.115***	.116	.008	.015		.005		.072	.055
Errors	Switching	.000	.002	.094	.008		.003		.014	.020
P300 ampl.	Flanker	.019	.019	.008	.111		.019		.117	.049
	Switching	.000	.029	**.471***	.100		.134		.055	.132
	Oddball	.050	.065	.139	.077		.043		.083	.076
P300 lat.	Flanker	**.266***	**.071***	.010	.080		.015		.110	.092
	Switching	.041	.222	.407	.149		.105		.357	.214
	Oddball	.012	.108	.004	.157		.002		.025	.051
N100 ampl.	Flanker	.006	.018	.077	.027		.047		**.334***	.085
	Oddball	.032	.052	.051	.036		.000		.119	.048
N100 lat.	Flanker	.099	.404	.063	.077		.031		.312	.164
	Oddball	.034	.034	.225	.064		.073		.188	.103
P200 ampl.	Flanker	.025	.099	.229	.078		.015		.035	.080
	Switching	.033	.151	.082	.010		.018		.231	.088
	Oddball	.009	.079	.171	.100		**.193***		.059	.102
P200 lat.	Flanker	.001	.008	.022	.032		.000		.072	.023
	Switching	.016	.106	.121	.076		.027		.389	.123
	Oddball	.008	.066	.503	.099		.031		.345	.175
Total (Mean η^2)		.039	.085	.139	.067		.041		.155	

4.4 Relationships between Fitness and Cognition

4.4.1 Endurance

In this chapter, the relationship between endurance and cognitive performance is described. The independent variables are maximal oxygen uptake (VO_{2max}) and the performance at the individual anaerobic threshold (IAT). The groups for VO_{2max} were built according to a normative table from the Cooper Institute for Aerobics Research, Texas, USA (printed in Heyward, 1998, p. 48). In the poor performance group, males have a VO_{2max} of $\leq$ 45 ml/kg/min and females $\leq$ 35 ml/kg/min. In the good performance group, males have a VO_{2max} of $\geq$ 45.1 ml/kg/min and females $\geq$ 35.1 ml/kg/min (table 89). The groups for the individual anaerobic threshold were built individually for the study sample due to a lack of reference or normative values. In the poor performance group, males have an IAT of $\leq$ 180 W and females $\leq$ 125 W. In the good performance group, males have an IAT of $\geq$ 180.1 W and females $\geq$ 125.1 W (table 90).

Table 89:   Number of participants per VO_{2max} group (M ± SD in ml/kg/min)

	Poor		Good		Total	
	n (%)	M (SD)	n (%)	M (SD)	n	p
♂	30 (30)	40.64 (4.10)	69 (70)	55.54 (7.14)	99	.00
♀	13 (31)	30.26 (3.36)	29 (69)	42.59 (7.57)	42	.00
Total	43 (31)	37.50 (6.17)	98 (69)	51.71 (9.36)	141	.00

Table 90:   Number of participants per IAT group (M ± SD in W)

	Poor		Good		Total	
	n (%)	M (SD)	n (%)	M (SD)	n	p
♂	52 (50)	148.74 (23.76)	52 (50)	217.71 (31.22)	104	.00
♀	22 (50)	97.56 (15.90)	22 (50)	147.26 (14.00)	44	.00
Total	74 (50)	133.52 (31.97)	74 (50)	196.77 (42.29)	148	.00

> *RQ 8*: Is aerobic endurance related to cognitive performance in young adulthood?

4.4.1.1 VO_{2max} - Results

Behavioral Measures

VO_{2max} did not have any effect on behavioral measures in the Flanker task. For response times during the Switching task, there was a significant condition x VO_{2max} interaction (F [1, 116] = 4.95, p = .03, η^2 = .041). Participants with good VO_{2max} had a larger difference in response times between the homogenous (527.74 ms; 95 % CI 512.98 - 542.50) and heterogeneous condition (887.50 ms; 95 % CI 861.25 - 913.75) than participants with poor VO_{2max} (homogenous: 533.65 ms; 95 % CI 511.37 - 555.92; heterogeneous: 852.63 ms; 95 % CI 813.01 - 892.25).

In addition, participants with good VO_{2max} had faster response times during homogenous but longer response times during heterogeneous conditions compared to participants with a poor VO_{2max}. For errors in the Switching task, there was a main effect (F [1, 120] = 4.77, p = .03, η^2 = .038; males: F [1, 83] = 3.86, p = .05, η^2 = .044; females: F [1, 35] = 1.06, p = .31, η^2 = .029) with fewer errors for participants with good VO_{2max}. This main effect is superseded by a condition x VO_{2max} interaction (F [1, 116] = 60.87, p = <.01, η^2 = .077). Participants with good VO_{2max} had a smaller difference in errors between the homogenous (2.58; 95 % CI 2.19 - 2.96) and heterogeneous condition (4.54; 95 % CI 3.91 - 5.17) than participants with poor VO_{2max} (homogenous: 2.72; 95 % CI 2.12 - 3.32; heterogeneous: 6.24; 95 % CI 5.26 - 7.21). No further effects were found (table 91 & 92, figure 76 & 77).

Table 91:   Mean response times for VO_{2max} groups in the Flanker, Switching and Oddball task

		Poor VO_{2max}		Good VO_{2max}			
		n	ms (95 % CI)	n	ms (95 % CI)	p	η^2
Flanker	♂	26	492.16 (475.83 - 508.50)	61	493.83 (483.16 - 504.49)	.87	.000
	♀	10	497.19 (467.26 - 527.12)	24	516.02 (496.70 - 535.34)	.29	.035
	Total	36	493.56 (479.15 - 507.97)	85	500.09 (490.72 - 509.47)	.45	.005
Switching	♂	26	693.46 (662.35 - 724.57)	57	707.56 (686.54 - 728.57)	.46	.007
	♀	10	692.32 (626.93 - 757.70)	25	707.76 (666.41 - 749.11)	.69	.005
	Total	36	693.14 (664.79 - 721.49)	82	707.62 (688.83 - 726.40)	.40	.006
Oddball	♂	26	482.34 (446.26 - 518.42)	59	512.70 (488.75 - 536.65)	.17	.023
	♀	10	552.28 (487.39 - 617.17)	27	520.34 (480.85 - 559.84)	.40	.020
	Total	36	501.77 (470.12 - 533.42)	86	515.10 (494.62 - 535.58)	.49	.004

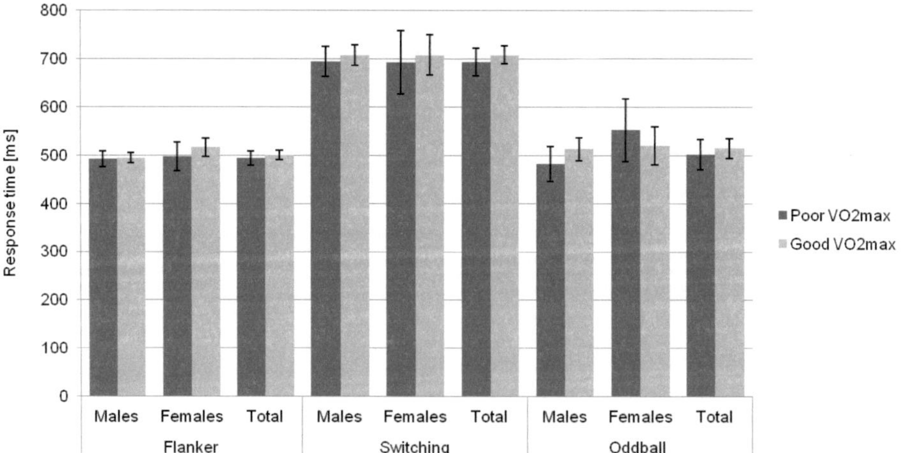

Figure 76:   Mean response times during Flanker, Switching and Oddball task for VO_{2max} groups; error bars represent 95% confidence intervals

Table 92:   Errors in the Switching task for VO_{2max} groups

		Poor VO_{2max}		Good VO_{2max}			
		n	No. (95 % CI)	n	No. (95 % CI)	p	η^2
Errors	♂	26	4.42 (3.63 - 5.21)	59	3.49 (2.96 - 4.01)	.05	.044
	♀	10	4.63 (3.09 - 6.16)	27	3.71 (2.78 - 4.65)	.31	.029
	Total	36	4.48 (3.78 - 5.18)	86	3.56 (3.10 - 4.01)	.03	.038

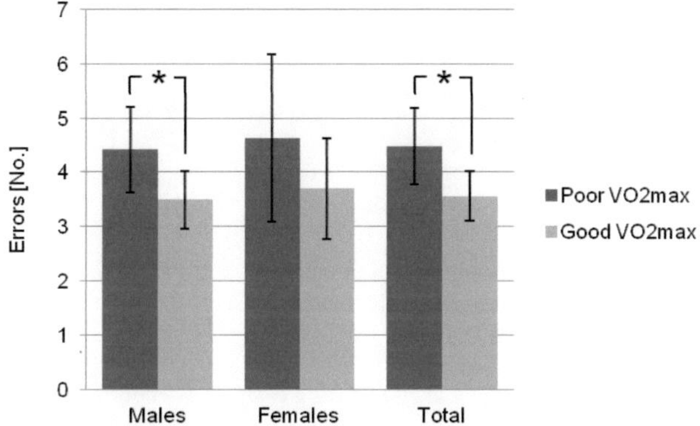

Figure 77:   Errors during the Switching task for VO_{2max} groups; error bars represent 95% confidence intervals

P300

VO_{2max} had no effect on the P300 component in the Flanker, Switching and Oddball task. The mean P300 amplitudes are presented in table 93 and figure 78, the latencies are given in table 94 and figure 79.

Table 93:   Mean P300 amplitudes for VO_{2max} groups during Flanker, Switching and Oddball task

		Poor VO_{2max}		Good VO_{2max}			
		n	μV (95 % CI)	n	μV (95 % CI)	p	η^2
Flanker	♂	26	4.53 (3.99 - 5.07)	59	4.79 (4.43 - 5.14)	.43	.008
	♀	8	4.58 (3.49 - 5.67)	22	4.41 (3.75 - 5.07)	.79	.003
	Total	34	4.54 (4.06 - 5.02)	81	4.68 (4.37 - 4.99)	.62	.002
Switching	♂	19	3.63 (3.20 - 4.05)	40	3.84 (3.54 - 4.13)	.42	.012
	♀	8	4.84 (3.97 - 5.72)	14	3.96 (3.30 - 4.62)	.11	.124
	Total	27	3.99 (3.59 - 4.39)	54	3.87 (3.59 - 4.15)	.64	.003
Oddball	♂	22	5.26 (4.25 - 6.26)	47	5.92 (5.24 - 6.61)	.28	.017
	♀	7	6.06 (4.31 - 7.81)	22	6.53 (5.54 - 7.51)	.64	.008
	Total	29	5.45 (4.59 - 6.31)	69	6.12 (5.56 - 6.67)	.20	.017

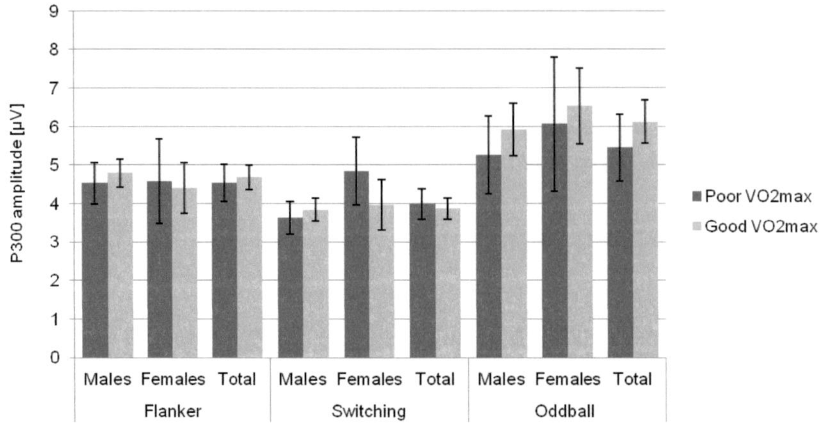

Figure 78:   Mean P300 amplitudes during Flanker, Switching and Oddball task for VO_{2max} groups; error bars represent 95% confidence intervals

Table 94:   Mean P300 latencies for VO_{2max} groups during Flanker, Switching and Oddball task

		Poor VO_{2max}		Good VO_{2max}			
		n	ms (95 % CI)	n	ms (95 % CI)	p	η^2
Flanker	♂	22	391.15 (378.02 - 404.27)	48	392.80 (383.92 - 401.69)	.84	.001
	♀	7	391.01 (346.08 - 435.93)	19	392.91 (365.64 - 420.18)	.94	.000
	Total	29	391.11 (376.67 - 405.56)	67	392.83 (383.33 - 402.34)	.84	.000
Switching	♂	10	381.54 (350.14 - 412.95)	25	385.64 (365.78 - 405.51)	.82	.002
	♀	6	416.86 (379.12 - 454.61)	9	370.13 (339.31 - 400.95)	.06	.248
	Total	16	394.79 (370.87 - 418.71)	34	381.53 (365.12 - 397.94)	.36	.017
Oddball	♂	19	418.35 (381.71 - 455.00)	46	444.45 (420.90 - 468.00)	.24	.022
	♀	5	383.85 (334.48 - 433.23)	21	392.08 (367.98 - 416.17)	.76	.004
	Total	24	411.17 (380.30 - 442.03)	67	428.03 (409.56 - 446.51)	.35	.010

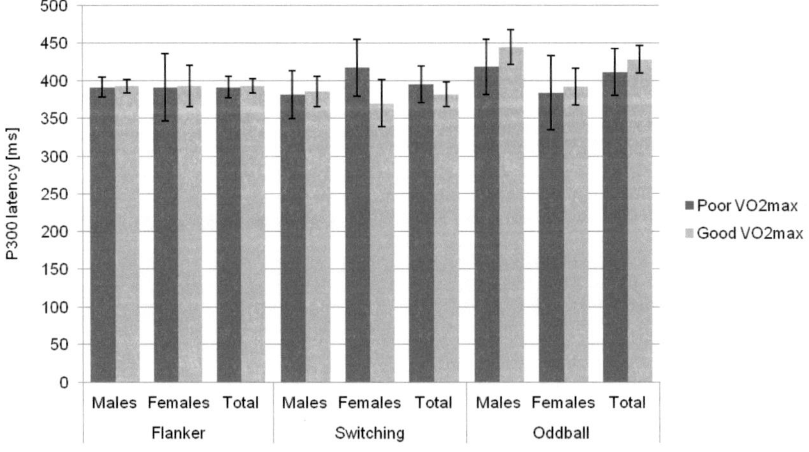

Figure 79:   Mean P300 latencies during Flanker, Switching and Oddball task for VO_{2max} groups; error bars represent 95% confidence intervals

N100 and P200

For N100 latency during the Flanker task, there was a condition x VO_{2max} interaction (F [1, 41] = 5.52, p = .02, η^2 = .119). Participants with a good VO_{2max} had a smaller latency in the accuracy (136.71 ms; 95 % CI 130.04 - 143.37) compared to the speed condition (143.47; 95 % CI 136.81 - 150.13) whereas participants with poor VO_{2max} had a faster latency in the speed condition (accuracy: 145.83; 95 % CI 136.73 - 154.94; speed: 136.72; 95 % CI 127.62 - 145.82). Participants with a good VO_{2max} further revealed smaller latencies in the accuracy but longer latencies in the speed condition when compared to participants with a poor VO_{2max}. No additional effects were observed (table 95 & 96, figure 80 - 83).

Table 95:   Mean N100 and P200 amplitudes for VO_{2max} groups in the Flanker, Switching and Oddball task

| | | | Poor VO_{2max} | | Good VO_{2max} | | | |
			n	µV (95 % CI)	n	µV (95 % CI)	p	η^2
Flanker	N100	♂	26	-1.56 (-1.92 - -1.21)	60	-1.73 (-1.97 - -1.50)	.43	.008
		♀	9	-2.29 (-2.87 - -1.71)	22	-1.78 (-2.15 - -1.40)	.14	.074
		Total	35	-1.75 (-2.05 - -1.44)	82	-1.74 (-1.94 - -1.55)	.98	.000
	P200	♂	26	3.15 (2.58 - 3.72)	60	2.80 (2.43 - 3.17)	.31	.012
		♀	9	4.20 (3.02 - 5.39)	22	3.78 (3.02 - 5.45)	.55	.013
		Total	35	3.42 (2.89 - 3.95)	82	3.06 (2.72 - 3.41)	.27	.011
Switching	P200	♂	20	3.56 (2.91 - 4.21)	42	4.14 (3.69 - 4.59)	.14	.035
		♀	8	4.53 (3.20 - 5.85)	14	4.53 (3.52 - 5.53)	.99	.000
		Total	28	3.84 (3.25 - 4.42)	56	4.24 (3.82 - 4.65)	.27	.015
Oddball	N100	♂	22	-4.92 (-5.86 - -3.98)	47	-4.56 (-5.20 - -3.91)	.53	.006
		♀	8	-5.57 (-7.03 - -4.12)	22	-5.45 (-6.33 - -4.57)	.88	.001
		Total	30	-5.09 (-5.88 - -4.31)	69	-4.84 (-5.36 - -4.32)	.59	.003
	P200	♂	22	2.89 (2.05 - 3.73)	47	3.06 (2.48 - 3.64)	.73	.002
		♀	8	2.51 (.75 - 4.27)	22	2.81 (1.74 - 3.87)	.77	.003
		Total	30	2.79 (2.02 - 3.55)	69	2.98 (2.48 - 3.48)	.68	.002

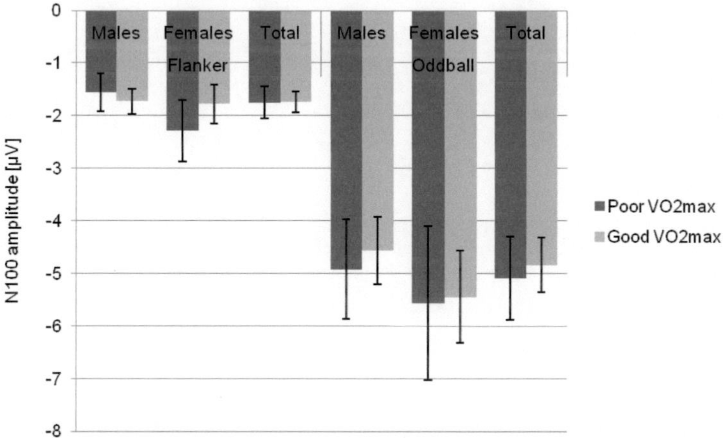

Figure 80:   Mean N100 amplitudes during Flanker and Oddball task for VO_{2max} groups; error bars represent 95% confidence intervals

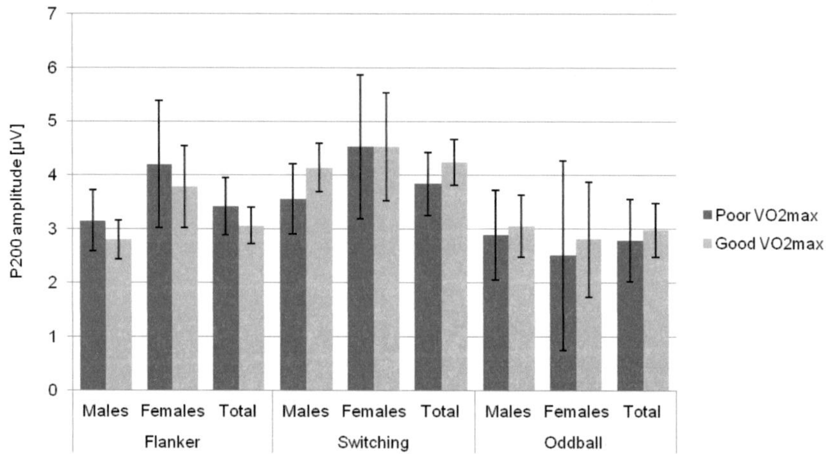

Figure 81: Mean P200 amplitudes during Flanker, Switching and Oddball task for VO_{2max} groups; error bars represent 95% confidence intervals

Table 96: Mean N100 and P200 latencies for VO_{2max} groups in the Flanker, Switching and Oddball task

				Poor VO_{2max}		Good VO_{2max}		
			n	ms (95 % CI)	n	ms (95 % CI)	p	η^2
Flanker	N 100	♂	8	143.14 (132.22 - 154.06)	22	138.82 (132.24 - 145.41)	.49	.017
		♀	7	139.15 (129.96 - 148.33)	6	144.73 (134.81 - 154.65)	.38	.070
		Total	15	141.28 (134.03 - 148.53)	28	140.09 (134.78 - 145.39)	.79	.002
	P 200	♂	21	227.46 (215.50 - 239.41)	44	220.72 (212.46 - 228.98)	.36	.013
		♀	8	228.59 (207.91 - 249.27)	15	226.56 (211.46 - 241.66)	.87	.001
		Total	29	227.77 (217.67 - 237.87)	59	222.21 (215.13 - 229.29)	.37	.009
Switching	P 200	♂	16	191.85 (182.39 - 201.30)	39	187.85 (181.79 - 193.90)	.48	.010
		♀	8	183.55 (167.07 - 200.02)	14	193.83 (181.38 - 206.29)	.31	.051
		Total	24	189.08 (181.01 - 197.15)	53	189.43 (184.00 - 194.86)	.94	.000
Oddball	N 100	♂	20	174.61 (166.96 - 182.26)	43	167.28 (162.06 - 172.50)	.12	.039
		♀	8	178.91 (170.43 - 187.38)	20	175.39 (170.03 - 180.75)	.48	.020
		Total	28	175.84 (169.88 - 181.80)	63	169.85 (165.88 - 173.83)	.10	.030
	P 200	♂	14	259.54 (248.61 - 270.47)	38	260.84 (254.20 - 267.47)	.84	.001

| | | ♀ | 4 | 267.77
(245.91 - 289.64) | 12 | 258.33
(245.71 - 270.96) | .44 | .044 |
| | | Total | 18 | 261.37
(251.89 - 270.86) | 50 | 260.23
(254.54 - 266.93) | .84 | .001 |

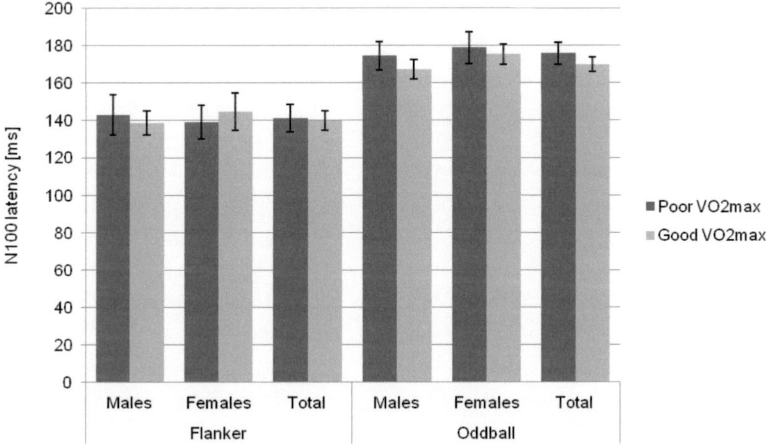

Figure 82:   Mean N100 latencies during Flanker and Oddball task for VO_{2max} groups; error bars represent 95% confidence intervals

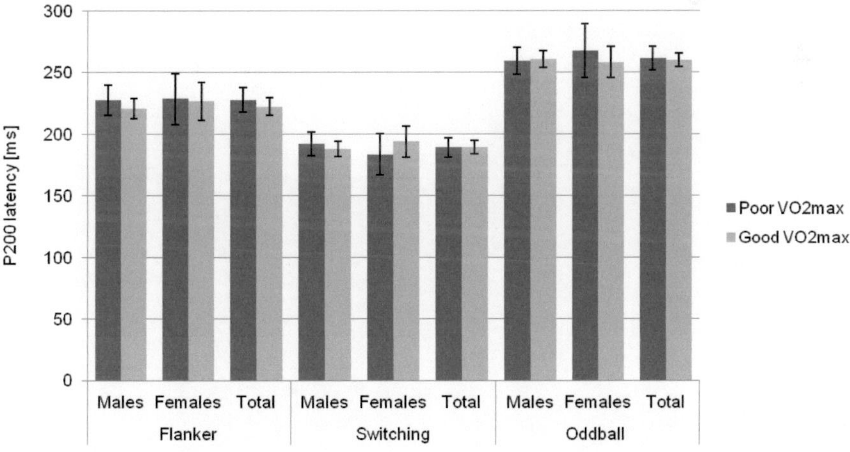

Figure 83:   Mean P200 latencies during Flanker, Switching and Oddball task for VO_{2max} groups; error bars represent 95% confidence intervals

4.4.1.2 Individual Anaerobic Threshold - Results

Behavioral Measures

For response times during the Switching task, a condition x IAT interaction (F [1, 122] = 4.13, p = .04, η^2 = .033) was observed. Participants with a poor performance at IAT had a smaller difference between the homogenous (524.42 ms; 95 % CI 507.58 - 541.27) and heterogeneous condition (854.51 ms; 95 % CI 823.35 - 885.68) compared to those with a good performance (homogenous: 531.29 ms; 95 % CI 514.71 - 547.86; heterogeneous: 896.91 ms; 95 % CI 866.24 - 927.58). No further effects were found on behavioral measures. Mean response times are given in table 97 and figure 84, errors in the Switching task in table 98 and figure 85.

Table 97:   Mean response times for IAT groups in the Flanker, Switching and Oddball task

		\multicolumn{2}{Poor performance at IAT}		\multicolumn{2}{Good performance at IAT}			
		n	ms (95 % CI)	n	ms (95 % CI)	p	η^2
Flanker	♂	45	491.83 (479.32- 504.35)	47	493.20 (480.96- 505.45)	.88	.000
	♀	19	508.20 (486.78- 529.62)	17	515.08 (492.44- 537.72)	.66	.006
	Total	64	496.69 (485.85- 507.53)	64	499.01 (488.17- 509.85)	.77	.001
Switching	♂	43	695.26 (670.79-719.73)	44	708.98 (684.79- 733.17)	.43	.007
	♀	18	675.62 (628.46- 722.79)	19	725.95 (680.04- 771.85)	.13	.064
	Total	61	689.47 (667.67- 711.27)	63	714.10 (692.65-735.54)	.11	.020
Oddball	♂	46	488.07 (461.39- 514.74)	44	514.61 (487.33- 541.89)	.17	.021
	♀	20	531.28 (485.94- 576.61)	19	527.81 (481.30- 574.32)	.91	.000
	Total	66	501.16 (478.20- 524.12)	63	518.59 (495.09- 542.09)	.30	.009

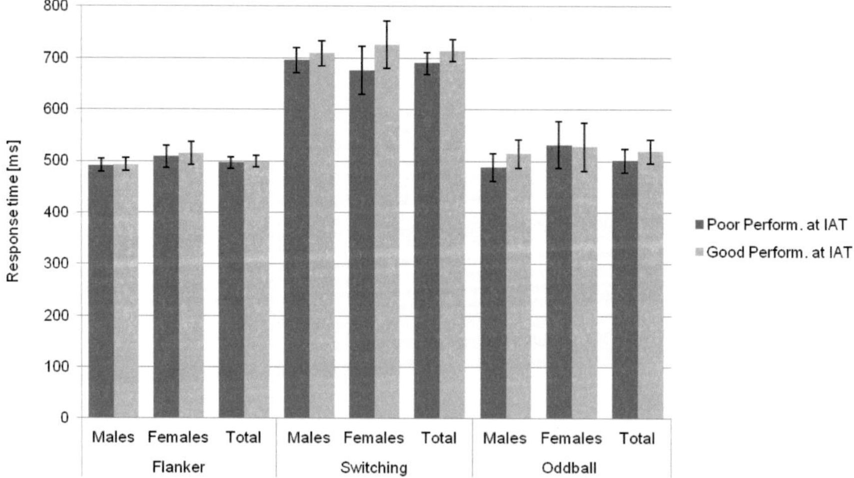

Figure 84:   Mean response times during Flanker, Switching and Oddball task for IAT groups; error bars represent 95% confidence intervals

Table 98:   Errors in the Switching task for IAT groups

		Poor performance at IAT		Good performance at IAT			
		n	No. (95 % CI)	n	No. (95 % CI)	p	η^2
Errors	♂	45	4.06 (3.44- 4.67)	45	3.44 (2.83- 4.05)	.16	.022
	♀	19	4.34 (3.10- 5.59)	19	3.96 (2.72- 5.20)	.66	.005
	Total	64	4.14 (3.58- 4.70)	64	3.59 (3.04- 4.15)	.17	.015

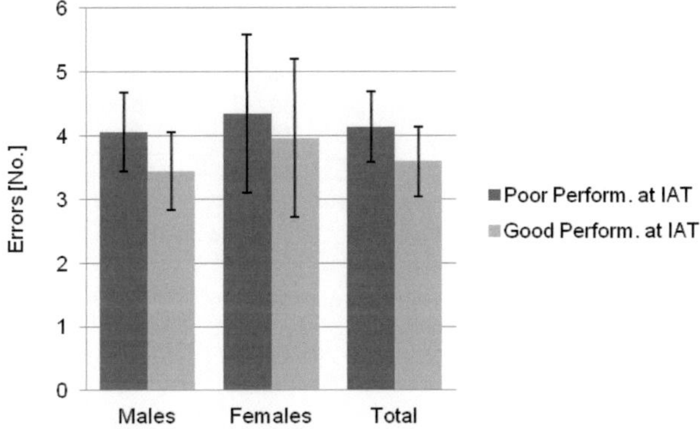

Figure 85:   Errors during the Switching task for IAT groups; error bars represent 95% confidence intervals

P300

Performance at IAT had no effect on P300 component in the Flanker, Switching and Oddball task. The mean P300 amplitudes are presented in table 99 and figure 86, the latencies are given in table 100 and figure 87.

Table 99:   Mean P300 amplitudes for IAT groups in the Flanker, Switching and Oddball task

		Poor performance at IAT		Good performance at IAT			
		n	µV (95 % CI)	n	µV (95 % CI)	p	η^2
Flanker	♂	44	4.57 (4.16- 4.98)	45	4.87 (4.46- 5.27)	.31	.012
	♀	16	4.35 (3.59- 5.12)	16	4.45 (3.68- 5.21)	.86	.001
	Total	60	4.51 (4.16- 4.87)	61	4.76 (4.40- 5.11)	.34	.008
Switching	♂	30	3.74 (3.40- 4.08)	31	3.85 (3.52- 4.19)	.64	.004
	♀	11	4.12 (3.33- 4.90)	11	4.45 (3,66- 5,24)	.54	.019
	Total	41	3.84 (3.52- 4.16)	42	4.01 (3.69- 4.33)	.47	.007
Oddball	♂	38	5.39 (4.61- 6.17)	34	6.21 (5.39- 7.04)	.16	.029
	♀	13	6.19 (4.90- 7.47)	16	6.60 (5.44- 7.76)	.63	.009
	Total	51	5.60 (4.94- 6.25)	50	6.33 (5.67-7.00)	.12	.024

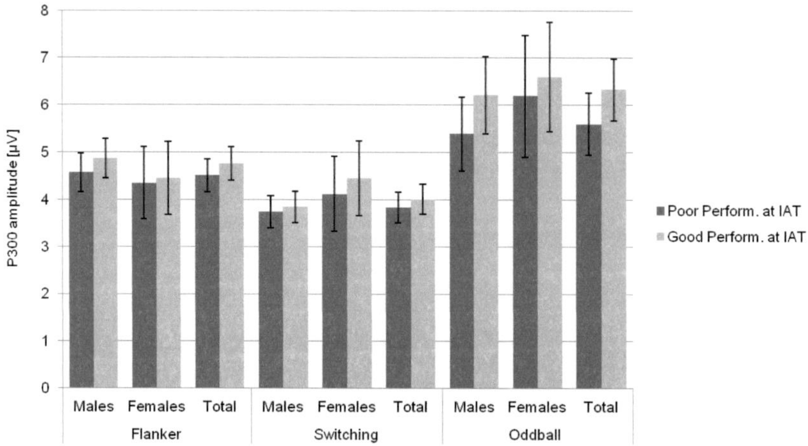

Figure 86:   Mean P300 amplitudes during Flanker, Switching and Oddball task for IAT groups; error bars represent 95% confidence intervals

Table 100:   Mean P300 latencies for IAT groups in the Flanker, Switching and Oddball task

			Poor performance at IAT		Good performance at IAT		
		n	ms (95 % CI)	n	ms (95 % CI)	p	η^2
Flanker	♂	36	392.87 (382.64- 403.11)	37	390.81 (380.72- 400.91)	.78	.001
	♀	15	381.16 (351.60- 410.73)	12	403.68 (370.62- 536.74)	.31	.042
	Total	51	389.43 (378.65- 400.21)	49	393.96 (382.96- 404.96)	.56	.003
Switching	♂	19	380.15 (357.46- 402.84)	16	389.60 (364.88- 414.33)	.57	.010
	♀	6	387.95 (344.42- 431.47)	9	389.41 (353.87- 424.95)	.96	.000
	Total	25	382.02 (362.77- 401.26)	25	389.53 (370.29- 408.78)	.58	.006
Oddball	♂	36	426.99 (400.67- 453.31)	32	443.64 (415.72- 471.56)	.39	.011
	♀	10	400.39 (365.81- 434.97)	16	384.31 (356.97- 411.65)	.46	.023
	Total	46	421.21 (399.12- 443.30)	48	423.86 (402.24- 445.48)	.87	.000

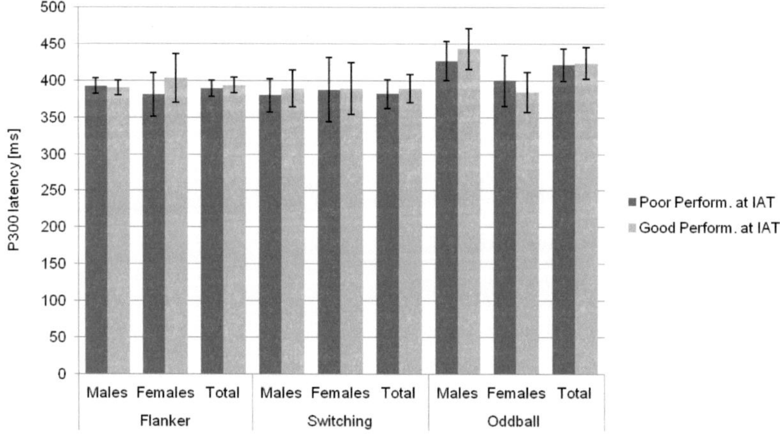

Figure 87:   Mean P300 latencies during Flanker, Switching and Oddball task for IAT groups; error bars represent 95% confidence intervals

N100 and P200

For N100 latency during the Flanker task, there was a condition x IAT interaction (F [1, 44] = 4.14, p = .05, η^2 = .086). Participants with a poor performance at IAT had a faster latency in the speed (137.81 ms; 95 % CI 130.83 - 144.79) compared to the accuracy condition (142.69; 95 % CI 135.61 - 149.77) whereas participants with a good performance at IAT had a faster latency in the accuracy condition (accuracy: 136.47; 95 % CI 129.08 - 143.87; speed: 144.11; 95 % CI 136.81 - 151.40). In addition, participants with a poor performance had faster latencies during the speed condition but longer latencies during the accuracy condition when compared to participants with a good performance.

There was a main effect for females on P200 amplitude during the Flanker task (F [1, 121] = 1.77, p = .19, η^2 = .014; males: F [1, 88] = .07, p = .79, η^2 = .001; females: F [1, 31] = 4.52, p = .04, η^2 = .127). For P200 latency during the Flanker task, a main effect for males was observed (F [1, 91] = .75, p = .39, η^2 = .008; males: F [1, 67] = 3.96, p = .05, η^2 = .056; females: F [1, 22] = 3.03, p = .10, η^2 = .121). No further effects were found for N100 and P200 amplitudes (table 101, figure 88 & 89) and latencies (table 102, figure 90 & 91)

Table 101:   Mean N100 and P200 amplitudes for IAT groups in the Flanker, Switching and Oddball task

			\multicolumn{2}{l}{Poor performance at IAT}		\multicolumn{2}{l}{Good performance at IAT}		p	η^2
			n	µV (95 % CI)	n	µV (95 % CI)	p	η^2
Flanker	N100	♂	44	-1.76 (-2.03- -1.49)	46	-1.62 (-1.89- -1.36)	.48	.006
		♀	17	-1.83 (-2.28- -1.37)	16	-1.98 (-2.45- -1.51)	.65	.007
		Total	61	-1.78 (-2.01- -1.55)	62	-1.71 (-1.94- -1.49)	.70	.001
	P200	♂	44	2.94 (2.49- 3.39)	46	3.02 (2.59- 3.46)	.79	.001
		♀	17	3.21 (2.37- 4.06)	16	4.48 (3.61- 5.36)	.04	.127
		Total	61	3.01 (2.61- 3.42)	62	3.40 (3.00- 3.80)	.19	.014
Switching	P200	♂	31	3.69 (3.17- 4.21)	33	4.24 (3.74- 4.74)	.13	.036
		♀	11	4.62 (3.49- 5.75)	11	4.43 (3.30- 5.56)	.81	.003
		Total	42	3.93 (3.46- 4.41)	44	4.29 (3.82- 4.75)	.29	.013
Oddball	N100	♂	38	-4.92 (-5.62- -4.22)	34	-4.31 (-5.05- -3.57)	.24	.020
		♀	14	-5.42 (-6.52- -4.32)	16	-5.54 (-6.57- -4.51)	.88	.001
		Total	52	-5.05 (-5.65- -4.46)	50	-4.70 (-5.31- -4.10)	.41	.007
	P200	♂	38	3.24 (2.62- 3.87)	34	2.76 (2.09- 3.42)	.29	.016
		♀	14	2.30 (.99- 3.62)	16	3.10 (1.87- 4.32)	.37	.028
		Total	52	2.99 (2.42- 3.56)	50	2.86 (2.28- 3.45)	.76	.001

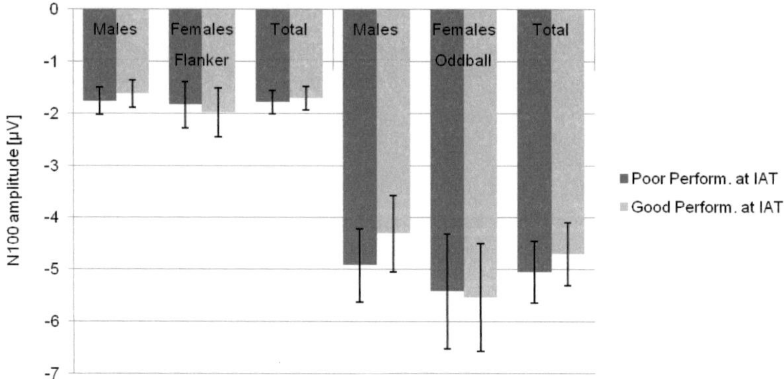

Figure 88:   Mean N100 amplitudes during Flanker and Oddball task for IAT groups; error bars represent 95% confidence intervals

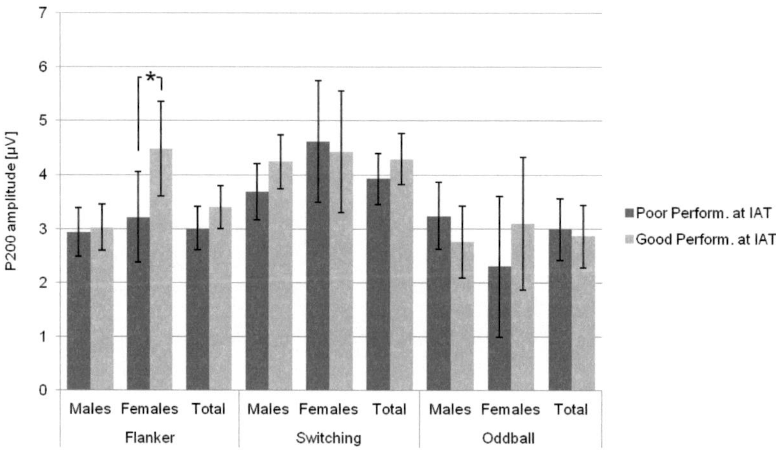

Figure 89:   Mean P200 amplitudes during Flanker, Switching and Oddball task for IAT groups; error bars represent 95% confidence intervals

Table 102:   Mean N100 and P200 latencies for IAT groups in the Flanker, Switching and Oddball task

			Poor performance at IAT		Good performance at IAT			
			n	ms (95 % CI)	n	ms (95 % CI)	p	η^2
Flanker	N100	♂	16	141.61 (134.16- 149.07)	16	138.44 (130.98- 145.90)	.54	.012
		♀	8	137.52 (129.25- 145.80)	6	145.22 (135.66- 154.78)	.21	.128
		Total	24	140.25 (134.68- 145.82)	22	140.29 (134.47- 146.11)	.99	.000
	P200	♂	32	229.47 (219.99- 238.96)	37	216.57 (207.75- 225.39)	.05	.056
		♀	12	217.56	12	235.95	.10	.121

				(202.06- 233.06)			(220.46- 251.45)		
	Total		44	226.23	49	221.32	.39	.008
				(218.05- 234.40)		(213.57- 229.06)		
Switching	P200	♂	26	191.98	31	186.18	.25	.024
				(184.68- 199.28)		(179.50- 192.87)		
		♀	11	184.32	11	195.86	.24	.070
				(170.41- 198.23)		(181.95- 209.78)		
		Total	37	189.70	42	188.72	.83	.001
				(183.24- 196.16)		(182.66-194.78)		
Oddball	N100	♂	35	170.54	30	168.49	.64	.004
				(164.71- 176.36)		(162.20- 174.78)		
		♀	13	175.00	15	177.60	.56	.013
				(168.33- 181.67)		(171.39- 183.82)		
		Total	48	171.75	45	171.53	.95	.000
				(167.16- 176.33)		(166.79- 176.27)		
	P200	♂	30	259.77	25	258.59	.83	.001
				(252.25- 267.28)		(250.36- 266.83)		
		♀	7	258.15	9	262.67	.67	.013
				(241.36- 274.94)		(247.87- 277.48)		
		Total	37	259.46	34	259.67	.97	.000
				(252.80- 266.12)		(252.73- 266.62)		

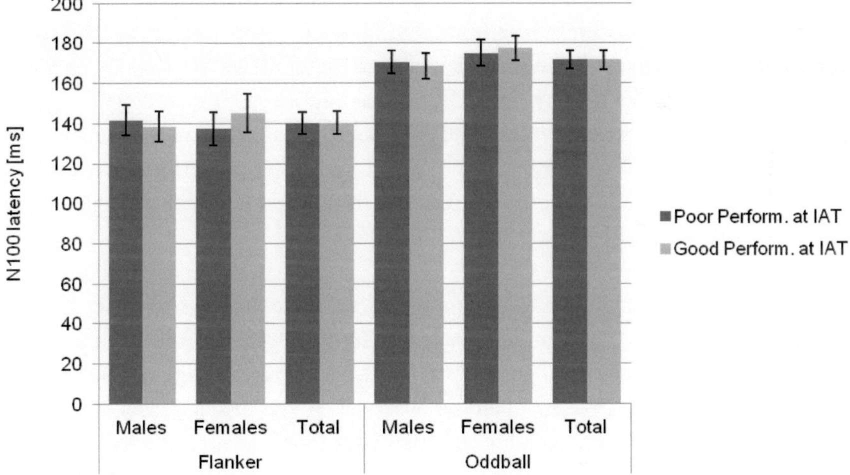

Figure 90: Mean N100 latencies during Flanker and Oddball task for IAT groups; error bars represent 95% confidence intervals

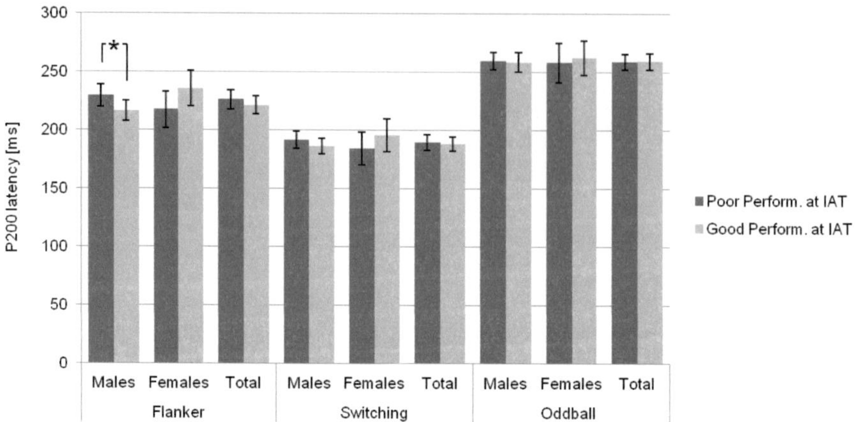

Figure 91:   Mean P200 latencies during Flanker, Switching and Oddball task for IAT groups; error bars represent 95% confidence intervals

4.4.1.3 Discussion

Interestingly, no constant effect of aerobic capacity on ERPs was found. Regarding behavioral measures, there was a significant interaction for the Switching task with faster response times during homogenous trials but longer response times during heterogeneous trials for the high compared to the poor VO_{2max} group. Since heterogeneous trials are more difficult to respond to than homogenous ones, this result indicates that a good VO_{2max} level might even impair cognitive performance. In contrast, participants with a good VO_{2max} had significantly less errors in the Switching task compared to those with a poor VO_{2max}. Participants with a good VO_{2max} further revealed faster N100 latencies for the accuracy condition in the Flanker task, but longer latencies during the speed condition.

For the performance at the individual anaerobic threshold, there was also a tendency for faster response times for the poor VO_{2max} group. With regard to event-related potentials, the high VO_{2max} group had larger P300 amplitudes but longer P300 latencies across the cognitive tasks. However, no significant effects for P300 component could be observed. A significant gender-specific effect was found for Flanker P200 component with higher amplitudes for females with good VO_{2max} compared to females with poor VO_{2max}. Males with good VO_{2max} revealed faster P200 latencies in the Flanker task than those with a poor VO_{2max}. Consistent with the finding in VO_{2max}, participants with good performance at IAT had faster Flanker N100 latencies in the accuracy but longer latencies in the speed condition.

Given that cognitive benefits have been reported for aerobic exercise in recent publications (Åberg et al., 2009; Stroth et al., 2010; Stroth et al., 2009), the so-called cardiovascular fitness hypothesis has received much attention. Research underlying this hypothesis is trying to establish a link between cardiopulmonary (aerobic) fitness and cognitive function with aerobic fitness as the primary independent variable typically being indicated by VO_{2max}. Many studies among young adults support this hypothesis (e.g. Kamijo et al., 2010; Themanson et al., 2008; Themanson & Hillman, 2006), others reporting no associations between cardiovascular fitness and cognition (e.g. Magnié et al., 2000; Scisco et al., 2008), and others finding evidence for age-related correlations (Dustman et al., 1990). Several potential mechanisms are discussed that might mediate effects of cardiovascular fitness on cognition such as increased cerebral blood flow due to better oxygen supply and enhanced capillary density as well as generation of new blood vessels in brain regions that respond to cardiovascular exercise and thus have an increased metabolic demand (Voelcker-Rehage & Windisch, 2013).

Indeed, there is strong evidence for older adults that aerobic exercise and cardiovascular fitness are related to an enhanced cognitive performance and increased executive functioning (Colcombe & Kramer, 2003). For young adults, the relationship seems to be more uncertain and, as described above, is not supported by the current study, too. In a meta-analysis from Nowell, Landers and Sibley (2006) the cardiovascular fitness hypothesis could also not be confirmed. Thus, further research and especially intervention studies are needed to clarify the relationship between cardiovascular fitness and cognitive performance in young adults.

4.4.2 Strength

In this chapter, the relationship between strength and cognitive performance is described. The independent variables are maximal isometric strength (N) and maximal dynamic strength (m). The groups were built individually for the study sample due to a lack of normative values. In the poor performance group for maximal isometric strength, males have ≤ 4000 N and females ≤ 3000 N. In the good performance group, males have ≥ 4000.1 N and females ≥ 3000.1 N (table 103). For maximal dynamic strength, the poor performance group is characterized by males having $\leq .35$ m and females $\leq .21$ m. In the good performance group, males have a height of $\geq .36$ m and females $\geq .22$ m (table 104).

Table 103:   Number of participants per maximal isometric strength group (M ± SD in N)

	Poor		Good		Total	
	n (%)	M (SD)	n (%)	M (SD)	n	p
♂	47 (48)	3468.23 (409.81)	52 (52)	5014.29 (826.46)	99	.00
♀	21 (54)	2341.90 (301.06)	18 (46)	3608.22 (446.88)	39	.00
Total	68 (49)	3120.40 (645.90)	70 (51)	4652.73 (968.08)	138	.00

Table 104:   Number of participants per maximal dynamic strength group (M ± SD in m)

	Poor		Good		Total	
	n (%)	M (SD)	n (%)	M (SD)	n	p
♂	46 (45)	.32 (.03)	57 (55)	.41 (.05)	103	.00
♀	20 (47)	.19 (.02)	23 (43)	.26 (.05)	43	.00
Total	66 (45)	.28 (.06)	80 (55)	.37 (.08)	146	.00

RQ 9: Is maximal strength related to cognitive performance in young adulthood?

4.4.2.1 Maximal Isometric Strength - Results

Behavioral Measures

Maximal isometric strength has no effect on behavioral measures in the Flanker, Switching or Oddball task except for response times during the Flanker task. There was a main effect for females (F [1, 117] = .34, p = .56, η^2 = .003; males: F [1, 86] = 1.06, p = .31, η^2 = .012; females: F [1, 29] = 5.15, p = .03, η^2 = .151) with faster response times for females with good maximal isometric strength (table 105 & figure 92). Mean errors are given in table 106 and figure 93.

Table 105:   Mean response times for maximal isometric strength groups in the Flanker, Switching and Oddball task

		\multicolumn{2}{c}{Poor isometric strength}		\multicolumn{2}{c}{Good isometric strength}			
		n	ms (95 % CI)	n	ms (95 % CI)	p	η^2
Flanker	♂	40	487.50 (474.66 - 500.33)	48	496.48 (484.77 - 508.20)	.31	.012
	♀	18	528.84 (507.87 - 549.82)	13	492.92 (468.24 - 517.60)	.03	.151
	Total	58	500.33 (489.06 - 511.60)	61	495.73 (484.74 - 506.71)	.56	.003
Switching	♂	36	701.41 (675.09 - 727.73)	47	704.92 (681.88 - 727.95)	.84	.000
	♀	16	703.87 (650.35 - 757.39)	16	714.40 (660.88 - 767.92)	.78	.003
	Total	52	702.17 (678.47 - 725.87)	63	707.33 (685.79 - 728.86)	.75	.001
Oddball	♂	39	506.98 (477.84 - 536.13)	47	500.43 (473.88 - 526.98)	.74	.001
	♀	19	549.66 (502.30 - 597.02)	15	521.79 (468.49 - 575.08)	.43	.019
	Total	58	520.96 (496.31 - 545.62)	62	505.60 (481.75 - 529.45)	.38	.007

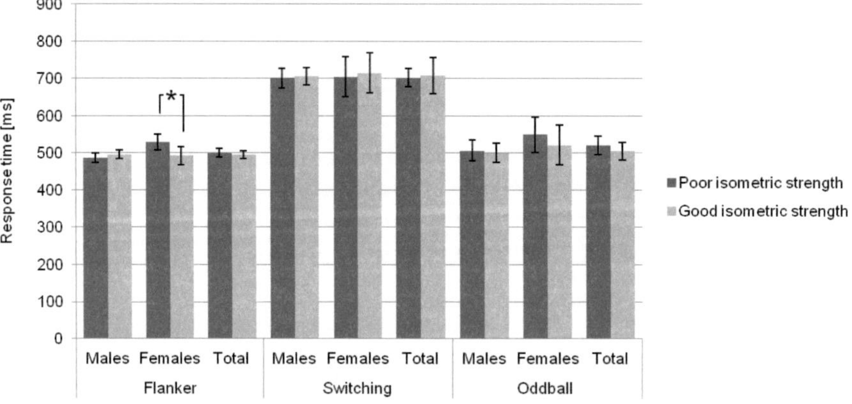

Figure 92:   Mean response times during Flanker, Switching and Oddball task for maximal isometric strength groups; error bars represent 95% confidence intervals

Table 106:   Errors in the Switching task for maximal isometric strength groups

		Poor isometric strength		Good isometric strength			
		n	No. (95 % CI)	n	No. (95 % CI)	p	η^2
Errors	♂	38	3.82 (3.11 - 4.52)	48	3.80 (3.17 - 4.43)	.97	.000
	♀	17	4.35 (2.97 - 5.74)	16	4.11 (2.69 - 5.53)	.80	.002
	Total	55	3.98 (3.35 - 4.61)	64	3.88 (3.29 - 4.46)	.81	.001

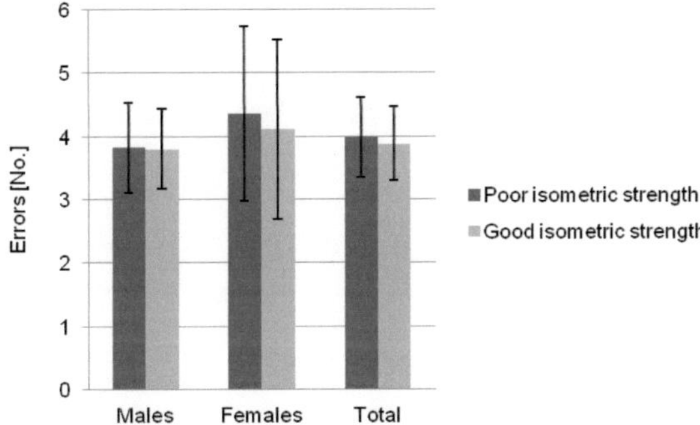

Figure 93:   Errors during the Switching task for maximal isometric strength groups; error bars represent 95% confidence intervals

P300

No effect was observed for P300 amplitudes (table 107 & figure 94).

Table 107:   Mean P300 amplitudes for maximal isometric strength groups in the Flanker, Switching and Oddball task

		Poor isometric strength		Good isometric strength			
		n	µV (95 % CI)	n	µV (95 % CI)	p	η^2
Flanker	♂	38	4.70 (4.27 - 5.12)	46	4.66 (4.27 - 5.05)	.91	.000
	♀	16	4.51 (3.70 - 5.32)	12	4.32 (3.39 - 5.26)	.75	.004
	Total	54	4.64 (4.27 - 5.01)	58	4.59 (4.23 - 4.95)	.85	.000
Switching	♂	22	3.86 (3.47 - 4.25)	34	3.71 (3.39 - 4.03)	.55	.007
	♀	8	4.43 (3.46 - 5.41)	11	4.39 (3.55 - 5.22)	.94	.000
	Total	30	4.01 (3.63 - 4.40)	45	3.87 (3.56 - 4.19)	.58	.004
Oddball	♂	30	5.94 (5.02 - 6.86)	39	5.64 (4.84 - 6.45)	.63	.004
	♀	11	5.59 (4.14 - 7.04)	14	6.83 (5.54 - 8.11)	.20	.071
	Total	41	5.85 (5.08 - 6.62)	53	5.96 (5.28 - 6.63)	.84	.000

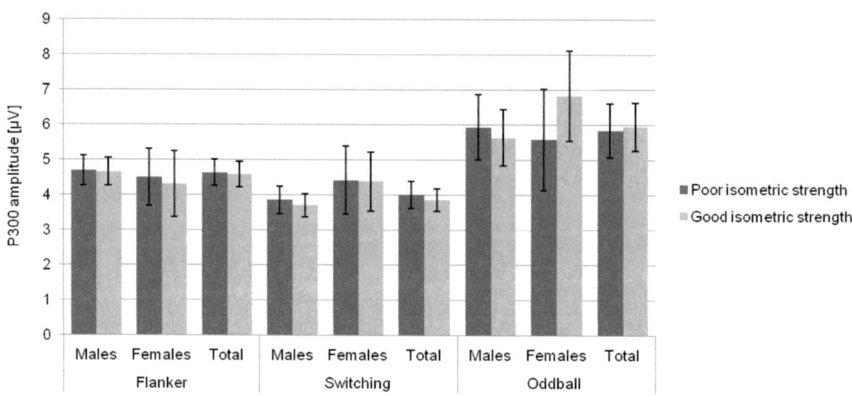

Figure 94:   Mean P300 amplitudes during Flanker, Switching and Oddball task for maximal isometric strength groups; error bars represent 95% confidence intervals

For P300 latencies during the Switching task, there was a main effect for males (F [1, 42] = 1.38, p = .25, η^2 = .032; males: F [1, 29] = 5.34, p = .03, η^2 = .156; females: F [1, 11] = 1.09, p = .32, η^2 = .090). For P300 latencies during the Oddball task, a main effect for females was observed (F [1, 85] = 1.35, p = .25, η^2 = .016; males: F [1, 62] = .10, p = .75, η^2 = .002; females: F [1, 21] = 5.27, p = .03, η^2 = .201). In both cases, participants with a good isometric strength had faster latencies. Mean P300 latencies are presented in table 108 and figure 95.

Table 108:   Mean P300 latencies for maximal isometric strength groups in the Flanker, Switching and Oddball task

		Poor isometric strength		Good isometric strength			
		n	ms (95 % CI)	n	ms (95 % CI)	p	η^2
Flanker	♂	31	392.36 (381.29 - 403.44)	38	388.82 (378.82 - 398.83)	.64	.003
	♀	13	403.12 (371.30 - 434.93)	10	396.48 (360.21 - 432.76)	.78	.004
	Total	44	395.54 (384.18 - 406.90)	48	390.42 (379.55 - 401.29)	.52	.005
Switching	♂	13	405.39 (381.82 - 428.97)	18	370.44 (350.41 - 390.47)	.03	.156
	♀	6	372.97 (329.18 - 416.76)	7	401.24 (360.69 - 441.78)	.32	.090
	Total	19	395.16 (374.29 - 416.02)	25	379.06 (360.87 - 397.25)	.25	.032
Oddball	♂	28	437.37 (406.83 - 467.91)	36	430.86 (403.92 - 457.80)	.75	.002
	♀	9	422.25 (386.93 - 457.57)	14	372.27 (343.95 - 400.58)	.03	.201
	Total	37	433.69 (408.72 - 458.66)	50	414.45 (392.98 - 435.93)	.25	.016

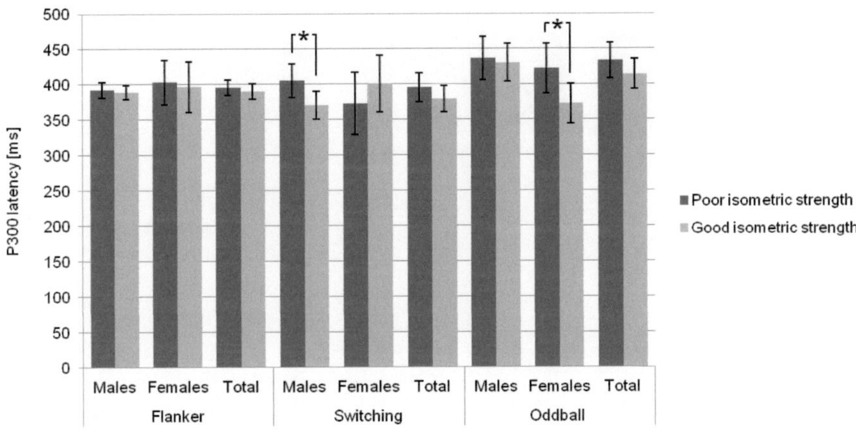

Figure 95: Mean P300 latencies during Flanker, Switching and Oddball task for maximal isometric strength groups; error bars represent 95% confidence intervals

N100 and P200

For P200 amplitudes during the Flanker task, there was a main effect for females (F [1, 113] = .41, p = .52, η^2 = .004; males: F [1, 84] = .12, p = .73, η^2 = .001; females: F [1, 27] = 4.36, p = .05, η^2 = .139) with larger amplitudes for females with good isometric strength. For P200 amplitude in the Oddball task, there was also a main effect for females (F [1, 93] = 2.12, p = .15, η^2 = .022; males: F [1, 67] = .26, p = .61, η^2 = .004; females: F [1, 24] = 13.16, p = <.01, η^2 = .354). P200 and N100 amplitudes are given in table 109 and figure 96 and 97.

Table 109: Mean N100 and P200 amplitudes for maximal isometric strength groups in the Flanker, Switching and Oddball task

			\multicolumn{2}{c}{Poor isometric strength}		\multicolumn{2}{c}{Good isometric strength}			
			n	μV (95 % CI)	n	μV (95 % CI)	p	η^2
Flanker	N100	♂	39	-1.70 (-1.98 - -1.42)	47	-1.63 (-1.88 - -1.37)	.70	.002
		♀	17	-1.78 (-2.27 - -1.29)	12	-2.02 (-2.60 - -1.44)	.52	.015
		Total	56	-1.72 (-1.96 - -1.49)	59	-1.71 (-1.94 - -1.47)	.92	.000
	P200	♂	39	3.12 (2.65 - 3.58)	47	3.01 (2.58 - 3.43)	.73	.001
		♀	17	3.18 (2.30 - 4.06)	12	4.57 (3.52 - 5.62)	.05	.139
		Total	56	3.13 (2.71 - 3.56)	59	3.32 (2.91 - 3.73)	.52	.004
Switching	P200	♂	23	4.25 (3.64 - 4.87)	36	3.78 (3.29 - 4.27)	.23	.025
		♀	8	4.26 (2.92 - 5.61)	11	5.06 (3.92 - 6.21)	.35	.051
		Total	31	4.26 (3.69 - 4.82)	47	4.08 (3.62 - 4.54)	.63	.003
Oddball	N100	♂	30	-4.77 (-5.58 - -3.97)	39	-4.53 (-5.23 - -3.82)	.65	.003
		♀	12	-5.25 (-6.49 - -4.00)	14	-5.42 (-6.58 - -4.27)	.83	.002
		Total	42	-4.91 (-5.58 - -4.24)	53	-4.76 (-5.36 - -4.17)	.75	.001
	P200	♂	30	3.14 (2.44 - 3.85)	39	2.90 (2.29 - 3.52)	.61	.004
		♀	12	1.34 (.16 - 2.53)	14	4.19 (3.09 - 5.29)	<.01	.354
		Total	42	2.63 (2.00 - 3.26)	53	3.24 (2.69 - 3.80)	.15	.022

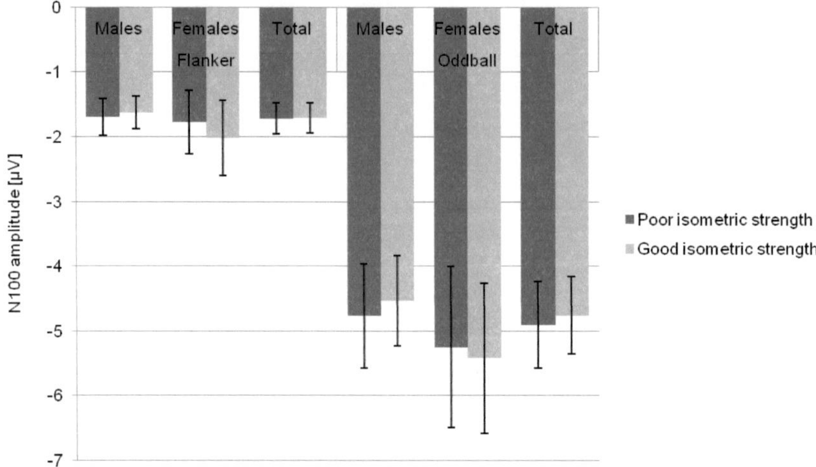

Figure 96:   Mean N100 amplitudes during Flanker and Oddball task for maximal isometric strength groups; error bars represent 95% confidence intervals

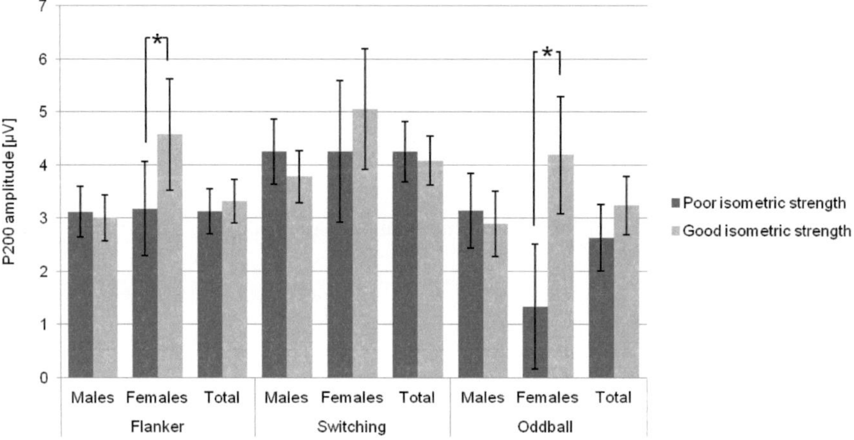

Figure 97:   Mean P200 amplitudes during Flanker, Switching and Oddball task for maximal isometric strength groups; error bars represent 95% confidence intervals

For N100 latency during the Flanker task, there was a congruency x maximal isometric strength interaction (F [1, 40] = 4.07, p = .05, η^2 = .092). Participants with a poor maximal isometric strength had a faster latency in incongruent trials (138.92 ms; 95 % CI 132.39 - 145.45) compared to congruent trials (143.44; 95 % CI 135.81 - 151.08) whereas participants with a good maximal isometric strength had a faster latency in congruent trials (congruent: 136.36; 95 % CI 129.42 - 143.30; incongruent: 141.37; 95 % CI 135.44 - 147.31).

There was further a main effect for P200 latencies during Flanker task (F [1, 86] = 5.09, p = .03, η^2 = .056; males: F [1, 66] = 4.66, p = .04, η^2 = .066; females: F [1, 18] = .54, p = .47, η^2 = .029) with faster latencies for the poor isometric strength group. P200 and N100 latencies are given in table 110 and figure 98 and 99.

Table 110:   Mean N100 and P200 latencies for maximal isometric strength groups in the Flanker, Switching and Oddball task

			\multicolumn{2}{l}{Poor isometric strength}		\multicolumn{2}{l}{Good isometric strength}		p	η^2
			n	ms (95 % CI)	n	ms (95 % CI)		p	η^2
Flanker	N100	♂	13	141.80 (133.23 - 150.37)	17	138.75 (131.26 - 146.25)		.59	.011
		♀	6	139.84 (132.33 - 147.36)	6	139.19 (131.68 - 146.71)		.89	.002
		Total	19	141.18 (135.02 - 147.35)	23	138.87 (133.26 - 144.47)		.58	.008
	P200	♂	31	215.58 (206.16 - 224.99)	37	229.37 (220.75 - 237.99)		.04	.066
		♀	10	219.34 (200.64 - 238.03)	10	228.61 (209.92 - 247.31)		.47	.029
		Total	41	216.49 (208.30 - 224.68)	47	229.21 (221.56 - 236.86)		.03	.056
Switching	P200	♂	22	187.74 (179.60 - 195.89)	30	190.33 (183.36 - 197.31)		.63	.005
		♀	8	197.71 (181.40 - 214.01)	11	180.86 (166.96 - 194.76)		.12	.139
		Total	30	190.40 (183.12 - 197.67)	41	187.79 (181.57 - 194.01)		.59	.004
Oddball	N100	♂	27	170.66 (163.90 - 177.42)	35	169.09 (163.15 - 175.02)		.73	.002
		♀	11	175.36 (167.68 - 183.03)	13	175.90 (168.84 - 182.96)		.91	.001
		Total	38	172.02 (166.75 - 177.29)	48	170.93 (166.24 - 175.62)		.76	.001
	P200	♂	23	254.82 (246.36 - 263.29)	29	260.26 (252.77 - 267.81)		.34	.018
		♀	3	270.05 (243.46 - 296.64)	11	259.16 (245.28 - 273.05)		.44	.050
		Total	26	256.58 (248.62 - 264.54)	40	259.96 (253.54 - 266.38)		.51	.007

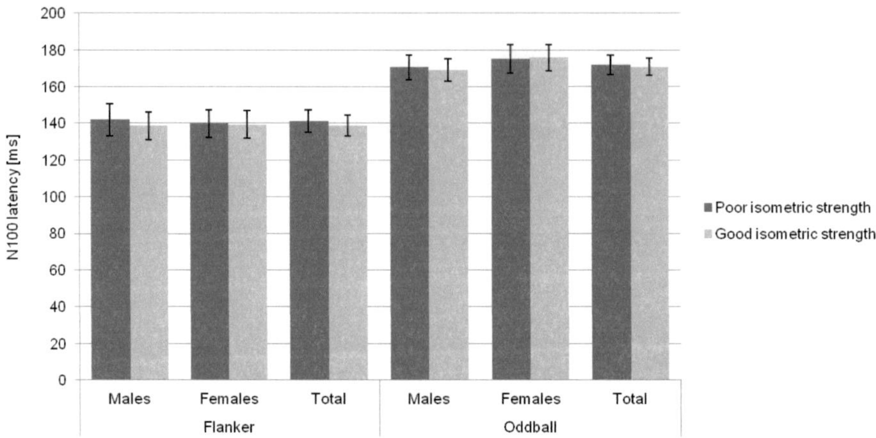

Figure 98:   Mean N100 latencies during Flanker and Oddball task for maximal isometric strength groups; error bars represent 95% confidence intervals

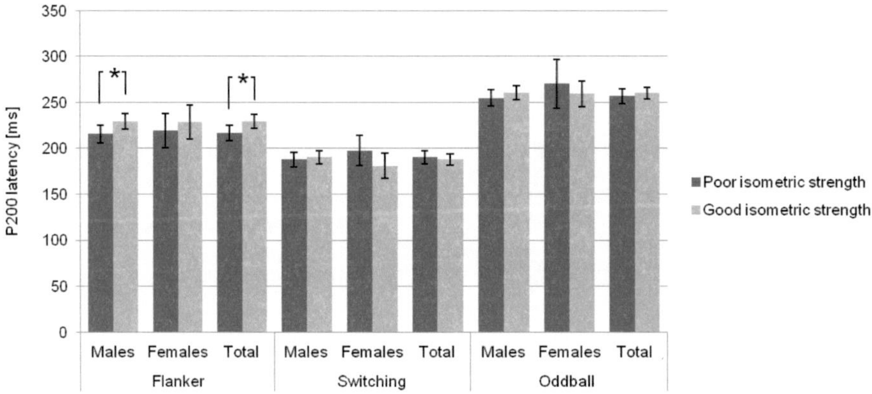

Figure 99:   Mean P200 latencies during Flanker, Switching and Oddball task for maximal isometric strength groups; error bars represent 95% confidence intervals

4.4.2.2 Maximal Dynamic Strength - Results

Behavioral Measures

There was a main effect on response times observed for males during the Flanker task (F [1, 125] = 1.49, p = .23, η^2 = .012; males: F [1, 90] = 3.90, p = .05, η^2 = .042; females: F [1, 33] = .79, p = .38, η^2 = .023). No other effects on behavioral measures were found (table 111 & 112, figure 100 & 101).

Table 111: Mean response times for maximal dynamic strength groups in the Flanker, Switching and Oddball task

		Poor dynamic strength		Good dynamic strength			
		n	ms (95 % CI)	n	ms (95 % CI)	p	η^2
Flanker	♂	43	482.63 (470.53 - 494.72)	49	499.09 (487.77 - 510.42)	.05	.042
	♀	14	519.83 (494.70 - 544.96)	21	505.70 (485.18 - 526.22)	.38	.023
	Total	57	491.76 (480.54 - 502.99)	70	501.07 (490.95 - 511.20)	.23	.012
Switching	♂	41	706.08 (681.66 - 730.51)	46	696.21 (673.15 - 719.26)	.56	.004
	♀	15	727.36 (674.46 - 780.25)	21	682.33 (637.62 - 727.03)	.20	.049
	Total	56	711.78 (689.23 - 734.33)	67	691.86 (671.24 - 712.47)	.20	.014
Oddball	♂	41	496.99 (468.79 - 525.18)	49	503.85 (478.05 - 529.64)	.72	.001
	♀	15	517.98 (465.11 - 570.85)	23	536.10 (493.41 - 578.80)	.59	.008
	Total	56	502.61 (477.74 - 527.48)	72	514.15 (492.22 - 536.08)	.49	.004

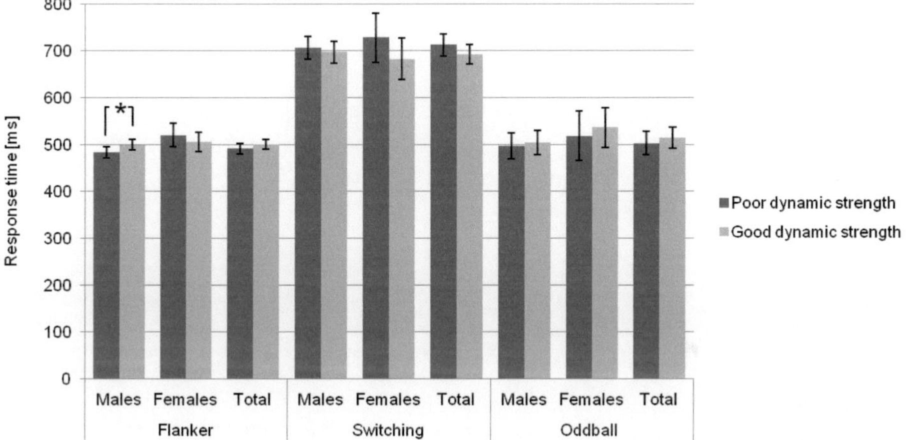

Figure 100: Mean response times during Flanker, Switching and Oddball task for maximal dynamic strength
groups; error bars represent 95% confidence intervals

Table 112: Errors in the Switching task for maximal dynamic strength groups

		Poor dynamic strength		Good dynamic strength			
		n	No. (95 % CI)	n	No. (95 % CI)	p	η^2
Errors	♂	41	3.66 (2.99 - 4.33)	49	4.02 (3.40 - 4.63)	.44	.007
	♀	16	4.27 (2.90 - 5.63)	21	4.17 (2.97 - 5.36)	.91	.000
	Total	57	3.83 (3.22 - 4.44)	70	4.06 (3.51 - 4.61)	.58	.003

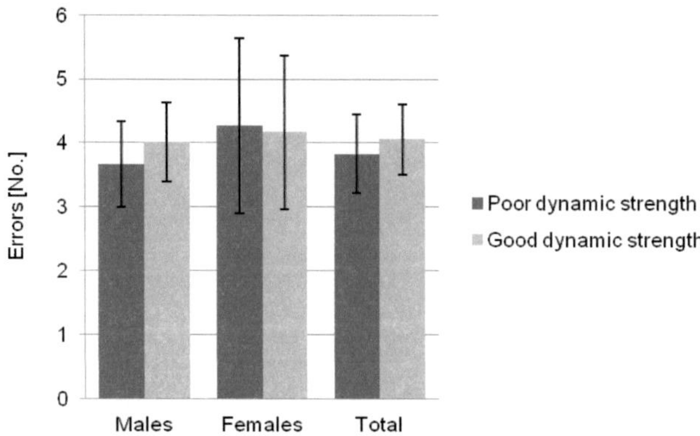

Figure 101: Errors during the Switching task for maximal dynamic strength groups; error bars represent 95% confidence intervals

P300

A single main effect was found for P300 amplitudes during Switching task for females (F [1, 79] = .40, p = .53, η^2 = .005; males: F [1, 57] = .35, p = .55, η^2 = .006; females: F [1, 20] = 4.82, p = .04, η^2 = .194). P300 amplitudes are given in table 113 and figure 102.

Table 113:   Mean P300 amplitudes for maximal dynamic strength groups in the Flanker, Switching and Oddball task

		Poor dynamic strength		Good dynamic strength			
		n	µV (95 % CI)	n	µV (95 % CI)	p	η^2
Flanker	♂	40	4.82 (4.39 - 5.26)	48	4.61 (4.21 - 5.00)	.46	.006
	♀	11	4.83 (3.93 - 5.72)	21	4.18 (3.53 - 4.82)	.24	.046
	Total	51	4.83 (4.44 - 5.21)	69	4.48 (4.14 - 4.81)	.18	.015
Switching	♂	32	3.73 (3.39 - 4.06)	27	3.88 (3.51 - 4.24)	.55	.006
	♀	10	4.86 (4.12 - 5.61)	12	3.80 (3.12 - 4.48)	.04	.194
	Total	42	4.00 (3.68 - 4.32)	39	3.85 (3.52 - 4.19)	.53	.005
Oddball	♂	34	5.28 (4.44 - 6.11)	38	6.17 (5.38 - 6.96)	.13	.033
	♀	9	6.68 (5.14 - 8.23)	20	6.29 (5.25 - 7.33)	.67	.007
	Total	43	5.57 (4.85 - 6.30)	58	6.21 (5.59 - 6.84)	.19	.018

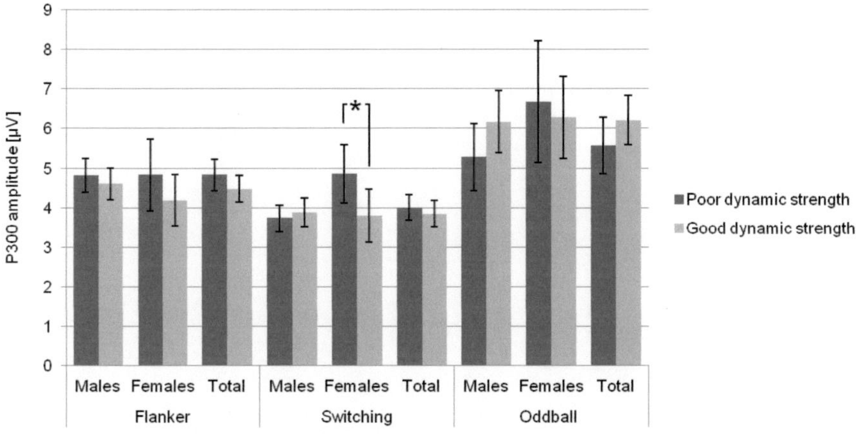

Figure 102: Mean P300 amplitudes during Flanker, Switching and Oddball task for maximal dynamic strength groups; error bars represent 95% confidence intervals

For P300 latencies during the Switching task, there was a condition x maximal dynamic strength interaction (F [1, 47] = 8.39, p = <.01, η^2 = .151). This interaction indicates that participants with poor dynamic strength had faster latencies during homogenous (390.96 ms; 95 % CI 370.53 - 411.40) compared to heterogeneous trials (399.25 ms; 95 % CI 376.48 - 422.02) whereas participants with a good dynamic strength revealed faster latencies during heterogeneous trials (homogenous: 394.78 ms; 95 % CI 373.92 - 415.64; heterogeneous: 360.76 ms; 95 % CI 337.53 - 384.00). When comparing both groups, participants with poor dynamic strength reveal faster latencies during homogenous but longer latencies during heterogeneous trials than participants with good performance.

A main effect was found for P300 latencies during the Oddball task (F [1, 91] = 7.56, p = <.01, η^2 = .077; males: F [1, 65] = 6.55, p = .01, η^2 = .092; females: F [1, 24] = .00, p = .99, η^2 = .000) with faster latencies for the good dynamic strength group (table 114 & figure 103).

Table 114: Mean P300 latencies for maximal dynamic strength groups in the Flanker, Switching and Oddball task

| | | Poor dynamic strength | | Good dynamic strength | | | |
		n	ms (95 % CI)	n	ms (95 % CI)	p	η^2
Flanker	♂	34	396.23 (385.82 - 406.64)	37	386.47 (376.49 - 396.44)	.18	.026
	♀	9	398.13 (359.29 - 436.97)	18	387.69 (360.23 - 415.16)	.66	.008
	Total	43	396.63 (384.91 - 408.35)	55	386.87 (376.50 - 397.23)	.22	.016
Switching	♂	18	397.31 (374.55 - 420.07)	16	372.51 (348.37 - 396.65)	.14	.068
	♀	7	389.43 (349.12 - 429.73)	8	388.30 (350.60 - 426.00)	.97	.000
	Total	25	395.10 (376.07 - 414.14)	24	377.77 (358.34 - 397.20)	.21	.034
Oddball	♂	30	461.72 (433.88 - 489.56)	37	413.70 (388.64 - 438.77)	.01	.092
	♀	7	390.59 (348.77 - 432.40)	19	390.46 (365.08 - 415.84)	.99	.000
	Total	37	448.26 (424.46 - 472.06)	56	405.82 (386.47 - 425.16)	<.01	.077

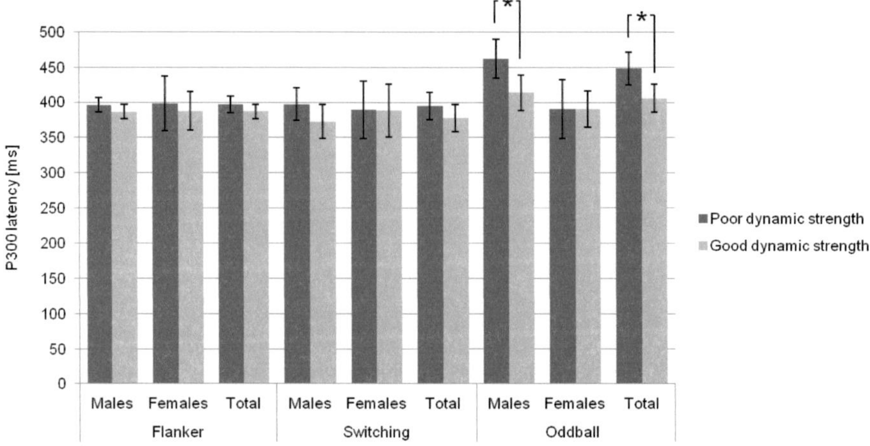

Figure 103: Mean P300 latencies during Flanker, Switching and Oddball task for maximal dynamic strength groups; error bars represent 95% confidence intervals

N100 and P200

A main effect was found for N100 amplitudes during the Oddball task for males (F [1, 100] = 1.89, p = .17, η^2 = .019; males: F [1, 70] = 5.27, p = .03, η^2 = .070; females: F [1, 28] = .42, p = .52, η^2 = .015) with larger amplitudes for the poor dynamic strength group. No further effects on N100 or P200 amplitudes were found (table 115, figure 104 & 105).

Table 115: Mean N100 and P200 amplitudes for maximal dynamic strength groups in the Flanker, Switching and Oddball task

			Poor dynamic strength		Good dynamic strength			
			n	μV (95 % CI)	n	μV (95 % CI)	p	η^2
Flanker	N100	♂	41	-1.53 (-1.79 - -1.26)	49	-1.77 (-2.01 - -1.53)	.18	.020
		♀	12	-2.20 (-2.72 - -1.67)	21	-1.73 (-2.13 - -1.33)	.16	.062
		Total	53	-1.68 (-1.92 - -1.44)	70	-1.76 (-1.97 - -1.55)	.62	.002
	P200	♂	41	3.01 (2.56 - 3.46)	49	3.06 (2.65 - 3.48)	.87	.000
		♀	12	3.42 (2.36 - 4.49)	21	4.06 (3.25 - 4.86)	.34	.030
		Total	53	3.11 (2.68 - 3.53)	70	3.63 (2.99 - 3.74)	.37	.007
Switching	P200	♂	34	4.00 (3.49 - 4.50)	28	3.93 (3.37 - 4.49)	.86	.001
		♀	10	4.91 (3.75 - 6.07)	12	4.20 (3.14 - 5.27)	.36	.042
		Total	44	4.21 (3.74 - 4.67)	40	4.01 (3.52 - 4.50)	.58	.004
Oddball	N100	♂	34	-5.23 (-5.95 - -4.51)	38	-4.09 (-4.77 - -3.40)	.03	.070
		♀	10	-5.15 (-6.44 - -3.86)	20	-5.65 (-6.56 - -4.74)	.52	.015
		Total	44	-5.21 (-5.85 - -4.57)	58	-4.63 (-5.18 - -4.07)	.17	.019
	P200	♂	34	2.94 (2.28 - 3.60)	38	3.14 (2.52 - 3.77)	.66	.003
		♀	10	2.32 (.76 - 3.89)	20	2.93 (1.82 - 4.03)	.52	.015
		Total	44	2.80 (2.18 - 3.42)	58	3.07 (2.53 - 3.61)	.52	.004

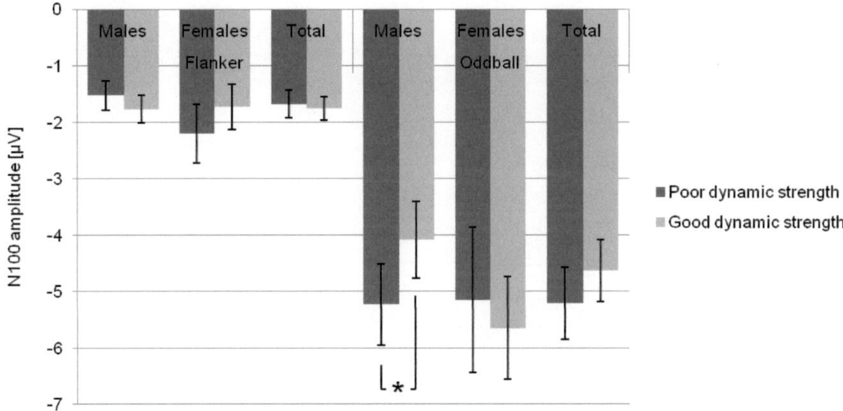

Figure 104: Mean N100 amplitudes during Flanker and Oddball task for maximal dynamic strength groups; error bars represent 95% confidence intervals

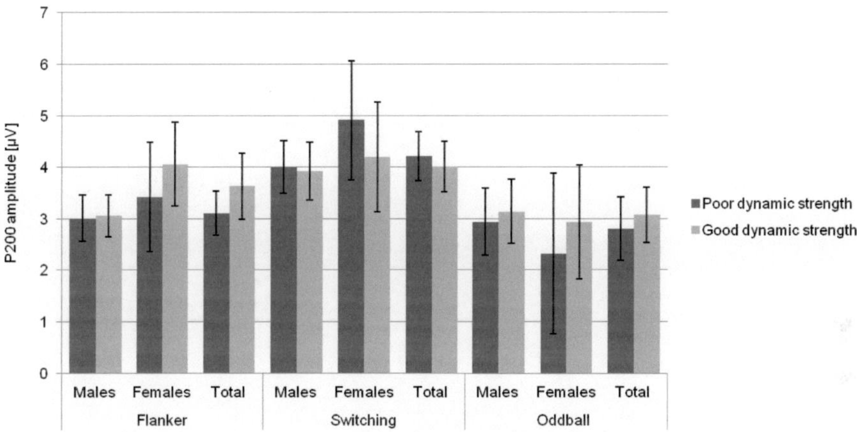

Figure 105: Mean P200 amplitudes during Flanker, Switching and Oddball task for maximal dynamic strength groups; error bars represent 95% confidence intervals

A main effect was found for P200 latencies during the Oddball task for females (F [1, 69] = 3.46, p = .07, η^2 = .048; males: F [1, 53] = .85, p = .36, η^2 = .016; females: F [1, 14] = 7.98, p = .01, η^2 = .363). However, only five females were in the poor strength group. Mean N100 and P200 latencies are given in table 116 and figure 106 and 107.

Table 116: Mean N100 and P200 latencies for maximal dynamic strength groups in the Flanker, Switching and Oddball task

			\multicolumn{2}{c|}{Poor dynamic strength}		\multicolumn{2}{c|}{Good dynamic strength}			
			n	ms (95 % CI)	n	ms (95 % CI)	p	η^2
Flanker	N100	♂	11	142.06 (132.91 - 151.21)	20	138.72 (131.93 - 145.51)	.55	.012
		♀	7	139.84	7	141.80	.76	.008

				(130.41 - 149.28)		(132.36 - 151.23)		
	Total		18	141.20	27	139.52	.69	.004
				(134.70 - 147.70)		(134.21 - 144.82)		
	P200	♂	33	219.18	38	225.27	.35	.013
				(209.75 - 228.61)		(216.48 - 234.06)		
		♀	8	222.24	16	229.02	.57	.015
				(202.14 - 242.34)		(214.81 - 243.23)		
	Total		41	219.78	54	226.38	.24	.015
				(211.42 - 228.14)		(219.09 - 233.67)		
Switching	P200	♂	30	185.06	25	193.67	.09	.052
				(178.24 - 191.88)		(186.20 - 201.14)		
		♀	10	186.91	12	192.74	.56	.018
				(171.92 - 201.91)		(179.05 - 206.43)		
	Total		40	185.52	37	193.37	.08	.040
				(179.36 - 191.69)		(186.96 - 199.78)		
Oddball	N100	♂	33	167.66	32	171.95	.32	.016
				(161.72 - 173.60)		(165.91 - 177.98)		
		♀	8	172.07	20	178.13	.22	.058
				(163.76 - 180.38)		(172.87 - 183.38)		
	Total		41	168.52	52	174.32	.08	.033
				(163.66 - 173.39)		(170.00 - 178.64)		
	P200	♂	25	261.72	30	256.64	.36	.016
				(253.56 - 269.88)		(249.19 - 264.09)		
		♀	5	278.13	11	252.77	.01	.363
				(262.16 - 294.09)		(242.01 - 263.53)		
	Total		30	264.45	41	255.60	.07	.048
				(257.24 - 271.66)		(249.43 - 261.77)		

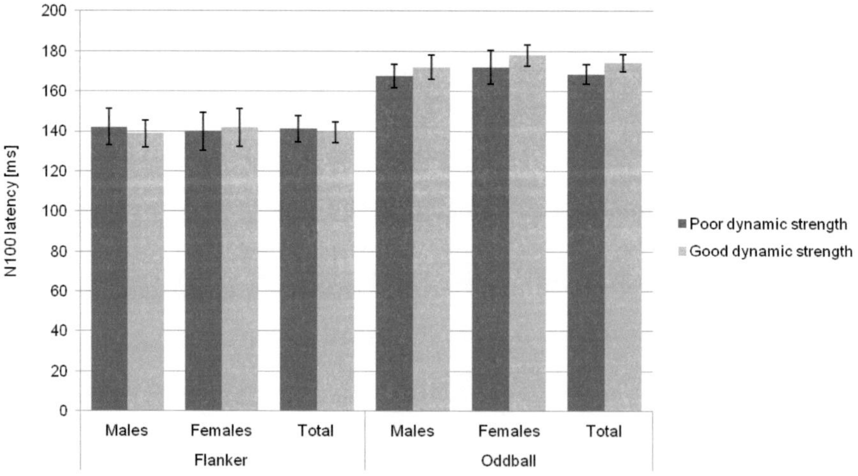

Figure 106: Mean N100 latencies during Flanker and Oddball task for maximal dynamic strength groups; error bars represent 95% confidence intervals

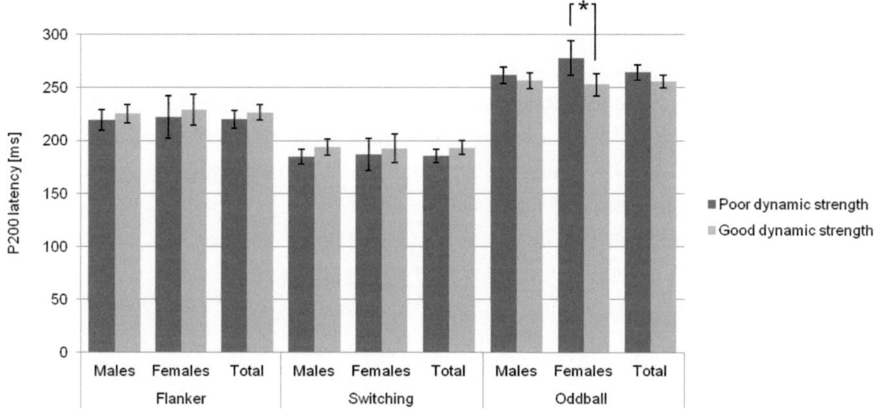

Figure 107: Mean P200 latencies during Flanker, Switching and Oddball task for maximal dynamic strength
groups; error bars represent 95% confidence intervals

4.4.2.3 Discussion

For maximal strength, many small gender-specific effects on cognitive performance were observed
in the present study.

Females with a good maximal isometric strength had faster response times in the Flanker task. In
contrast to that, males with a poor dynamic strength responded significantly faster in the Flanker
task than males with a good dynamic strength. This indicates that maximal isometric strength is
beneficial for response speed during the Flanker task for females but dynamic strength might impair
response speed in males.
A good maximal isometric strength was beneficial for P300 latencies during the Switching (only for
males) and Oddball task (only for females). In the Switching task, females even had a tendency to
longer latencies for the good isometric strength group whereas males in the Oddball task also had a
tendency for faster P300 latencies for the good strength group. Furthermore, male participants with
a good dynamic strength had faster P300 latencies during the Oddball task. This could be predictors
of faster cognitive processing speed during tasks requiring a high amount of top-down executive
control. Females with a good dynamic strength had smaller P300 amplitudes in the Switching task.

For P200 amplitudes during the Flanker and Oddball task, females with a high isometric strength
revealed significantly larger amplitudes. However, good isometric strength was related to longer
P200 latencies during the Flanker task for males and the total sample which might indicate an impaired stimulus classification. In contrast to this, females with good dynamic strength had faster
P200 latencies during the Oddball task.
It can be concluded, that the results are very inconsistent and are mainly gender-specific. Liu-
Ambrose, Nagamatsu, Graf, Beattie, Ashe et al. (2010) conducted a 12 month randomized con-
trolled trial in older women and found that resistance training lead to an enhanced selective attention and conflict resolution, but cognitive abilities associated with shifting between task sets or in-

structions and manipulating verbal information in working memory were not improved. In addition, Goekint et al. (2010) also found no improvements of resistance training on working memory in young adults. In the current study, effects of maximal strength were found for all three cognitive tasks. However, based on literature and the current study, strength-related increases of cognitive performance seem to be very selective. To explain strength-related benefits on cognition, it is often suggested that resistance exercise leads to increased levels of neurotrophins such as BDNF and IGF-I which induce the generation of new neurons and thus might mediate benefits on cognitive performance. However, the study by Goekint et al. (2010) failed to detect any effects of a 10 week strength training on resting BDNF or IGF-I levels in young adults.

4.4.3 Coordination

In this chapter, the relationship between coordination and cognitive performance is described. The independent variables are static and dynamic performances on the Posturomed, number of steps in the balancing backwards test and number of jumps in the jumping side-to-side test. The groups were built individually for the study sample due to a lack of normative values.
In the poor performance group for the static Posturomed test, males and females have ≤ 410 points and in the good performance group, they have ≥ 411 points (table 117).

Table 117:   Number of participants per static Posturomed group (M ± SD in points)

	Poor		Good		Total	
	n (%)	M (SD)	n (%)	M (SD)	n	p
♂	45 (52)	174.52 (151.01)	42 (48)	641.15 (138.98)	87	.00
♀	19 (46)	156.30 (118.12)	22 (54)	682.93 (134.17)	41	.00
Total	64 (50)	169.11 (141.36)	64 (50)	655.51 (137.73)	128	.00

In the poor performance group for the dynamic Posturomed test, males and females have ≤ 300 points and in the good performance group, they have ≥ 301 points (table 118).

Table 118:   Number of participants per dynamic Posturomed group (M ± SD in points)

	Poor		Good		Total	
	n (%)	M (SD)	n (%)	M (SD)	n	p
♂	43 (49)	119.82 (111.21)	44 (51)	486.90 (140.60)	87	.00
♀	24 (59)	121.36 (110.64)	17 (41)	485.80 (90.69)	41	.00
Total	67 (52)	120.37 (110.17)	61 (48)	486.59 (127.91)	128	.00

In the poor performance group for the balancing backwards test, males and females have ≤ 43 steps and in the good performance group, they have ≥ 44 steps (table 119).

Table 119: Number of participants per balancing backwards group (M ± SD in steps)

		Poor		Good		Total	
		n (%)	M (SD)	n (%)	M (SD)	n	p
♂		54 (52)	36.39 (6.37)	49 (48)	47.06 (1.44)	103	.00
♀		18 (42)	38.06 (5.88)	25 (58)	46.64 (1.73)	43	.00
Total		72 (49)	36.81 (6.25)	74 (51)	46.92 (1.54)	146	.00

In the poor performance group for the jumping side to side test, males have ≤ 43.5 jumps and females have ≤ 42 jumps. In the good performance group, males have ≥ 44 jumps and females ≥ 42.5 jumps (table 120).

Table 120: Number of participants per jumping group (M ± SD in jumps)

		Poor		Good		Total	
		n (%)	M (SD)	n (%)	M (SD)	n	p
♂		52 (51)	38.92 (3.55)	51 (19)	49.30 (3.89)	103	.00
♀		21 (49)	37.95 (3.29)	22 (51)	46.25 (3.56)	43	.00
Total		73 (50)	38.64 (3.48)	73 (50)	48.38 (4.02)	146	.00

> *RQ 10*: Is coordinative performance related to cognitive performance in young adulthood?

4.4.3.1 Static Posturomed Performance - Results

Behavioral Measures

No effects were observed for behavioral measures in the Flanker, Switching and Oddball task (table 121 & 122, figure 108 & 109).

Table 121: Mean response times for static Posturomed groups in the Flanker, Switching and Oddball task

		Poor static performance		Good static performance			
		n	ms (95 % CI)	n	ms (95 % CI)	p	η^2
Flanker	♂	43	487.98 (475.97 - 499.98)	36	489.65 (476.53 - 502.77)	.85	.000
	♀	16	516.94 (493.30 - 540.57)	19	506.64 (484.95 - 528.33)	.52	.013
	Total	59	495.83 (484.79 - 506.86)	55	495.52 (484.09 - 506.95)	.97	.000
Switching	♂	41	688.91 (665.25 - 712.58)	34	705.77 (679.78 - 731.75)	.34	.012
	♀	17	723.00 (675.16 - 770.85)	18	671.01 (624.51 - 717.50)	.12	.071
	Total	58	698.91 (677.07 - 720.74)	52	693.73 (670.67 - 716.80)	.75	.001
Oddball	♂	42	503.21 (476.04 - 530.38)	35	495.51 (465.75 - 525.27)	.71	.002
	♀	18	533.11 (484.69 - 581.53)	20	525.20 (479.27 - 571.14)	.81	.002
	Total	60	512.18 (488.39 - 535.97)	55	506.31 (481.46 - 531.16)	.74	.001

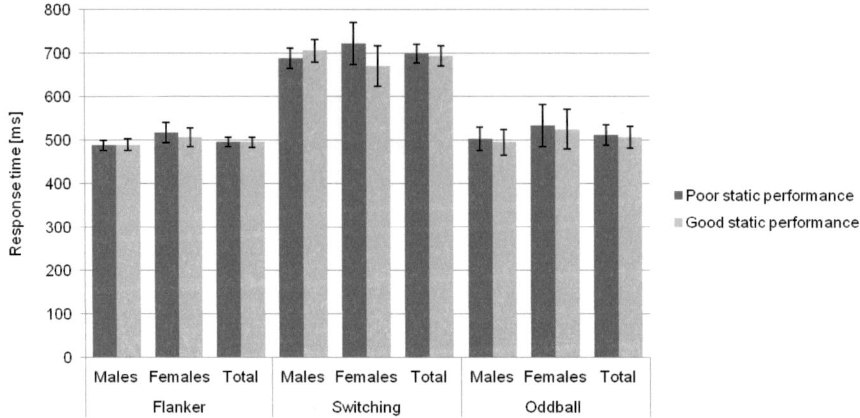

Figure 108: Mean response times during Flanker, Switching and Oddball task for static Posturomed groups; error bars represent 95% confidence intervals

Table 122:   Errors in the Switching task for static Posturomed groups

		\multicolumn{2}{c}{Poor static performance}	\multicolumn{2}{c}{Good static performance}		
		n	No. (95 % CI)	n	No. (95 % CI)	p	η^2
Errors	♂	42	3.88 (3.18 - 4.57)	35	4.22 (3.46 - 4.98)	.51	.006
	♀	17	4.22 (2.88 - 5.56)	19	4.26 (2.99 - 5.53)	.96	.000
	Total	56	3.98 (3.36 - 4.59)	54	4.24 (3.59 - 4.88)	.56	.003

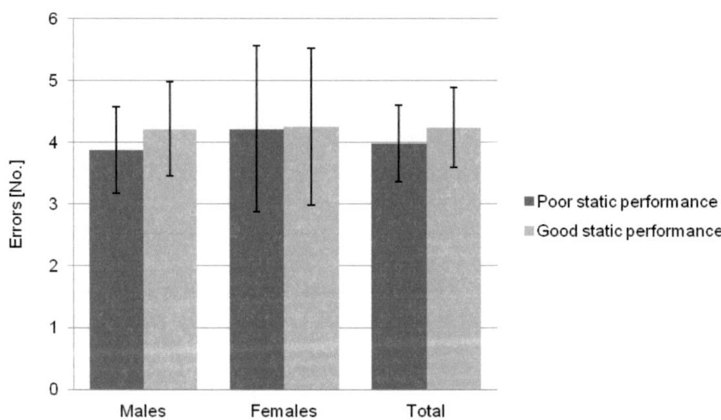

Figure 109: Errors during the Switching task for static Posturomed groups; error bars represent 95% confidence intervals

P300

No effects were observed for P300 amplitudes (table 123 & figure 110) and latencies (table 124 & figure 111) in the Flanker, Switching and Oddball task.

Table 123:   Mean P300 amplitudes for static Posturomed groups in the Flanker, Switching and Oddball task

		Poor static performance		Good static performance			
		n	µV (95 % CI)	n	µV (95 % CI)	p	η^2
Flanker	♂	42	4.69 (4.25 - 5.13)	33	4.81 (4.32 - 5.30)	.72	.002
	♀	13	3.81 (3.01 - 4.60)	19	4.80 (4.15 - 5.46)	.06	.115
	Total	55	4.48 (4.10 - 4.87)	52	4.81 (4.42 - 5.20)	.24	.013
Switching	♂	26	3.86 (3.47 - 4.25)	25	3.88 (3.48 - 4.27)	.95	.000
	♀	9	3.65 (2.84 - 4.46)	12	4.69 (3.99 - 5.39)	.06	.180
	Total	35	3.81 (3.45 - 4.16)	37	4.14 (3.79 - 4.49)	.18	.025
Oddball	♂	31	5.52 (4.60 - 6.43)	29	5.74 (4.79 - 6.69)	.74	.002
	♀	11	5.52 (4.19 - 6.85)	18	6.96 (5.92 - 8.00)	.09	.101
	Total	42	5.52 (4.77 - 6.27)	47	6.21 (5.50 - 6.91)	.19	.020

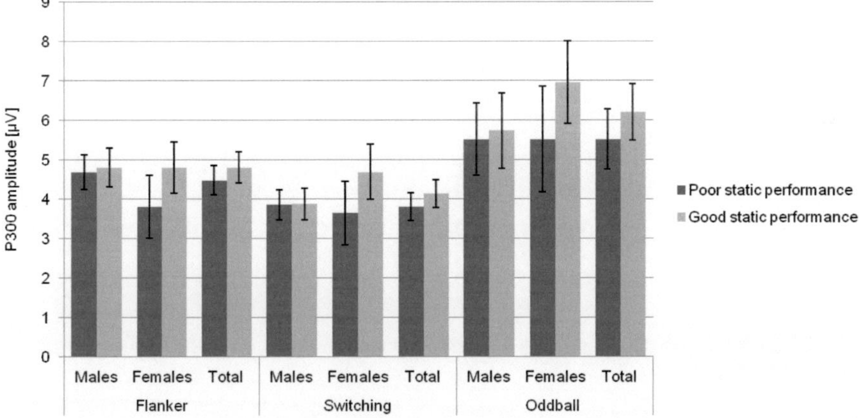

Figure 110:   Mean P300 amplitudes during Flanker, Switching and Oddball task for static Posturomed groups; error bars represent 95% confidence intervals

Table 124:   Mean P300 latencies for static Posturomed groups in the Flanker, Switching and Oddball task

		Poor static performance		Good static performance			
		n	ms (95 % CI)	n	ms (95 % CI)	p	η^2
Flanker	♂	31	391.67 (380.77 - 402.58)	27	391.39 (379.70 - 403.07)	.97	.000
	♀	10	379.88 (343.35 - 416.41)	17	397.81 (369.79 - 425.83)	.43	.025
	Total	41	388.80 (376.43 - 401.16)	44	393.87 (381.93 - 405.81)	.56	.004
Switching	♂	15	380.06 (354.35 - 405.76)	14	385.61 (359.00 - 412.22)	.76	.004
	♀	5	379.49 (330.78 - 428.20)	9	390.10 (353.79 - 426.40)	.71	.012
	Total	20	379.92 (358.33 - 401.50)	23	387.37 (367.24 - 407.49)	.61	.006
Oddball	♂	29	436.85 (406.88 - 466.83)	26	440.22 (408.57 - 471.88)	.88	.000
	♀	9	371.62 (336.07 - 407.16)	17	400.49 (374.63 - 426.35)	.19	.071
	Total	38	421.40 (396.96 - 445.85)	43	424.52 (401.54 - 447.50)	.85	.000

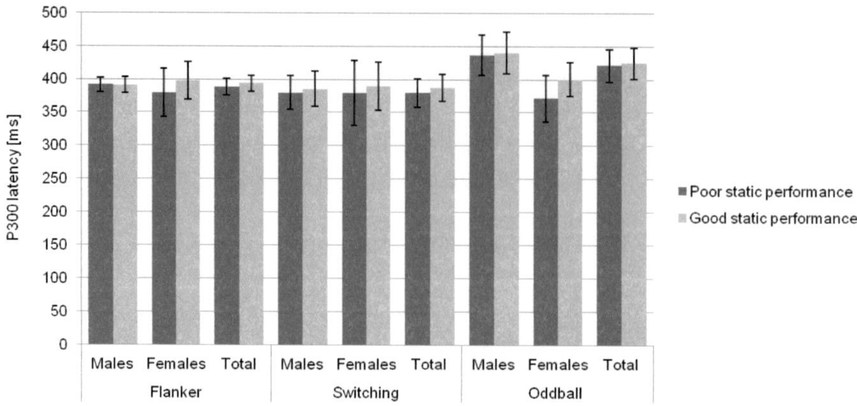

Figure 111: Mean P300 latencies during Flanker, Switching and Oddball task for static Posturomed groups; error bars represent 95% confidence intervals

N100 and P200

There was a main effect for P200 latencies during the Flanker task (F [1, 81] = 6.10, p = .02, η^2 = .070; males: F [1, 57] = 7.02, p = .01, η^2 = .110; females: F [1, 22] = .08, p = .78, η^2 = .004) with faster latencies for the poor performance group. No further effects were observed (table 125 & 126, figure 112 - 115).

Table 125: Mean N100 and P200 amplitudes for static Posturomed groups in the Flanker, Switching and Oddball task

			Poor static performance		Good static performance			
			n	µV (95 % CI)	n	µV (95 % CI)	p	η^2
Flanker	N100	♂	42	-1.49 (-1.72 - -1.26)	35	-1.64 (-1.90 - -1.39)	.38	.010
		♀	14	-2.00 (-2.50 - -1.49)	19	-1.83 (-2.26 - -1.40)	.61	.008
		Total	56	-1.62 (-1.83 - -1.40)	54	-1.71 (-1.93 - -1.49)	.55	.003
	P200	♂	42	3.16 (2.69 - 3.62)	35	2.86 (2.35 - 3.38)	.41	.009
		♀	14	3.23 (2.27 - 4.19)	19	4.27 (3.45 - 5.09)	.10	.084
		Total	56	3.17 (2.74 - 3.61)	54	3.36 (2.91 - 3.80)	.56	.003
Switching	P200	♂	28	4.05 (3.49 - 4.61)	25	3.93 (3.33 - 4.52)	.76	.002
		♀	9	4.37 (3.16 - 5.57)	12	4.41 (3.37 - 5.46)	.95	.000
		Total	37	4.13 (3.62 - 4.63)	37	4.08 (3.58 - 4.59)	.90	.000
Oddball	N100	♂	31	-4.78 (-5.59 - -3.97)	29	-4.38 (-5.21 - -3.54)	.49	.008
		♀	12	-5.01 (-6.18 - -3.85)	18	-5.80 (-6.75 - -4.84)	.30	.039
		Total	43	-4.85 (-5.51 - -4.18)	47	-4.92 (-5.56 - -4.29)	.87	.000
	P200	♂	31	2.93 (2.22 - 3.64)	29	3.36 (2.62 - 4.10)	.41	.012
		♀	12	2.90 (1.47 - 4.34)	18	2.61 (1.44 - 3.78)	.75	.004
		Total	43	2.92 (2.28 - 3.57)	47	3.07 (2.46 - 3.69)	.74	.001

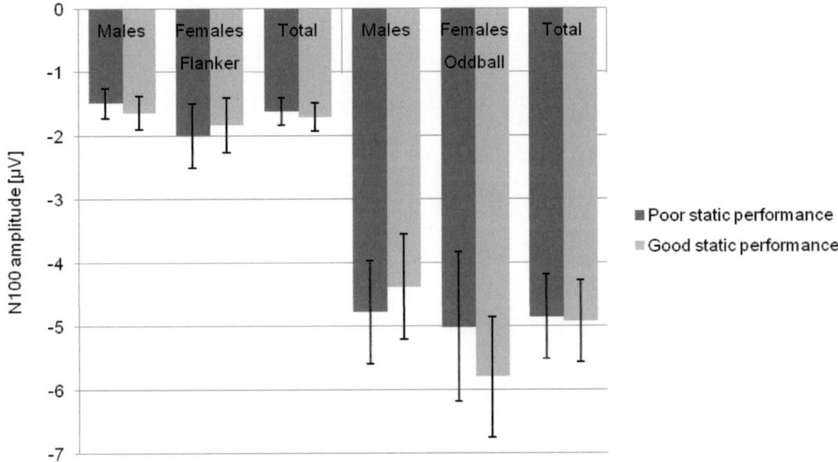

Figure 112: Mean N100 amplitudes during Flanker and Oddball task for static Posturomed groups; error bars represent 95% confidence intervals

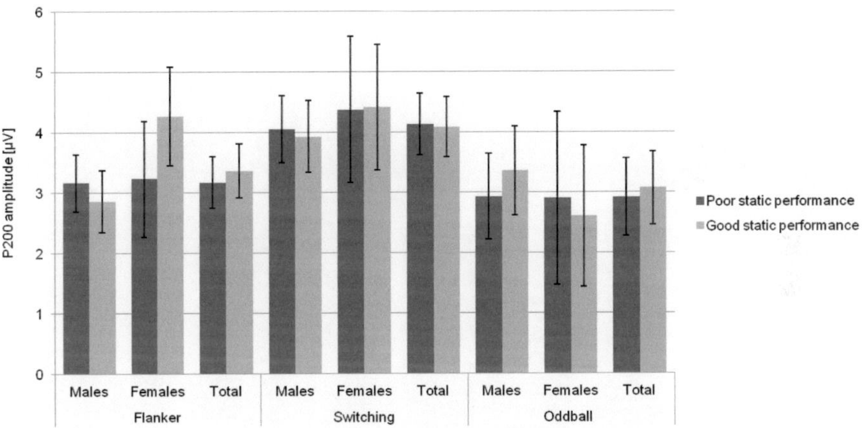

Figure 113: Mean P200 amplitudes during Flanker, Switching and Oddball task for static Posturomed groups; error bars represent 95% confidence intervals

Table 126:   Mean N100 and P200 latencies for static Posturomed groups in the Flanker, Switching and Oddball task

			\multicolumn{2}{l}{Poor static performance}		\multicolumn{2}{l}{Good static performance}			
			n	ms (95 % CI)	n	ms (95 % CI)	p	η^2
Flanker	N100	♂	10	141.70 (131.71 - 151.68)	13	137.44 (128.68 - 146.20)	.51	.021
		♀	7	141.80 (132.36 - 151.23)	7	139.84 (130.41 - 149.28)	.76	.008
		Total	17	141.74 (135.06 - 148.42)	20	138.28 (132.12 - 144.40)	.45	.017
	P200	♂	33	214.36	26	233.11	.01	.110

				Poor		Good		
				(204.95 - 223.77)		(222.51 - 243.70)		
		♀	9	224.70	15	228.00	.78	.004
				(205.64 - 243.75)		(213.24 - 242.75)		
		Total	42	216.57	41	231.24	.02	.070
				(208.27 - 224.88)		(222.83 - 239.64)		
Switching	P200	♂	22	185.12	24	190.02	.38	.018
				(177.11 - 193.13)		(182.35 - 197.68)		
		♀	9	185.63	12	193.39	.46	.029
				(169.46 - 201.81)		(179.39 - 207.40)		
		Total	31	185.27	36	191.14	.23	.022
				(178.15 - 192.39)		(184.54 - 197.74)		
Oddball	N100	♂	28	168.72	26	168.39	.95	.000
				(161.98 - 175.46)		(161.40 - 175.38)		
		♀	10	179.69	18	174.57	.27	.047
				(172.21 - 187.16)		(168.99 - 180.14)		
		Total	38	171.61	44	170.92	.85	.000
				(166.36 - 176.86)		(166.04 - 175.80)		
	P200	♂	23	263.49	23	253.64	.11	.056
				(254.81 - 272.16)		(244.96 - 262.31)		
		♀	6	258.33	10	262.11	.73	.009
				(240.16 - 276.51)		(248.03 - 276.19)		
		Total	29	262.42	33	256.20	.24	.023
				(254.78 - 270.06)		(249.04 - 263.37)		

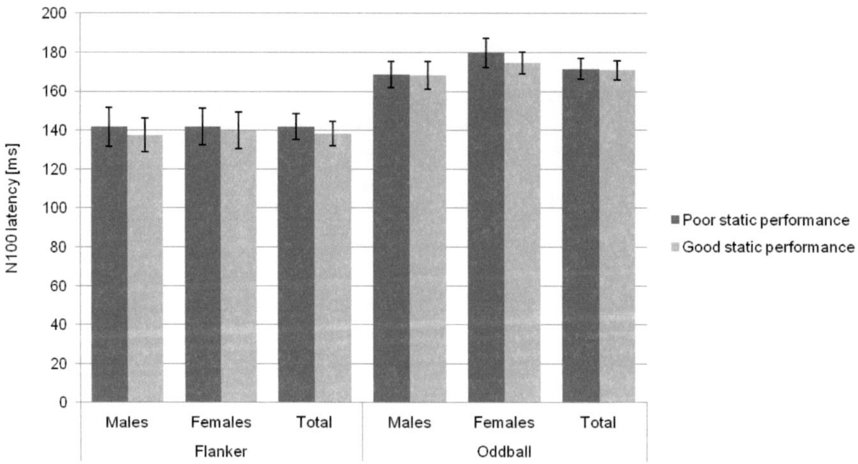

Figure 114: Mean N100 latencies during Flanker and Oddball task for static Posturomed groups; error bars represent 95% confidence intervals

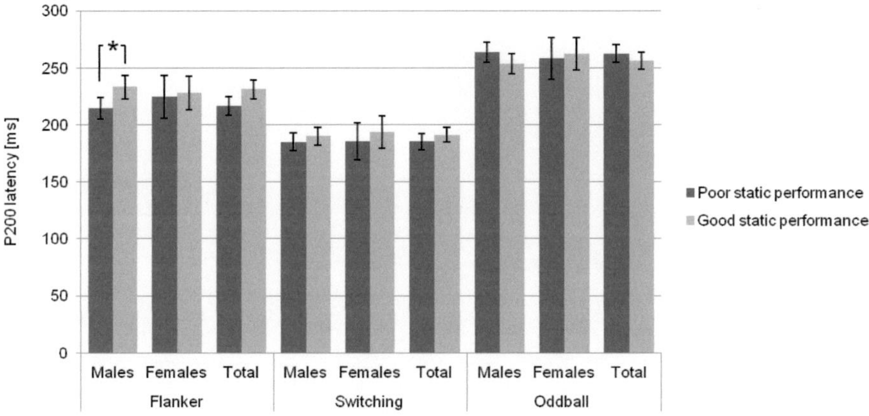

Figure 115: Mean P200 latencies during Flanker, Switching and Oddball task for static Posturomed groups; error bars represent 95% confidence intervals

4.4.3.2 Dynamic Posturomed Performance - Results

Behavioral Measures

A main effect was observed for Flanker response times for females (F [1, 112] = 3.50, p = .06, η^2 = .030; males: F [1, 77] = .46, p = .50, η^2 = .006; females: F [1, 33] = 4.81, p = .04, η^2 = .127). No further effects were observed for behavioral measures (table 127 & 128, figure 116 & 117).

Table 127: Mean response times for dynamic Posturomed groups in the Flanker, Switching and Oddball task

		Poor dynamic performance		Good dynamic performance			
		n	ms (95 % CI)	n	ms (95 % CI)	p	η^2
Flanker	♂	40	491.70 (479.28 - 504.11)	39	485.70 (473.13 - 498.27)	.50	.006
	♀	19	526.21 (505.82 - 546.60)	16	493.70 (471.48 - 515.92)	.04	.127
	Total	59	502.81 (491.94 - 513.68)	55	488.03 (476.78 - 499.28)	.06	.030
Switching	♂	37	703.34 (678.37 - 728.31)	38	689.94 (665.31 - 714.58)	.45	.008
	♀	20	710.71 (665.62 - 755.80)	15	677.00 (624.93 - 729.06)	.33	.029
	Total	57	705.93 (684.04 - 727.81)	53	686.28 (663.58 - 708.98)	.22	.014
Oddball	♂	39	502.92 (474.71 - 531.12)	38	496.42 (467.84 - 524.99)	.75	.001
	♀	22	544.82 (501.77 - 587.86)	16	507.13 (456.65 - 557.61)	.26	.036
	Total	61	518.03 (494.54 - 541.52)	54	499.59 (474.62 - 524.56)	.29	.010

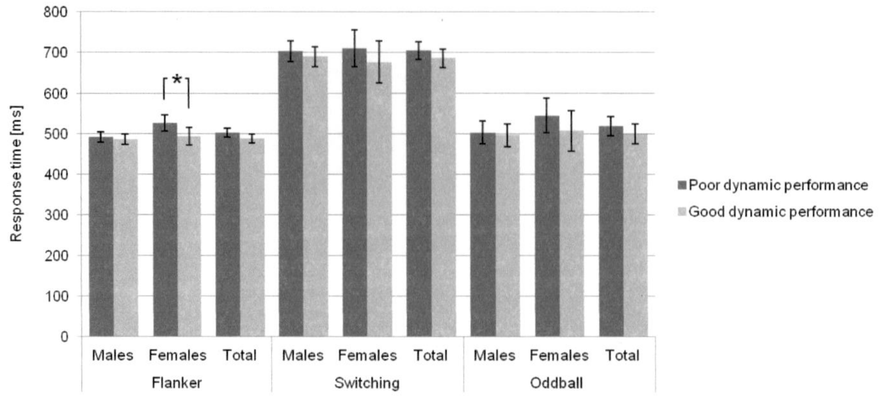

Figure 116: Mean response times during Flanker, Switching and Oddball task for dynamic Posturomed groups; error bars represent 95% confidence intervals

Table 128: Errors in the Switching task for dynamic Posturomed groups

		Poor dynamic performance		Good dynamic performance			
		n	No. (95 % CI)	n	No. (95 % CI)	p	η^2
Errors	♂	39	4.10 (3.37 - 4.82)	38	3.97 (3.23 - 4.70)	.80	.001
	♀	20	4.23 (2.99 - 5.46)	16	4.27 (2.88 - 5.65)	.97	.000
	Total	59	4.14 (3.52 - 4.76)	54	4.06 (3.41 - 4.70)	.85	.000

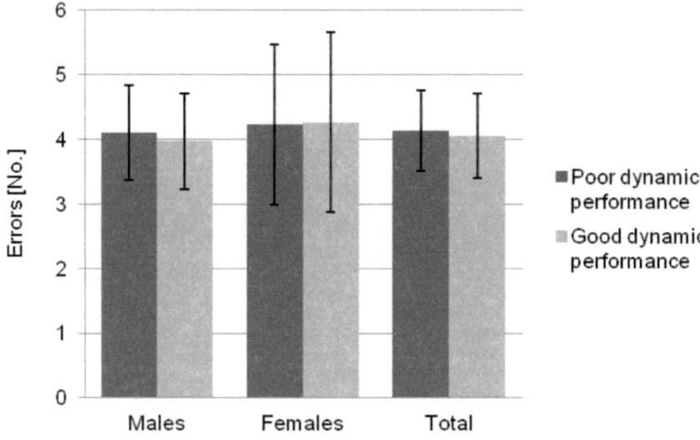

Figure 117: Errors during the Switching task for dynamic Posturomed groups; error bars represent 95% confidence intervals

P300

For P300 amplitudes (table 129 & figure 118) during the Switching task, there was a condition x dynamic Posturomed interaction (F [1, 70] = 4.42, p = .04, η^2 = .059). This interaction indicates that participants with poor dynamic Posturomed performance had a smaller difference between amplitudes (homogenous: 3.88 µV; 95 % CI 3.45 - 4.31; heterogeneous: 3.70 µV; 95 % CI 3.30 - 4.09) compared to participants with a good dynamic Posturomed performance (homogenous: 4.50 µV; 95 % CI 4.10 - 4.89; heterogeneous: 3.78 µV; 95 % CI 3.42 - 4.14). For P300 latencies (table 130 &

figure 119) during the Flanker task, there was a main effect for males (F [1, 83] = 1.22, p = .27, η^2 = .015; males: F [1, 56] = 5.19, p = .03, η^2 = .085; females: F [1, 25] = .11, p = .74, η^2 = .005). No further effects were observed.

Table 129: Mean P300 amplitudes for dynamic Posturomed groups in the Flanker, Switching and Oddball task

		Poor dynamic performance		Good dynamic performance			
		n	μV (95 % CI)	n	μV (95 % CI)	p	η^2
Flanker	♂	38	4.71 (4.25 - 5.17)	37	4.78 (4.32 - 5.25)	.82	.001
	♀	16	4.03 (3.29 - 4.76)	16	4.77 (4.04 - 5.51)	.15	.067
	Total	54	4.50 (4.12 - 4.89)	53	4.78 (4.39 - 5.17)	.32	.009
Switching	♂	27	3.76 (3.36 - 4.16)	27	3.96 (3.59 - 4.34)	.46	.011
	♀	9	3.86 (3.00 - 4.72)	12	4.53 (3.79 - 5.28)	.23	.075
	Total	33	3.79 (3.42 - 4.15)	39	4.14 (3.80 - 4.48)	.16	.027
Oddball	♂	32	5.34 (4.44 - 5.24)	28	5.95 (4.99 - 6.91)	.36	.015
	♀	15	5.89 (4.72 - 7.05)	14	6.98 (5.77 - 8.18)	.19	.062
	Total	47	5.52 (4.81 - 6.22)	42	6.29 (5.54 - 7.04)	.14	.025

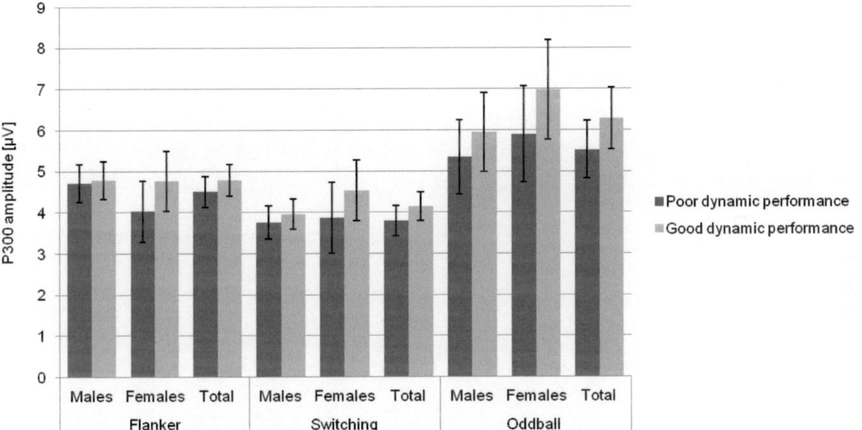

Figure 118: Mean P300 amplitudes during Flanker, Switching and Oddball task for dynamic Posturomed groups; error bars represent 95% confidence intervals

Table 130: Mean P300 latencies for dynamic Posturomed groups in the Flanker, Switching and Oddball task

		Poor dynamic performance		Good dynamic performance			
		n	ms (95 % CI)	n	ms (95 % CI)	p	η^2
Flanker	♂	29	382.86 (372.08 - 393.65)	29	400.22 (389.43 - 411.00)	.03	.085
	♀	12	395.30 (361.60 - 428.99)	15	387.87 (357.73 - 418.01)	.74	.005
	Total	41	386.50 (374.20 - 398.80)	44	396.01 (384.13 - 407.88)	.27	.015
Switching	♂	14	374.17 (347.94 - 400.41)	15	390.73 (365.39 - 416.08)	.36	.031
	♀	6	372.54 (329.30 - 415.78)	8	396.64 (359.19 - 434.08)	.38	.066
	Total	20	373.68 (352.48 - 394.88)	23	392.79 (373.02 - 412.56)	.19	.041

Oddball	♂	28	424.44 (394.43 - 454.45)	27	452.97 (422.41 - 483.53)	.19	.033
	♀	14	390.50 (360.93 - 420.06)	12	390.50 (358.56 - 422.43)	.99	.000
	Total	42	413.13 (390.09 - 436.16)	39	433.75 (409.84 - 457.65)	.22	.019

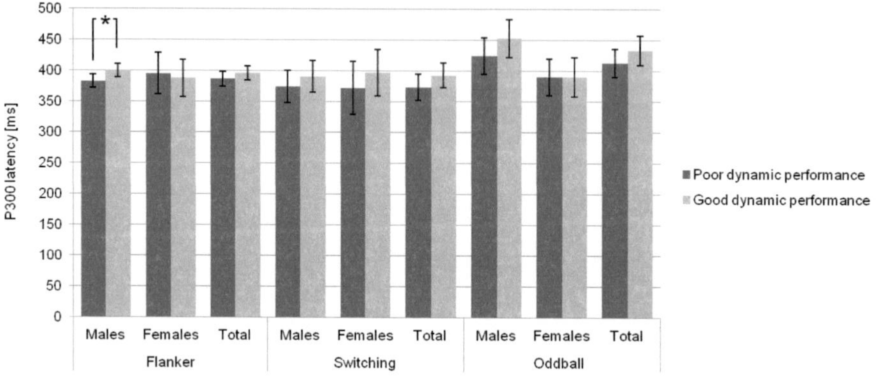

Figure 119: Mean P300 latencies during Flanker, Switching and Oddball task for dynamic Posturomed groups; error bars represent 95% confidence intervals

N100 and P200

There was a main effect for N100 amplitudes during the Oddball task for males (F [1, 88] = 2.67, p = .11, η^2 = .029; males: F [1, 58] = 4.15, p = .05, η^2 = .067; females: F [1, 28] = .01, p = .94, η^2 = .000). No further effects were observed (table 131 & 132, figure 120 - 123).

Table 131: Mean N100 and P200 amplitudes for dynamic Posturomed groups in the Flanker, Switching and Oddball task

| | | | \multicolumn{2}{c}{Poor dynamic performance} | | \multicolumn{2}{c}{Good dynamic performance} | | | |
			n	μV (95 % CI)	n	μV (95 % CI)	p	η^2
Flanker	N100	♂	39	-1.49 (-1.74 - -1.25)	38	-1.63 (-1.87 - -1.38)	.45	.008
		♀	17	-1.82 (-2.28 - -1.37)	16	-1.98 (-2.45 - -1.51)	.63	.008
		Total	56	-1.59 (-1.81 - -1.38)	54	-1.73 (-1.95 - -1.51)	.38	.007
	P200	♂	39	3.13 (2.64 - 3.62)	38	2.91 (2.42 - 3.41)	.54	.005
		♀	17	3.63 (2.73 - 4.53)	16	4.04 (3.11 - 4.97)	.52	.014
		Total	56	3.28 (2.84 - 3.72)	54	3.25 (2.80 - 3.69)	.92	.000
Switching	P200	♂	26	4.07 (3.49 - 4.65)	27	3.92 (3.35 - 4.49)	.71	.003
		♀	9	4.51 (3.31 - 5.71)	12	4.30 (3.26 - 5.35)	.79	.004
		Total	35	4.18 (3.67 - 4.70)	39	4.04 (3.55 - 4.53)	.68	.002
Oddball	N100	♂	32	-5.12 (-5.89 - -4.35)	28	-3.97 (-4.80 - -3.15)	.05	.067
		♀	16	-5.46 (-6.49 - -4.43)	14	-5.51 (-6.61 - -4.41)	.94	.000
		Total	48	-5.23 (-5.85 - -4.61)	42	-4.49 (-5.15 - -3.82)	.11	.029
	P200	♂	32	2.89 (2.19 - 3.59)	28	3.42 (2.68 - 4.17)	.30	.019
		♀	16	3.30 (2.10 - 4.50)	14	2.07 (.78 - 3.35)	.16	.069
		Total	48	3.03 (2.42 - 3.64)	42	2.97 (2.32 - 3.63)	.90	.000

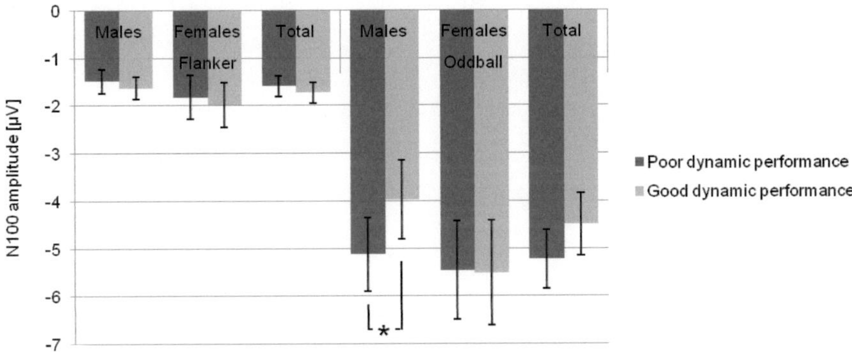

Figure 120: Mean N100 amplitudes during Flanker and Oddball task for dynamic Posturomed groups; error bars represent 95% confidence intervals

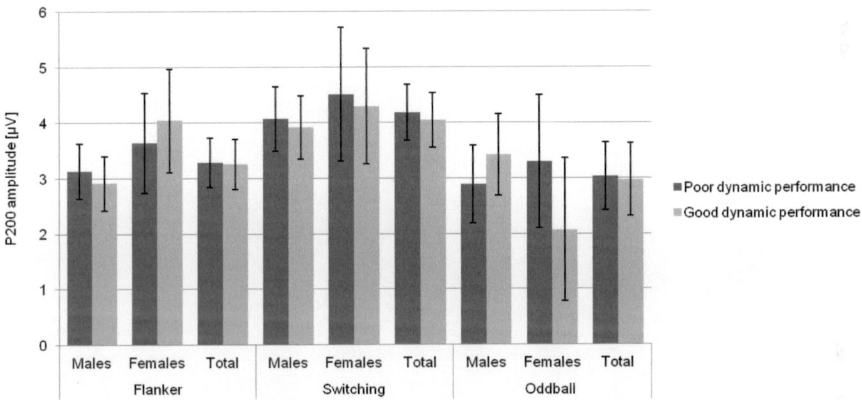

Figure 121: Mean P200 amplitudes during Flanker, Switching and Oddball task for dynamic Posturomed groups; error bars represent 95% confidence intervals

Table 132: Mean N100 and P200 latencies for dynamic Posturomed groups in the Flanker, Switching and Oddball task

			Poor dynamic performance		Good dynamic performance			
			n	ms (95 % CI)	n	ms (95 % CI)	p	η^2
Flanker	N100	♂	9	143.53 (133.19 - 153.88)	14	136.57 (128.27 - 144.86)	.29	.054
		♀	6	140.66 (130.42 - 150.89)	8	140.94 (132.08 - 149.81)	.96	.000
		Total	15	142.38 (135.30 - 149.47)	22	138.16 (132.31 - 144.01)	.36	.024
	P200	♂	31	216.80 (206.76 - 226.84)	28	229.06 (218.50 - 239.62)	.10	.047
		♀	11	228.98 (211.76 - 246.19)	13	224.88 (209.05 - 240.71)	.72	.006
		Total	42	219.99	41	227.73	.21	.020

					(211.47 - 228.52)		(219.11 - 236.36)		
Switching	P200	♂	20	187.60 (179.12 - 196.07)	26	187.73 (180.30 - 195.17)	.98	.000
		♀	9	190.95 (174.55 - 207.36)	12	189.40 (175.20 - 203.61)	.88	.001
		Total	29	188.64 (181.20 - 196.08)	38	188.26 (181.76 - 194.76)	.94	.000
Oddball	N100	♂	30	172.14 (165.80 - 178.47)	24	164.10 (157.01 - 171.18)	.10	.052
		♀	15	178.39 (172.25 - 184.53)	13	174.10 (167.50 - 180.69)	.34	.035
		Total	45	174.22 (169.50 - 178.94)	37	167.61 (162.40 - 172.82)	.07	.042
	P200	♂	24	260.29 (251.58 - 268.99)	22	256.68 (247.58 - 265.77)	.57	.008
		♀	9	258.77 (243.95 - 273.58)	7	263.17 (246.37 - 279.97)	.68	.013
		Total	33	259.87 (252.63 - 267.11)	29	258.24 (250.52 - 265.97)	.76	.002

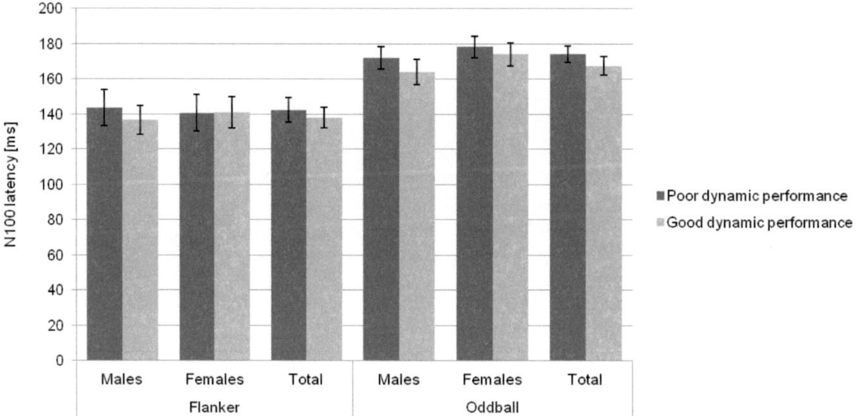

Figure 122: Mean N100 latencies during Flanker and Oddball task for dynamic Posturomed groups; error bars represent 95% confidence intervals

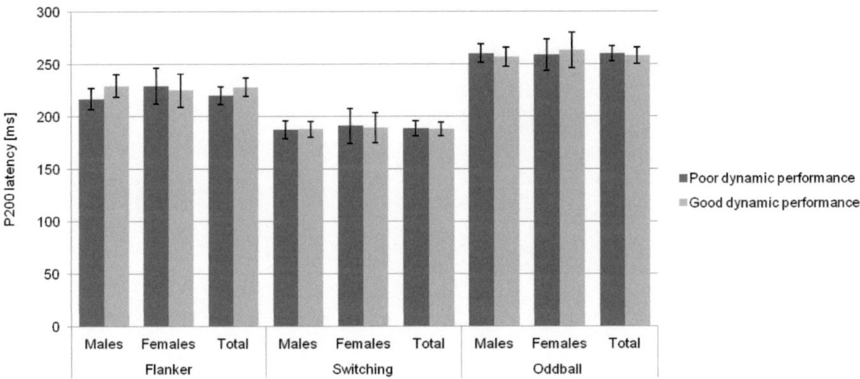

Figure 123: Mean P200 latencies during Flanker, Switching and Oddball task for dynamic Posturomed groups; error bars represent 95% confidence intervals

4.4.3.3 Balancing Backwards - Results

Behavioral Measures

No effects were observed for response times (table 133 & figure 124). For errors during the Switching task, there was a type of trials x balancing backwards interaction (F [1, 125] = 4.79, p = .03, η^2 = .037). This interaction indicates that participants with poor balancing backwards performance had more errors in odd/even trials (4.83; 95 % CI 4.12 - 5.54) but less in </> 5 trials (3.89; 95 % CI 3.32 - 4.46) than participants with good performance (odd/even: 4.27; 95 % CI 3.57 - 4.97; </> 5: 3.52; 95 % CI 2.96 - 4.07). No main effects on errors were found (table 134 & figure 125).

Table 133: Mean response times for balancing backwards groups in the Flanker, Switching and Oddball task

		Poor balancing performance		Good balancing performance			
		n	ms (95 % CI)	n	ms (95 % CI)	p	η^2
Flanker	♂	49	493.00 (481.44 - 504.56)	43	489.57 (477.23 - 501.91)	.69	.002
	♀	13	511.65 (485.26 - 538.04)	22	511.17 (490.89 - 531.45)	.98	.000
	Total	62	496.91 (486.09 - 507.73)	65	496.88 (486.31 - 507.45)	.99	.000
Switching	♂	47	700.16 (677.31 - 723.02)	40	701.68 (676.91 - 726.46)	.93	.000
	♀	15	714.20 (660.30 - 768.11)	21	691.72 (646.16 - 737.28)	.52	.012
	Total	62	703.56 (681.99 - 725.13)	61	698.25 (676.51 - 720.00)	.73	.001
Oddball	♂	49	510.25 (484.61 - 535.89)	41	489.33 (461.31 - 517.36)	.28	.013
	♀	14	510.77 (456.37 - 565.17)	24	539.55 (498.00 - 581.10)	.40	.020
	Total	63	510.37 (486.88 - 533.86)	65	507.88 (484.75 - 531.00)	.88	.000

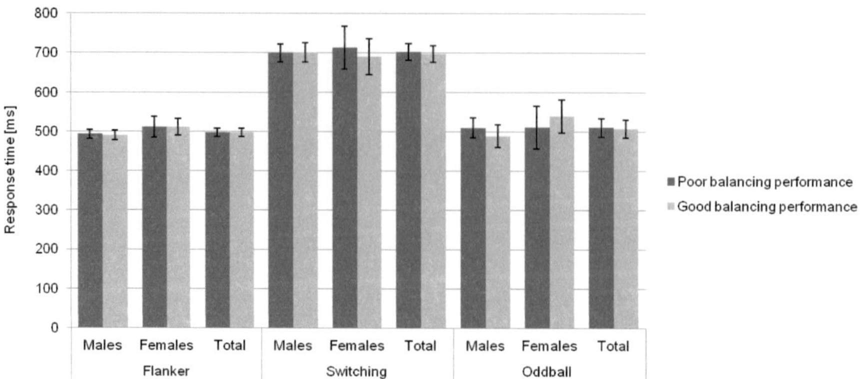

Figure 124: Mean response times during Flanker, Switching and Oddball task for balancing backwards groups; error bars represent 95% confidence intervals

Table 134:   Errors in the Switching task for balancing backwards groups

		Poor balancing performance		Good balancing performance			
		n	No. (95 % CI)	n	No. (95 % CI)	p	η^2
Errors	♂	48	4.01 (3.39 - 4.63)	42	3.67 (3.01 - 4.34)	.46	.006
	♀	14	4.07 (2.61 - 5.53)	23	4.29 (3.16 - 5.43)	.81	.002
	Total	62	4.02 (3.44 - 4.61)	65	3.89 (3.32 - 4.46)	.75	.001

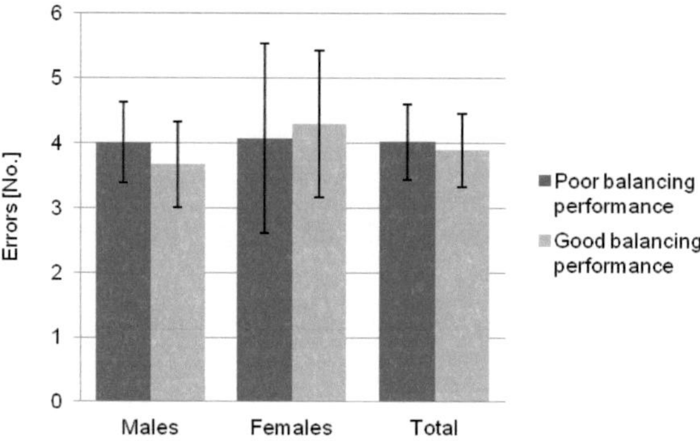

Figure 125: Errors during the Switching task for balancing backwards groups; error bars represent 95% confidence intervals

P300

For P300 amplitudes (table 135 & figure 126), a condition x balancing backwards interaction (F [1, 79] = 4.73, p = .03, η^2 = .056) and a type of trials x balancing backwards interaction (F [1, 79] = 4.67, p = .03, η^2 = .056) were found during the Switching task. These interactions indicate that participants with poor balancing backwards performance had a smaller amplitude in the homogenous condition (3.90 µV; 95 % CI 3.49 - 4.30) than participants with a good performance (4.35 µV; 95 % CI 3.98 - 4.71). However, participants with a good performance had a smaller amplitude during the heterogeneous condition (3.69 µV; 95 % CI 3.36 - 4.02) compared to participants with a poor performance (3.74 µV; 95 % CI 3.37 - 4.11). No effect was found for P300 latencies (table 136 & figure 127).

Table 135: Mean P300 amplitudes for balancing backwards groups in the Flanker, Switching and Oddball task

		Poor balancing performance		Good balancing performance			
		n	µV (95 % CI)	n	µV (95 % CI)	p	η^2
Flanker	♂	46	4.52 (4.12 - 4.92)	42	4.91 (4.49 - 5.32)	.19	.020
	♀	11	4.43 (3.51 - 5.35)	21	4.38 (3.72 - 5.05)	.93	.000
	Total	57	4.51 (4.14 - 4.87)	63	4.73 (4.38 - 5.08)	.38	.007
Switching	♂	27	3.68 (3.32 - 4.05)	32	3.89 (3.56 - 4.23)	.40	.012
	♀	9	4.22 (3.35 - 5.10)	13	4.32 (3.59 - 5.05)	.86	.002
	Total	36	3.82 (3.47 - 4.17)	45	4.02 (3.71 - 4.33)	.40	.009
Oddball	♂	40	5.70 (4.92 - 6.49)	32	5.81 (4.94 - 6.69)	.85	.000
	♀	9	6.68 (5.13 - 8.22)	20	6.29 (5.26 - 7.33)	.68	.007
	Total	49	5.88 (5.20 - 6.57)	52	6.00 (5.33 - 6.66)	.81	.001

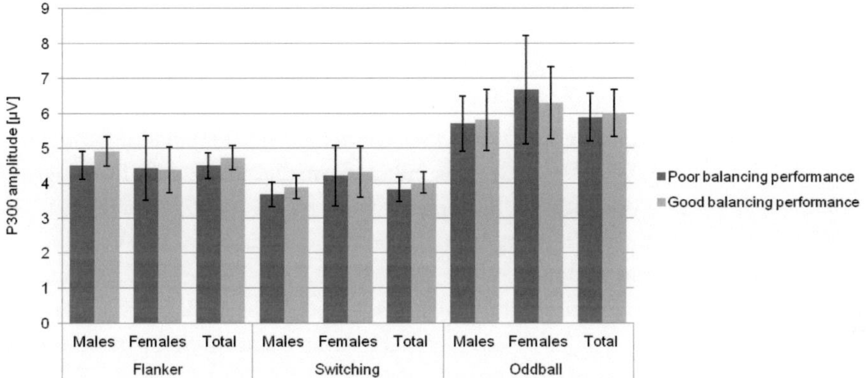

Figure 126: Mean P300 amplitudes during Flanker, Switching and Oddball task for balancing backwards groups; error bars represent 95% confidence intervals

Table 136: Mean P300 latencies for balancing backwards groups in the Flanker, Switching and Oddball task

		Poor balancing performance		Good balancing performance			
		n	ms (95 % CI)	n	ms (95 % CI)	p	η^2
Flanker	♂	38	387.57 (377.68 - 397.47)	33	395.25 (384.63 - 405.87)	.30	.016
	♀	9	382.97 (344.19 - 421.75)	18	395.27 (367.85 - 422.69)	.60	.011
	Total	47	386.69 (375.46 - 397.92)	51	395.26 (384.48 - 406.04)	.28	.012
Switching	♂	19	381.30 (358.48 - 404.11)	15	391.15 (365.46 - 416.83)	.56	.011
	♀	6	397.87 (354.91 - 440.84)	9	382.79 (347.71 - 417.87)	.57	.026
	Total	25	385.27 (365.91 - 404.63)	24	388.01 (368.25 - 407.77)	.84	.001
Oddball	♂	37	433.59 (407.30 - 459.88)	30	437.20 (408.00 - 466.40)	.86	.001
	♀	8	371.62 (333.69 - 409.54)	18	398.89 (373.60 - 424.17)	.23	.060
	Total	45	422.57 (400.11 - 445.03)	48	422.83 (401.09 - 444.57)	.99	.000

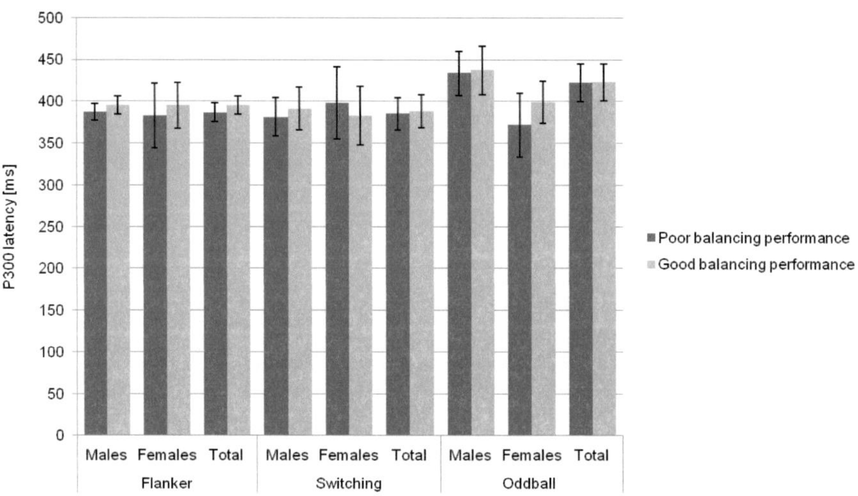

Figure 127: Mean P300 latencies during Flanker, Switching and Oddball task for balancing backwards groups; error bars represent 95% confidence intervals

N100 and P200

There was a congruency x balancing backwards interaction for N100 amplitudes during the Oddball task (F [1, 121] = 5.78, p = .02, η^2 = .046). This indicates smaller amplitudes for congruent (-1.72 µV; 95 % CI -1.98 - -1.46) compared to incongruent trials (-1.92 µV; 95 % CI -2.15 - -1.69) for participants with poor performance but larger amplitudes for congruent (-1.70 µV; 95 % CI -1.95 - -1.46) compared to incongruent trials (-1.57 µV; 95 % CI -1.79 - -1.35) for participants with good performance. For P200 latencies during the Flanker task, there was a condition x balancing backwards interaction (F [1, 93] = 5.11, p = .03, η^2 = .052). This indicates that participants with poor balancing backwards performance had a longer latency in the accuracy condition (222.47 ms; 95 % CI 213.65 - 231.29) than participants with a good performance (220.64 ms; 95 % CI 212.09 - 229.19). Participants with a good performance had a longer latency during the speed condition (228.89 ms; 95 % CI 221.16 - 236.62) compared to participants with a poor performance (221.96 ms; 95 % CI 213.98 - 229.94).

For P200 latency during the Switching task, there was a main effect observed for females (F [1, 75] =1.66, p = .20, η^2 = .022; males: F [1, 53] = .02, p = .88, η^2 = .000; females: F [1, 20] = 7.12, p = .02, η^2 = .263). Mean amplitude presented in table 137 and figure 128 and 129, latencies in table 138 and figure 130 and 131.

Table 137: Mean N100 and P200 amplitudes for balancing backwards groups in the Flanker, Switching and Oddball task

			Poor balancing performance		Good balancing performance			
			n	µV (95 % CI)	n	µV (95 % CI)	p	η^2
Flanker	N100	♂	47	-1.71 (-1.96 - -1.46)	43	-1.61 (-1.87 - -1.34)	.57	.004
		♀	11	-2.30 (-2.83 - -1.76)	22	-1.70 (-2.08 - -1.32)	.08	.097
		Total	58	-1.82 (-2.05 - -1.59)	65	-1.64 (-1.85 - -1.42)	.25	.011

	P200	♂	47	3.19 (2.78 - 3.61)	43	2.87 (2.44 - 3.31)	.30	.012
		♀	11	3.56 (2.43 - 4.68)	22	3.96 (3.17 - 4.76)	.55	.012
		Total	58	3.26 (2.85 - 3.67)	65	3.24 (2.85 - 3.63)	.95	.000
Switching	P200	♂	29	3.93 (3.38 - 4.48)	33	4.00 (3.49 - 4.52)	.85	.001
		♀	9	4.76 (3.51 - 6.00)	13	4.37 (3.33 - 5.40)	.62	.012
		Total	38	4.12 (3.62 - 4.63)	46	4.10 (3.65 - 4.56)	.96	.000
Oddball	N100	♂	40	-4.65 (-5.34 - -3.96)	32	-4.60 (-5.37 - -3.83)	.92	.000
		♀	10	-5.66 (-6.96 - -4.35)	20	-5.40 (-6.32 - -4.48)	.74	.004
		Total	50	-4.85 (-5.46 - -4.25)	52	-4.91 (-5.50 - -4.31)	.90	.000
	P200	♂	40	2.97 (2.36 - 3.58)	32	3.14 (2.46 - 3.82)	.72	.002
		♀	10	3.43 (1.89 - 4.97)	20	2.38 (1.29 - 3.47)	.27	.044
		Total	50	3.06 (2.48 - 3.64)	52	2.85 (2.28 - 3.42)	.60	.003

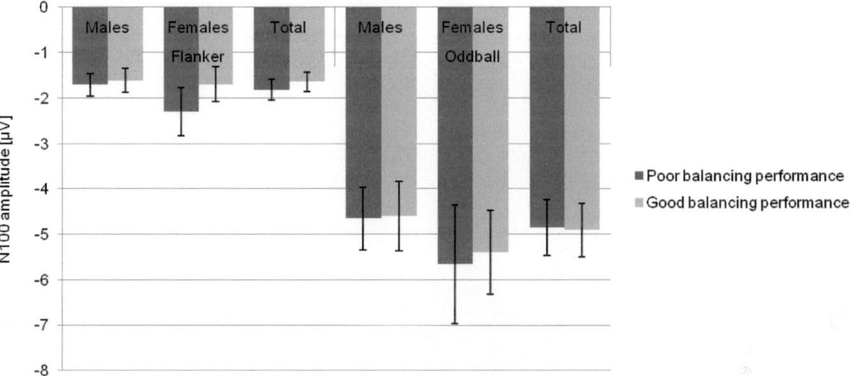

Figure 128: Mean N100 amplitudes during Flanker and Oddball task for balancing backwards groups; error bars represent 95% confidence intervals

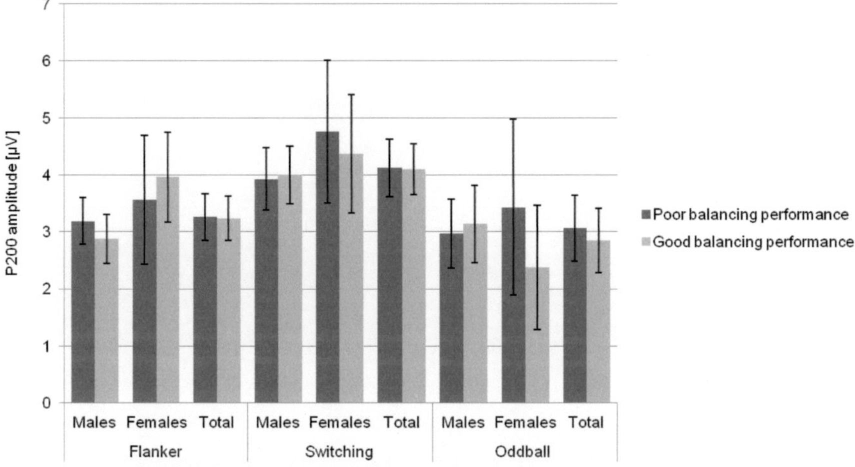

Figure 129: Mean P200 amplitudes during Flanker, Switching and Oddball task for balancing backwards groups; error bars represent 95% confidence intervals

Table 138:   Mean N100 and P200 latencies for balancing backwards groups in the Flanker, Switching and Oddball task

			Poor balancing performance		Good balancing performance			
			n	ms (95 % CI)	n	ms (95 % CI)	p	η^2
Flanker	N100	♂	14	142.84 (134.82 - 150.87)	17	137.49 (130.21 - 144.77)	.32	.034
		♀	7	137.33 (128.38 - 146.29)	7	144.31 (135.35 - 153.26)	.25	.107
		Total	21	141.01 (134.99 - 147.02)	24	139.48 (133.85 - 145.11)	.71	.003
	P200	♂	39	223.10 (214.37 - 231.83)	32	221.63 (212.00 - 231.27)	.82	.001
		♀	7	217.27 (196.21 - 238.33)	17	230.66 (217.15 - 244.18)	.28	.053
		Total	46	222.22 (214.27 - 230.16)	49	224.77 (217.07 - 232.46)	.65	.002
Switching	P200	♂	25	189.41 (181.74 - 197.09)	30	188.61 (181.60 - 195.61)	.88	.000
		♀	9	176.63 (162.93 - 190.32)	13	199.41 (188.02 - 210.81)	.02	.263
		Total	34	186.03 (179.28 - 192.78)	43	191.87 (185.87 - 197.88)	.20	.022
Oddball	N100	♂	37	169.30 (163.64 - 174.95)	28	170.40 (163.90 - 176.90)	.80	.001
		♀	9	177.17 (169.11 - 185.23)	19	176.03 (170.48 - 181.58)	.81	.002
		Total	46	170.84 (166.18 - 175.50)	47	172.67 (168.06 - 177.29)	.58	.003
	P200	♂	31	262.83 (255.62 - 270.03)	24	253.94 (245.75 - 262.13)	.11	.048
		♀	6	258.33 (240.16 - 276.51)	10	262.11 (248.03 - 276.19)	.73	.009
		Total	37	262.10 (255.51 - 268.68)	34	256.34 (249.47 - 263.21)	.23	.021

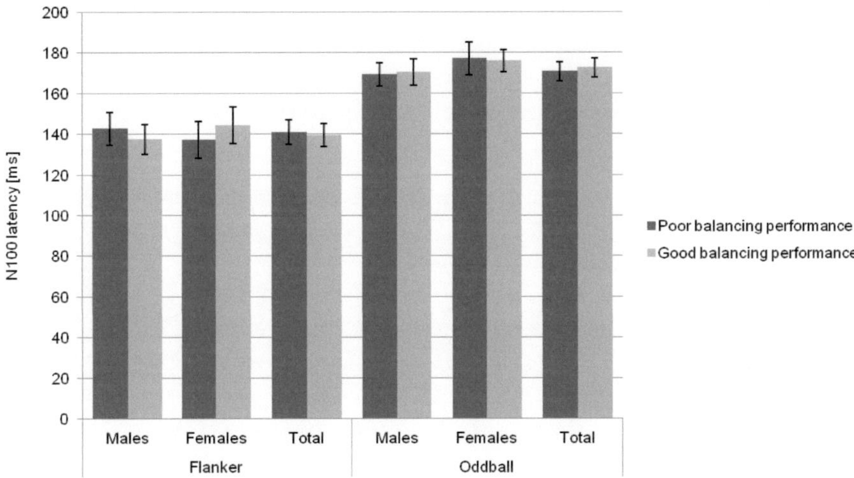

Figure 130: Mean N100 latencies during Flanker and Oddball task for balancing backwards groups; error bars represent 95% confidence intervals

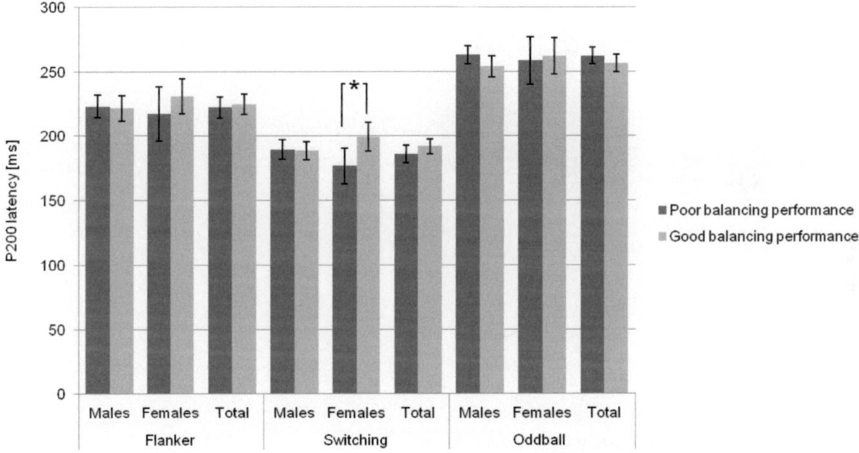

Figure 131: Mean P200 latencies during Flanker, Switching and Oddball task for balancing backwards groups; error bars represent 95% confidence intervals

4.4.3.4 Jumping Side-to-Side - Results

Behavioral Measures

For response times during the Flanker task, there was a main effect (F [1, 125] =4.47, p = .04, η^2 = .034; males: F [1, 90] = 4.38, p = .04, η^2 = .046; females: F [1, 33] = .35, p = .56, η^2 = .010) with faster times for the poor performance group (table 139 & figure 132). For response times during the Switching task, there was a condition x jumping interaction (F [1, 125] = 9.82, p = <.01, η^2 = .073). This indicates that participants with poor jumping performance had faster response times during the accuracy condition (517.33 ms; 95 % CI 502.79 - 531.88) than participants with good performance (547.52 ms; 95 % CI 532.39 - 562.66), whereas there was no difference in speed condition (poor performance: 461.20 ms; 95 % CI 452.29 - 470.10; good performance: 462.78 ms; 95 % CI 453.51 - 472.04). No effects on errors were found (table 140 & figure 133).

Table 139: Mean response times for jumping groups in the Flanker, Switching and Oddball task

		\multicolumn{2}{Poor jump performance}		\multicolumn{2}{Good jump performance}			
		n	ms (95 % CI)	n	ms (95 % CI)	p	η^2
Flanker	♂	49	483.26 (471.96 - 494.56)	43	500.66 (488.60 - 512.73)	.04	.046
	♀	19	506.57 (483.61 - 529.52)	18	515.86 (493.56 - 538.17)	.56	.010
	Total	66	489.27 (478.96 - 499.47)	61	505.15 (494.43 - 515.87)	.04	.034
Switching	♂	46	709.67 (686.74 - 732.61)	41	690.98 (666.68 - 715.27)	.27	.014
	♀	17	694.69 (643.84 - 745.55)	19	706.81 (658.71 - 754.91)	.73	.004
	Total	63	705.63 (684.26 - 727.00)	60	695.99 (674.09 - 717.89)	.53	.003
Oddball	♂	49	496.25 (470.48 - 522.02)	41	506.07 (477.89 - 534.24)	.61	.003
	♀	17	544.40 (495.03 - 593.77)	21	516.44 (472.02 - 560.86)	.40	.020
	Total	66	508.65 (485.70 - 531.60)	62	509.58 (485.90 - 533.26)	.96	.000

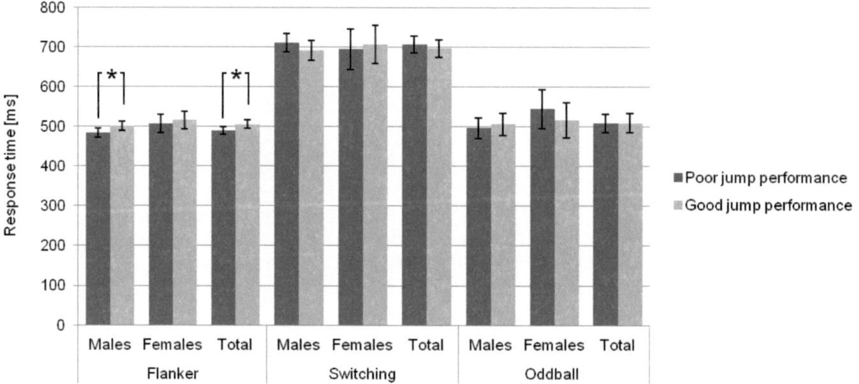

Figure 132: Mean response times during Flanker, Switching and Oddball task for jumping groups; error bars represent 95% confidence intervals

Table 140: Errors in the Switching task for jumping groups

		\multicolumn{2}{Poor jump performance}		\multicolumn{2}{Good jump performance}			
		n	No. (95 % CI)	n	No. (95 % CI)	p	η^2
Errors	♂	48	3.83 (3.21 - 4.45)	42	3.88 (3.22 - 4.55)	.91	.000
	♀	16	4.56 (3.21 - 5.92)	21	3.94 (2.76 - 5.13)	.49	.014
	Total	64	4.01 (3.44 - 4.59)	63	3.90 (3.32 - 4.48)	.79	.001

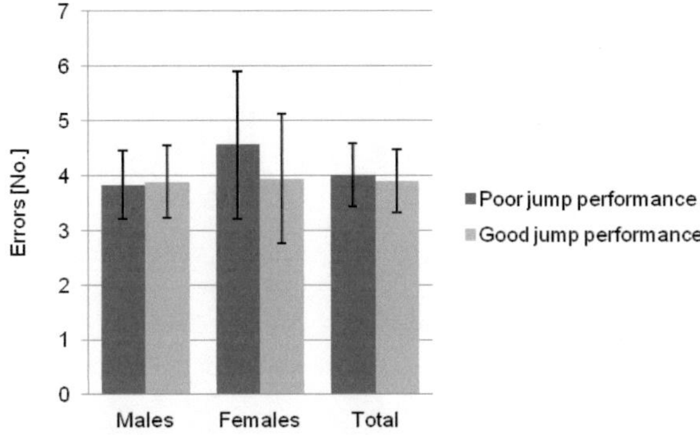

Figure 133: Errors during the Switching task for jumping groups; error bars represent 95% confidence intervals

P300

For P300 amplitudes during all three tasks, significant main effects were found: Flanker task (F [1, 118] = 3.89, p = .05, η^2 = .032; males: F [1, 86] = 4.73, p = .03, η^2 = .052; females: F [1, 30] = .20, p = .66, η^2 = .007), Switching task (F [1, 79] = 13.73, p = .00, η^2 = .148; males: F [1, 57] = 13.80, p = .00, η^2 = .195; females: F [1, 20] = 1.85, p = .19, η^2 = .085) and Oddball task (F [1, 99] = 4.28, p = .04, η^2 = .041; males: F [1, 70] = 2.36, p = .13, η^2 = .033; females: F [1, 27] = 1.51, p = .23, η^2 = .053). In all three tasks, the good jump performance group revealed larger P300 amplitudes than the poor jump performance group (table 141 & figure 134). No effects were observed for P300 latencies (table 142 & figure 135). However, there was a tendency for faster P300 latencies for the good jump performance group in all tasks.

Table 141: Mean P300 amplitudes for jumping groups in the Flanker, Switching and Oddball task

		Poor jump performance		Good jump performance			
		n	μV (95 % CI)	n	μV (95 % CI)	p	η^2
Flanker	♂	46	4.41 (4.01 - 4.80)	42	5.03 (4.62 - 5.44)	.03	.052
	♀	14	4.27 (3.46 - 5.08)	18	4.50 (3.79 - 5.22)	.66	.007
	Total	60	4.38 (4.02 - 4.73)	60	4.87 (4.52 - 5.23)	.05	.032
Switching	♂	28	3.36 (3.04 - 3.69)	31	4.19 (3.88 - 4.49)	.00	.195
	♀	10	3.90 (3.10 - 4.70)	12	4.60 (3.88 - 5.33)	.19	.085
	Total	38	3.50 (3.19 - 3.82)	43	4.30 (4.01 - 4.60)	.00	.148
Oddball	♂	38	5.33 (4.54 - 6.12)	34	6.22 (5.38 - 7.05)	.13	.033
	♀	12	5.81 (4.51 - 7.12)	17	6.84 (5.74 - 7.93)	.23	.053
	Total	50	5.45 (4.78 - 6.11)	51	6.42 (5.77 - 7.08)	.04	.041

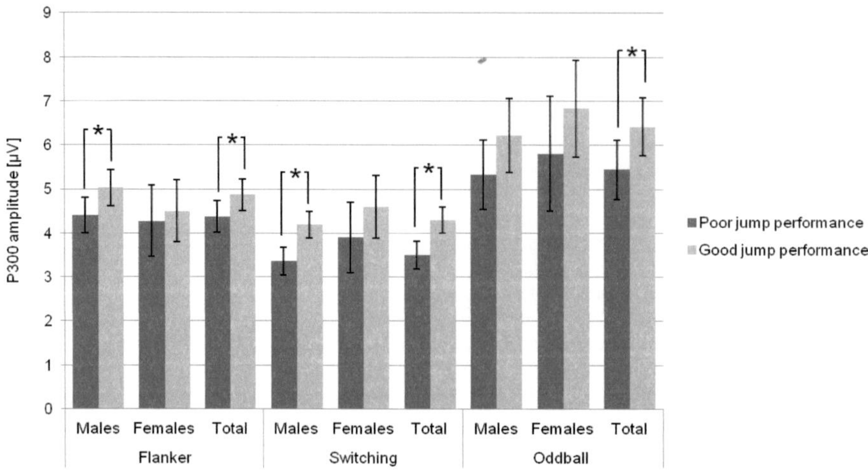

Figure 134: Mean P300 amplitudes during Flanker, Switching and Oddball task for jumping groups; error bars represent 95% confidence intervals

Table 142:   Mean P300 latencies for jumping groups in the Flanker, Switching and Oddball task

		Poor jump performance		Good jump performance			
		n	ms (95 % CI)	n	ms (95 % CI)	p	η^2
Flanker	♂	33	395.28 (384.66 - 405.90)	38	387.55 (377.66 - 397.44)	.29	.016
	♀	12	406.26 (373.53 - 438.98)	15	379.10 (349.83 - 408.38)	.22	.061
	Total	45	398.21 (386.82 - 409.59)	53	385.16 (374.67 - 395.65)	.10	.028
Switching	♂	14	389.98 (363.34 - 416.63)	20	382.60 (360.31 - 404.90)	.67	.006
	♀	7	404.49 (366.29 - 442.69)	8	375.11 (339.38 - 410.85)	.25	.102
	Total	21	394.82 (373.93 - 415.71)	28	380.46 (362.37 - 398.55)	.30	.023
Oddball	♂	37	450.37 (424.68 - 476.07)	30	416.49 (387.96 - 445.03)	.08	.046
	♀	9	385.36 (348.58 - 422.14)	17	393.21 (366.45 - 419.98)	.73	.005
	Total	46	437.65 (415.88 - 459.43)	47	408.07 (386.53 - 429.62)	.06	.039

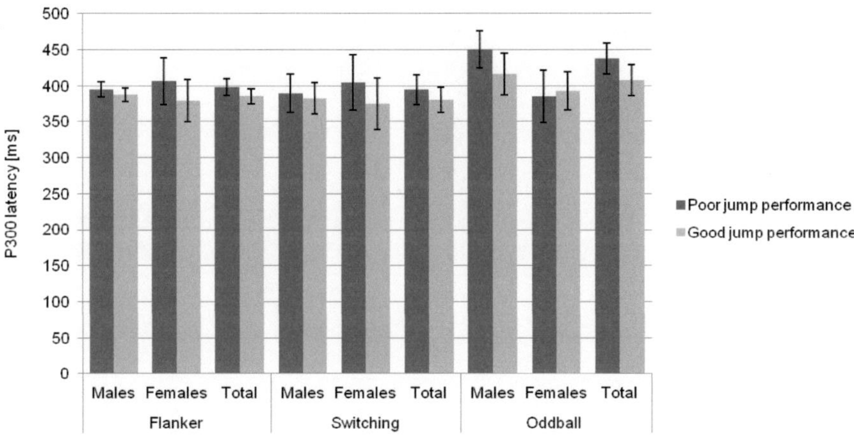

Figure 135: Mean P300 latencies during Flanker, Switching and Oddball task for jumping groups; error bars represent 95% confidence intervals

N100 and P200

For P200 amplitudes during the Switching task, there was a main effect (F [1, 82] =5.55, p = .02, η^2 = .063; males: F [1, 60] = 2.92, p = .09, η^2 = .046; females: F [1, 20] = 2.71, p = .12, η^2 = .119) with larger amplitudes for the good jump performance group (table 143 & figure 137). For P200 latency during the Oddball task, there was also a main effect found (F [1, 69] =4.36, p = .04, η^2 = .059; males: F [1, 53] = 3.84, p = .06, η^2 = .068; females: F [1, 14] = .69, p = .42, η^2 = .047) with faster latencies for the good performance group (table 144 & 139). No further effects were found.

Table 143: Mean N100 and P200 amplitudes for jumping groups in the Flanker, Switching and Oddball task

			Poor jump performance		Good jump performance			
			n	µV (95 % CI)	n	µV (95 % CI)	p	η^2
Flanker	N100	♂	47	-1.67 (-1.92 - -1.42)	43	-1.65 (-1.91 - -1.39)	.93	.000
		♀	15	-2.13 (-2.60 - -1.66)	18	-1.71 (-2.14 - -1.28)	.19	.055
		Total	62	-1.78 (-2.00 - -1.56)	61	-1.67 (-1.89 - -1.45)	.48	.004
	P200	♂	47	2.97 (2.55 - 3.39)	43	3.12 (2.68 - 3.56)	.63	.003
		♀	15	3.51 (2.55 - 4.46)	18	4.10 (3.23 - 4.97)	.36	.027
		Total	62	3.10 (2.70 - 3.50)	61	3.41 (3.01 - 3.81)	.28	.009
Switching	P200	♂	29	3.63 (3.10 - 4.17)	33	4.26 (3.76 - 4.76)	.09	.046
		♀	10	3.87 (2.76 - 4.99)	12	5.07 (4.05 - 6.09)	.12	.119
		Total	39	3.70 (3.21 - 4.18)	45	4.48 (4.03 - 4.92)	.02	.063
Oddball	N100	♂	38	-4.59 (-5.30 - -3.88)	34	-4.67 (-5.42 - -3.92)	.87	.000
		♀	13	-5.67 (-6.81 - -4.53)	17	-5.34 (-6.34 - -4.35)	.66	.007
		Total	51	-4.86 (-5.46 - -4.26)	51	-4.89 (-5.49 - -4.30)	.94	.000
	P200	♂	38	2.83 (2.21 - 3.45)	34	3.29 (2.63 - 3.95)	.31	.015
		♀	13	2.60 (1.22 - 3.98)	17	2.82 (1.62 - 4.03)	.81	.002
		Total	51	2.77 (2.20 - 3.34)	51	3.13 (2.56 - 3.71)	.38	.008

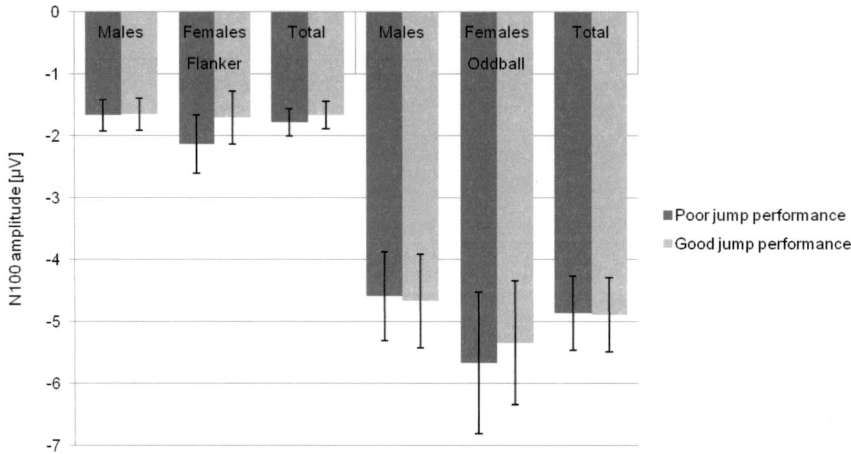

Figure 136: Mean N100 amplitudes during Flanker and Oddball task for jumping groups; error bars represent 95% confidence intervals

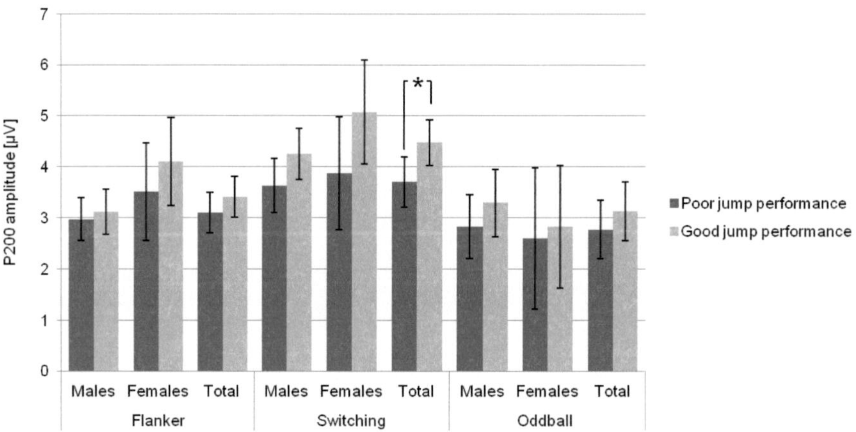

Figure 137: Mean P200 amplitudes during Flanker, Switching and Oddball task for jumping groups; error bars represent 95% confidence intervals

Table 144: Mean N100 and P200 latencies for jumping groups in the Flanker, Switching and Oddball task

			Poor jump performance		Good jump performance			
			n	ms (95 % CI)	n	ms (95 % CI)	p	η^2
Flanker	N100	♂	15	140.63 (132.75 - 148.50)	16	139.23 (131.61 - 146.86)	.80	.002
		♀	9	136.81 (129.60 - 144.02)	5	148.05 (138.37 - 157.72)	.07	.256
		Total	24	139.19 (133.57 - 144.81)	21	141.33 (135.33 - 147.34)	.60	.006
	P200	♂	37	221.98 (213.02 - 230.94)	34	222.94 (213.59 - 232.29)	.88	.000

				Poor jump performance		Good jump performance			
		♀	10	225.98 (207.87 - 244.08)	14	227.32 (212.02 - 242.62)		.91	.001
		Total	47	222.83 (214.96 - 230.70)	48	224.22 (216.43 - 232.00)		.80	.001
Switching	P200	♂	25	189.81 (182.14 - 197.47)	30	188.28 (181.28 - 195.28)		.77	.002
		♀	10	187.70 (172.64 - 202.75)	12	192.09 (178.35 - 205.83)		.66	.010
		Total	35	189.20 (182.48 - 195.93)	42	189.37 (183.23 - 195.51)		.97	.000
Oddball	N100	♂	35	167.97 (162.19 - 173.75)	30	171.88 (165.63 - 178.12)		.36	.013
		♀	12	177.93 (170.99 - 184.87)	16	175.24 (169.23 - 181.26)		.55	.014
		Total	47	170.51 (165.91 - 175.12)	46	173.05 (168.39 - 177.70)		.44	.006
	P200	♂	29	263.90 (256.53 - 271.27)	26	253.43 (245.64 - 261.21)		.06	.068
		♀	6	266.15 (248.32 - 283.97)	10	257.42 (243.62 - 271.23)		.42	.047
		Total	35	264.29 (257.65 - 270.92)	36	254.54 (247.99 - 261.08)		.04	.059

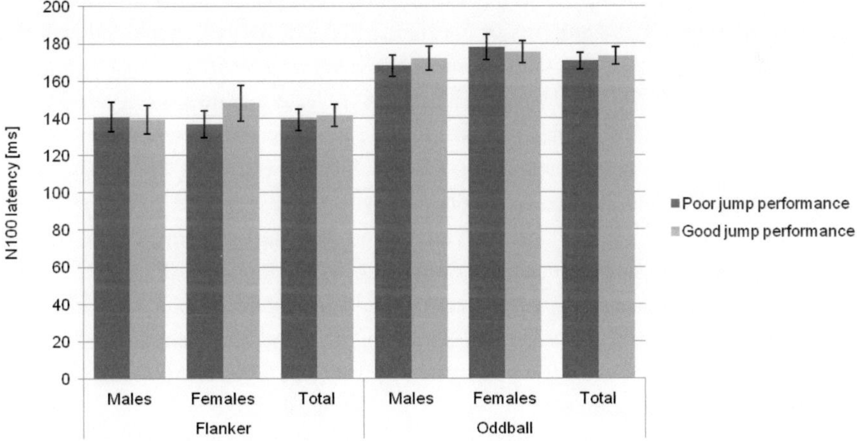

Figure 138: Mean N100 latencies during Flanker and Oddball task for jumping groups; error bars represent 95% confidence intervals

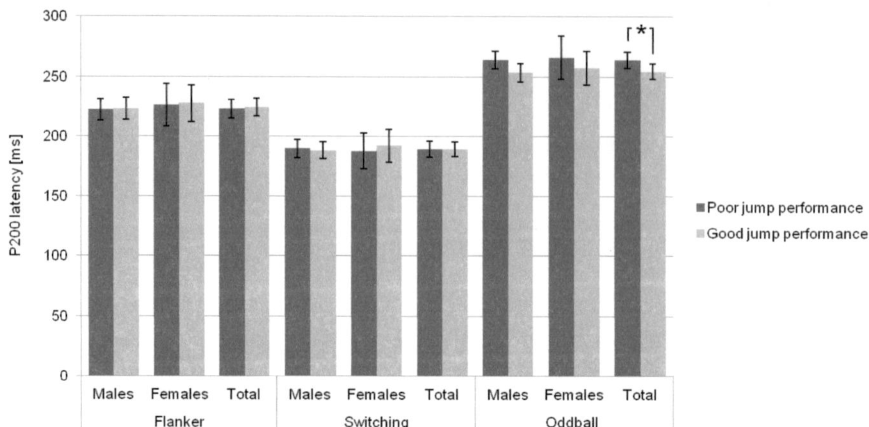

Figure 139: Mean P200 latencies during Flanker, Switching and Oddball task for jumping groups; error bars represent 95% confidence intervals

4.4.3.5 Discussion

Performance in the jumping side-to-side task was strongly related to cognitive performance as assessed by event-related potentials in young adults. The high performance group consistently revealed significantly higher P300 amplitudes during the Flanker, Switching and Oddball task. However, effects were higher for males (significant for Flanker and Switching task) compared to females with only non-significant tendencies. This might indicate a gender-specific effect of coordinative performance under time pressure on attentional resources and processes across a variety of cognitive task requiring top-down executive control, inhibition and information processing, selective attention and conflict monitoring. In accordance with the results on P300 amplitude, P300 latencies were faster by trend for participants with a good jumping performance but the differences did not become significant. For P200 amplitudes during the Switching task and for P200 latencies during the Oddball task, significantly higher amplitudes and faster latencies were observed for the good performance group. For the Flanker and Switching task, the same tendencies for P200 amplitude were observed. However, the good performance group had significantly longer response times during the Flanker task when compared to the poor performance group.

Interestingly, only few effects were found for the Posturomed and balancing backwards performance. A good static as well as dynamic Posturomed performance was related to larger P300 amplitudes during all cognitive tasks; however the tendencies were not significant. In contrast, P300 latencies tended to be longer for the good static and dynamic performance group. This tendency was significant for males in the Flanker task. These findings are contradictory since larger amplitudes indicate better attention processes but longer latencies impaired cognitive processing speed.

In addition, males with a good performance had significantly longer P200 latencies during the Flanker task. Females with a good dynamic Posturomed performance responded significantly faster during the Flanker task. On the other hand, males with a good dynamic Posturomed performance had significantly smaller N100 amplitudes in the Oddball task but tended to have shorter N100 latencies during the Flanker and the Switching task. One problem of the Posturomed testing were the large standard deviations for performances that limit the chance to detect any significant differences between groups. More effects were found for the balancing backward task, although performance in this test was limited to a maximal score of 48 points. The good performance group had larger P300 amplitudes in the homogenous condition of the Switching task but revealed smaller amplitudes in the heterogeneous condition. This indicates that during task conditions requiring less executive functioning, participants with a good performance had a higher attention. But during conditions requiring a greater involvement of executive functions, they had a smaller attention performance than the poor balancing backwards performance group. Females with a good balancing performance had a significantly longer P200 latency during the Switching task.

One might suggest that coordinative exercise and a good coordinative performance are beneficial for cognitive performance and thus the results of the current study are not surprising. However, there is only limited research investigating this relationship. In a study from 2010, Voelcker-Rehage et al. found strong relationships between motor fitness including balance and fine coordination and cognitive performance especially regarding visuo-spatial processing and action initiation. These cross-sectional data were confirmed by an intervention study in which participants conducted a 12 month coordination training and could also increase their cognitive performance during a Flanker task (Voelcker-Rehage et al., 2011). However, in contrast to the current study, the results have been observed in a sample of older adults which might have a considerably greater cognitive benefit from coordination training than young adults who are on the peak of their cognitive performance. A study assessing the effects of a 6-week coordination training on cognitive performance in young adults revealed significant improvements in coordinative performance for the intervention compared to a control group, but failed to detect any significant improvements on cognitive performance related to the increased coordinative abilities (Werner & Jansen, 2010). In general, it is speculated that coordinative training requires repeated functional brain activation and that this induces changes in the brain that might be underlying positive effects on cognitive performance. Additionally, the growth of neurons or an increased synaptogenesis could be responsible for coordination-related cognitive benefits (Voelcker-Rehage & Windisch, 2013).

4.4.4 Body Composition

In this chapter, the relationship between body composition and cognitive performance is described. The independent variables are body fat (%) and body mass index (BMI; kg/m^2). The groups for BMI (table 145) were built according to WHO guidelines (Underweight: < 18.5; Normal weight: 18.6 - 25; Overweight: > 25.1). Since there are only two participants in the underweight group, they were not considered for data analysis.

Table 145:   Number of participants per BMI group (M ± SD in kg/m^2)

	Normal weight		Overweight		Total	
	n (%)	M (SD)	n (%)	M (SD)	n	p
♂	92 (89)	22.4 (1.51)	11 (11)	26.5 (1.61)	103	.00
♀	36 (84)	21.53 (1.73)	7 (16)	26.06 (.77)	43	.00
Total	128 (88)	22.19 (1.62)	18 (12)	26.31 (1.33)	146	.00

The groups for body fat (table 146) are based on the BOD POD manufacturer's manual and were classified as follows: Males (Low BF: < 8 %; Normal BF: 8.1 % - 20 %; High BF: > 20.1 %), Females (Low BF: < 18 %; Normal BF: 18.1 % - 30 %; High BF: > 30.1 %).

Table 146:   Number of participants per Body fat group (M ± SD in %)

	Low BF		Normal BF		High BF		Total	
	n (%)	M (SD)	n (%)	M (SD)	n (%)	M (SD)	n	p
♂	6 (6)	5.65 (1.67)	82 (79)	13.99 (3.18)	16 (15)	24.70 (3.01)	104	.00
♀	2 (4)	14.45 (2.90)	24 (55)	24.31 (3.29)	18 (41)	35.04 (3.10)	44	.00
Total	8 (5)	7.85 (4.45)	106 (72)	16.33 (5.39)	34 (23)	30.17 (6.04)	148	.00

RQ 11: Is body composition related to cognitive performance in young adulthood?

4.4.4.1 BMI - Results

Behavioral Measures
No effects of BMI on behavioral measures in the Flanker, Switching, and Oddball task were observed (table 147 & 148, figure 140 & 141).

Table 147:   Mean response times for BMI groups in the Flanker, Switching and Oddball task

		Normal Weight		Overweight			
		n	ms (95 % CI)	n	ms (95 % CI)	p	η^2
Flanker	♂	80	493.67 (484.28 - 503.05)	11	487.91 (462.61 - 513.22)	.67	.002
	♀	29	512.88 (495.32 - 530.44)	6	501.84 (463.23 - 540.45)	.60	.008
	Total	109	498.78 (490.46 - 507.10)	17	492.83 (471.76 - 513.89)	.60	.002
Switching	♂	76	700.67 (682.19 - 719.15)	10	705.99 (655.04 - 756.93)	.85	.000

		n	No. (95 % CI)	n	No. (95 % CI)	p	η²
	♀	30	706.47 (668.71 - 744.23)	6	693.17 (608.73 - 777.61)	.77	.003
	Total	106	702.31 (685.61 - 719.01)	16	701.18 (658.20 - 744.16)	.96	.000
Oddball	♂	79	498.58 (477.98 - 519.17)	10	513.27 (455.38 - 571.16)	.64	.003
	♀	32	524.39 (488.27 - 560.50)	6	550.00 (466.59 - 633.40)	.57	.009
	Total	111	506.02 (488.21 - 523.83)	16	527.04 (480.13 - 573.95)	.41	.005

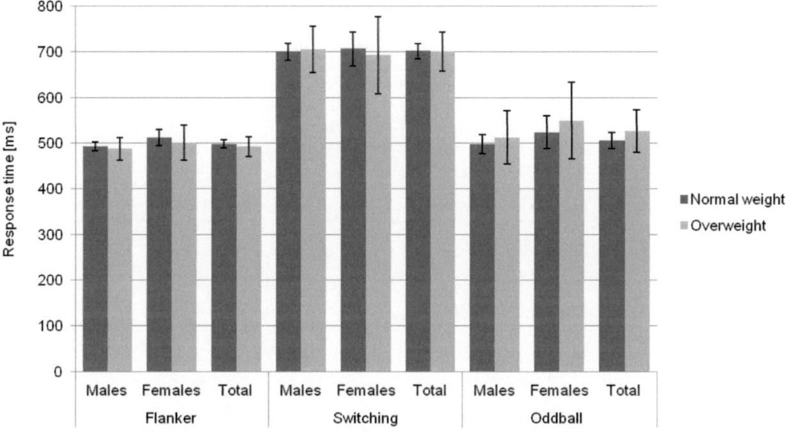

Figure 140: Mean response times during Flanker, Switching and Oddball task for BMI groups; error bars represent 95% confidence intervals

Table 148:   Errors in the Switching task for body fat groups

		Normal Weight		Overweight			
		n	No. (95 % CI)	n	No. (95 % CI)	p	η²
Errors	♂	79	3.72 (3.25 - 4.19)	10	3.88 (2.55 - 5.20)	.83	.001
	♀	31	3.90 (2.93 - 4.87)	6	5.33 (3.13 - 7.54)	.23	.040
	Total	110	3.77 (3.34 - 4.20)	16	4.42 (3.30 - 5.55)	.29	.009

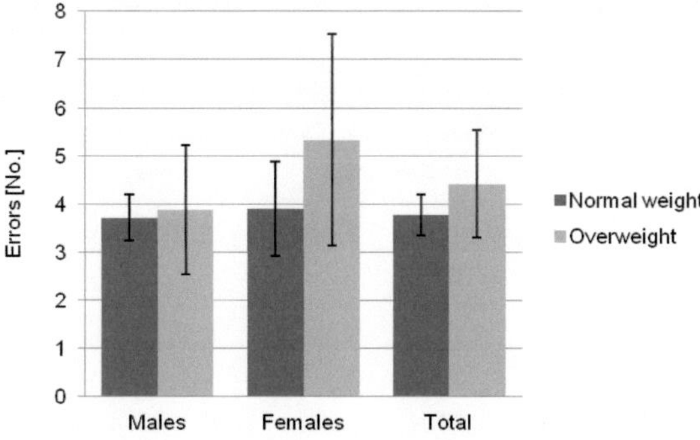

Figure 141: Errors during the Switching task for BMI groups; error bars represent 95% confidence intervals

P300

There was an effect on P300 amplitudes (table 149 & figure 142) in the Flanker task for males (F [1, 117] = 3.06, p = .08, η^2 = .025; males: F [1, 86] = 4.61, p = .04, η^2 = .051; females: F [1, 29] = .00, p = .96, η^2 = .000) with normal weighted participants having larger amplitudes than overweight participants.

Table 149:   Mean P300 amplitudes for BMI groups during Flanker, Switching and Oddball task

		Normal Weight		Overweight			
		n	µV (95 % CI)	n	µV (95 % CI)	p	η^2
Flanker	♂	77	4.84 (4.54 - 5.14)	11	3.92 (3.11 - 4.72)	.04	.051
	♀	27	4.39 (3.79 - 4.98)	4	4.43 (2.88 - 5.98)	.96	.000
	Total	104	4.72 (4.45 - 4.99)	15	4.05 (3.34 - 4.76)	.08	.025
Switching	♂	53	3.78 (3.53 - 4.04)	7	3.90 (3.19 - 4.61)	.76	.002
	♀	17	4.05 (3.44 - 4.66)	4	4.99 (3.72 - 6.25)	.18	.093
	Total	70	3.85 (3.61 - 4.09)	11	4.30 (3.68 - 4.91)	.18	.022
Oddball	♂	62	5.96 (5.36 - 6.57)	9	4.88 (3.29 - 6.48)	.21	.023
	♀	24	6.33 (5.37 - 7.30)	4	6.99 (4.63 - 9.35)	.60	.011
	Total	86	6.07 (5.56 - 6.58)	13	5.53 (4.22 - 6.84)	.45	.006

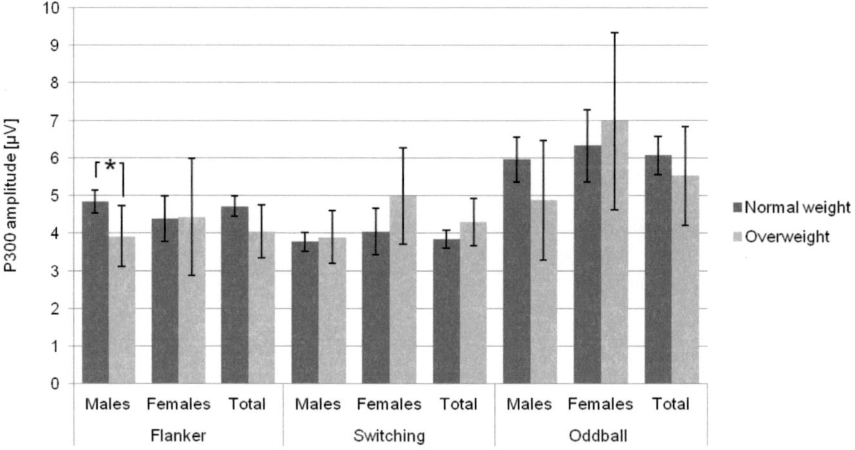

Figure 142: Mean P300 amplitudes during Flanker, Switching and Oddball task for BMI groups; error bars represent 95% confidence intervals

For P300 latencies (table 150 & figure 143), an effect in the Switching task for females was observed (F [1, 47] = .91, p = .35, η^2 = .019; males: F [1, 33] = .07, p = .79, η^2 = .002; females: F [1, 12] = 4.59, p = .05, η^2 = .276) with normal weighted participants having faster latencies than overweight participants. However, only three females were in the overweight group.

Table 150: Mean P300 latencies for BMI groups during Flanker, Switching and Oddball task

| | | Normal Weight | | Overweight | | | |
		n	ms (95 % CI)	n	ms (95 % CI)	p	η^2
Flanker	♂	64	390.56 (382.94 - 398.19)	8	404.78 (383.21 - 426.34)	.22	.021
	♀	23	389.36 (364.55 - 414.17)	3	395.70 (327.01 - 464.39)	.86	.001
	Total	87	390.24 (381.96 - 398.53)	11	402.30 (379.01 - 425.60)	.34	.010
Switching	♂	30	385.38 (367.26 - 403.51)	5	378.97 (334.57 - 423.37)	.79	.002
	♀	11	378.72 (350.68 - 406.75)	3	438.24 (384.56 - 491.92)	.05	.276
	Total	41	383.60 (368.58 - 398.61)	8	401.20 (367.20 - 435.19)	.35	.019
Oddball	♂	60	433.53 (413.05 - 454.01)	8	444.53 (388.43 - 500.63)	.71	.002
	♀	21	387.36 (364.04 - 410.69)	4	385.61 (332.18 - 439.05)	.95	.000
	Total	81	421.56 (404.87 - 438.25)	12	424.89 (381.52 - 468.26)	.89	.000

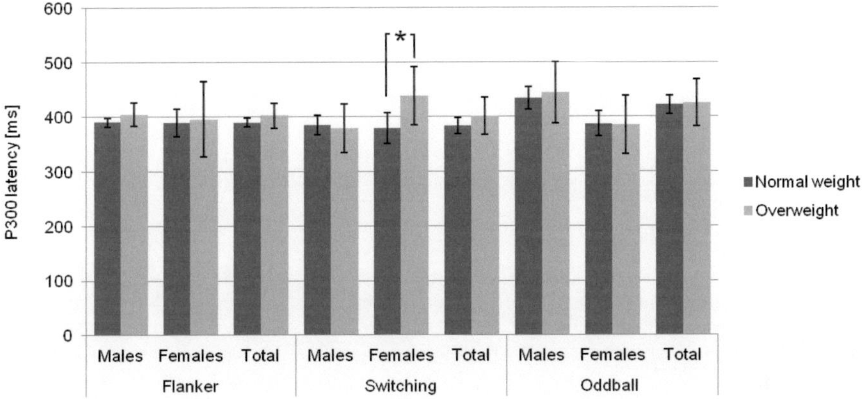

Figure 143: Mean P300 latencies during Flanker, Switching and Oddball task for BMI groups; error bars represent 95% confidence intervals

N100 and P200

No effect was observed for N100 and P200 amplitudes in the Flanker, Switching or Oddball task (table 151, figure 144 & 145).

Table 151: Mean N100 and P200 amplitudes for BMI groups in the Flanker, Switching and Oddball task

| | | | Normal Weight | | Overweight | | | |
			n	µV (95 % CI)	n	µV (95 % CI)	p	η^2
Flanker	N100	♂	78	-1.74 (-1.94 - -1.54)	11	-1.33 (-1.87 - -.79)	.15	.023
		♀	28	-1.85 (-2.21 - -1.50)	4	-2.28 (-3.23 - -1.34)	.40	.024
		Total	106	-1.77 (-1.95 - -1.60)	15	-1.58 (-2.05 - -1.20)	.45	.005
	P200	♂	78	3.03 (2.69 - 3.37)	11	2.65 (1.76 - 3.55)	.44	.007
		♀	28	3.82 (3.11 - 4.53)	4	4.27 (2.39 - 6.14)	.65	.007
		Total	106	3.24 (2.93 - 3.55)	15	3.08 (2.25 - 3.92)	.73	.001
Switching	P200	♂	54	4.09 (3.70 - 4.49)	9	3.33 (2.36 - 4.30)	.15	.034
		♀	17	4.57 (3.63 - 5.50)	4	4.44 (2.50 - 6.37)	.90	.001

			71	4.21 (3.84 - 4.58)	13	3.67 (2.81 - 4.53)	.26	.016
		Total	71	4.21 (3.84 - 4.58)	13	3.67 (2.81 - 4.53)	.26	.016
Oddball	N100	♂	62	-4.54 (-5.08 - -4.00)	9	-4.76 (-6.18 - -3.33)	.78	.001
		♀	24	-5.22 (-6.04 - -4.41)	5	-6.87 (-8.65 - -5.09)	.10	.100
		Total	86	-4.73 (-5.18 - -4.28)	14	-5.51 (-6.64 - -4.39)	.20	.016
	P200	♂	62	3.11 (2.62 - 3.60)	9	2.18 (.90 - 3.46)	.18	.026
		♀	24	2.44 (1.44 - 3.45)	5	3.80 (1.60 - 6.01)	.26	.047
		Total	86	2.92 (2.48 - 3.37)	14	2.76 (1.65 - 3.87)	.78	.001

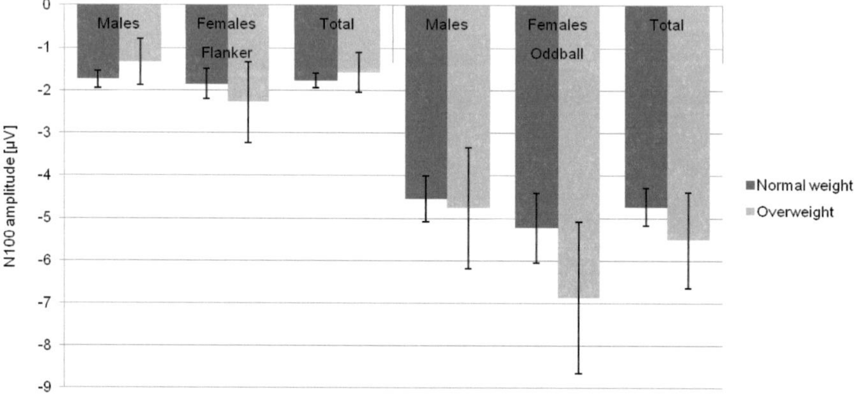

Figure 144: Mean N100 amplitudes during Flanker and Oddball task for BMI groups; error bars represent 95% confidence intervals

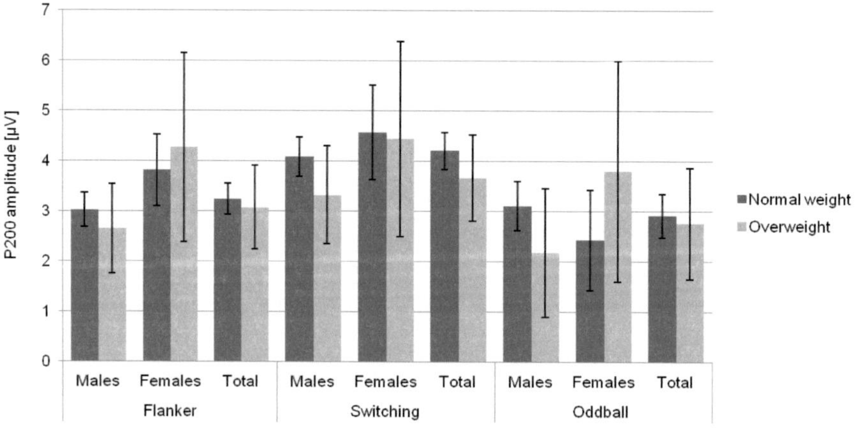

Figure 145: Mean P200 amplitudes during Flanker, Switching and Oddball task for BMI groups; error bars represent 95% confidence intervals

For the Flanker task, an effect on N100 latencies for males was observed (F [1, 44] = .66, p = .42, η^2 = .015; males: F [1, 30] = 5.87, p = .02, η^2 = .164; females: F [1, 12] = 1.03, p = .33, η^2 = .079) indicating faster latencies for normal weighted compared to overweight participants. It should be considered though that, for this analysis, only one men was in the overweight group. For P200 latencies during the Switching task, a main effect was observed (F [1, 75] = 4.00, p = .05, η^2 = .051; males: F [1, 54] = .20, p = .65, η^2 = .004; females: F [1, 19] = 6.51, p = .02, η^2 = .255). Further, a condition x BMI interaction (F [1, 75] = 4.06, p = .05, η^2 = .051) was revealed with faster latencies for homogenous compared to heterogeneous trials for normal weighted but faster latencies for heterogeneous compared to homogenous trials for overweight participants. Mean latencies are presented in table 152 and figure 146 and 147.

Table 152:   Mean N100 and P200 latencies for BMI groups in the Flanker, Switching and Oddball task

			Normal Weight		Overweight			
			n	ms (95 % CI)	n	ms (95 % CI)	p	η^2
Flanker	N 100	♂	31	138.99 (134.06 - 143.92)	1	172.07 (144.62 - 199.52)	.02	.164
		♀	12	142.04 (135.10 - 148.99)	2	133.50 (116.48 - 150.51)	.33	.079
		Total	43	139.84 (135.71 - 143.98)	3	146.35 (130.71 - 162.00)	.42	.015
	P 200	♂	61	224.13 (217.11 - 231.16)	7	212.11 (191.38 - 232.84)	.28	.018
		♀	21	228.90 (216.69 - 241.10)	3	211.78 (179.49 - 244.08)	.32	.046
		Total	82	225.35 (219.40 - 231.30)	10	212.01 (194.98 - 229.05)	.15	.023
Switching	P 200	♂	51	189.51 (184.25 - 194.78)	5	185.55 (168.72 - 202.37)	.65	.004
		♀	17	195.91 (185.63 - 206.19)	4	167.19 (146.00 - 188.38)	.02	.255
		Total	68	191.11 (186.44 - 195.79)	9	177.39 (164.54 - 190.23)	.05	.051
Oddball	N 100	♂	55	169.39 (164.70 - 174.08)	9	170.23 (158.64 - 181.82)	.89	.000
		♀	22	177.31 (172.38 - 182.24)	5	176.56 (166.22 - 186.90)	.89	.001
		Total	77	171.65 (168.01 - 175.29)	14	172.49 (163.95 - 181.03)	.86	.000
	P 200	♂	49	259.18 (253.25 - 265.12)	5	260.94 (242.36 - 279.51)	.86	.001
		♀	12	258.66 (245.39 - 271.93)	3	266.15 (239.61 - 292.68)	.60	.022
		Total	61	259.08 (253.84 - 264.33)	8	262.89 (248.41 - 277.37)	.62	.004

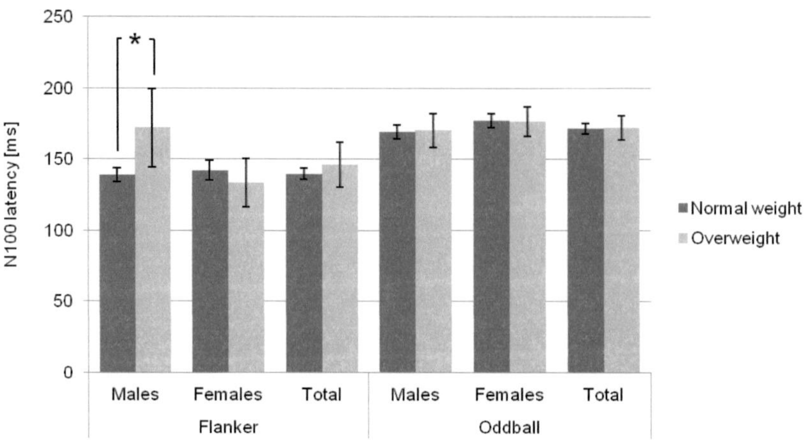

Figure 146: Mean N100 latencies during Flanker and Oddball task for BMI groups; error bars represent 95% confidence intervals

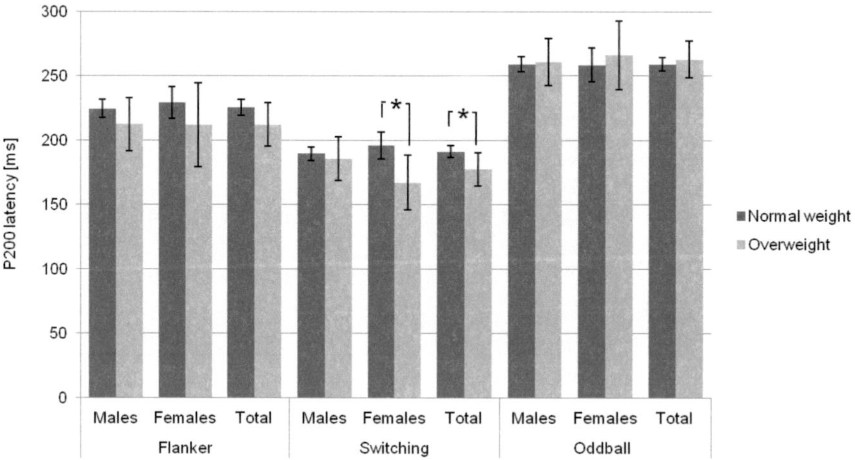

Figure 147: Mean P200 latencies during Flanker, Switching and Oddball task for BMI groups; error bars represent 95% confidence intervals

4.4.4.2 Body Fat - Results

Behavioral Measures

Analyses revealed a condition x body fat interaction for response times during the Flanker task (F [2, 125] = 6.26, p = <.01, η^2 = .091) indicating a large difference between accuracy (587.28 ms; 95 % CI 541.94 - 632.62) and speed condition (454.71 ms; 95 % CI 426.69 - 482.73) for the low body fat group compared to little differences in normal and high body fat groups. No further effects were observed for body fat (table 153 & 154, figure 148 & 149).

Table 153: Mean response times for body fat groups in the Flanker, Switching and Oddball task

			Low BF		Normal BF		High BF			
		n	ms (95 % CI)	n	ms (95 % CI)	n	ms (95 % CI)	p	η^2
Flanker	♂	5	518.79 (481.52 - 556.06)	73	489.94 (480.18 - 499.69)	14	496.69 (474.42 - 518.96	.31	.026
	♀	2	526.52 (459.52 - 593.52)	20	508.98 (487.79 - 530.16)	14	512.83 (487.50 - 538.15)	.87	.008
	Tot.	7	521.00 (488.51 - 553.48)	93	494.03 (485.12 - 502.94)	28	504.76 (488.51 - 521.00)	.19	.027
Switching	♂	4	752.71 (672.60 - 832.82)	70	697.75 (678.60 -716.90)	13	710.60 (666.16 - 755.04)	.39	.022
	♀	2	629.69 (488.48 - 770.90)	19	681.43 (635.61 - 727.24)	16	734.24 (684.31 - 784.16)	.18	.097
	Tot.	6	711.71 (641.93 -781.48)	89	694.27 (676.15-712.38)	29	723.64 (691.90-755.38)	.28	.021
Oddball	♂	4	568.75 (478.09- 459.40)	72	500.12 (478.76- 521.49)	14	486.43 (437.98- 534.89)	.28	.029
	♀	2	462.37 (318.77- 605.98)	22	532.08 (488.78- 575.38)	15	534.89 (482.46- 587.33)	.62	.026
	Tot.	6	533.29 (456.64-609.94)	94	507.60 (488.24-526.97)	29	511.50 (476.63-546.37)	.81	.003

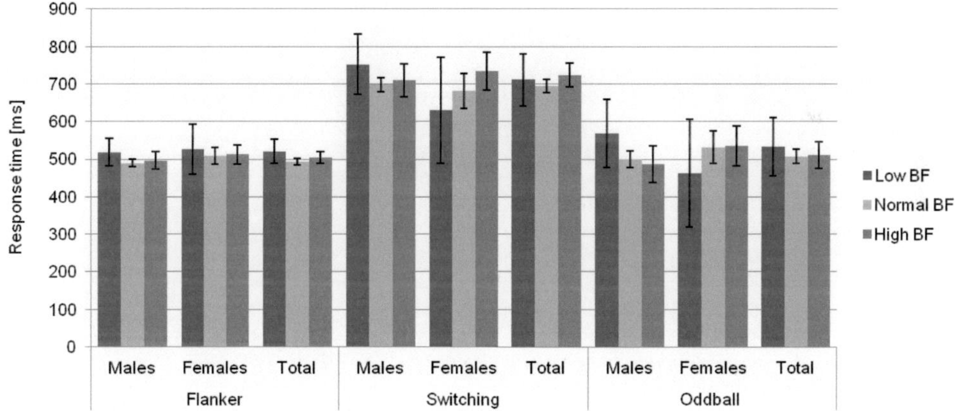

Figure 148: Mean response times during Flanker, Switching and Oddball task for body fat groups; error bars represent 95% confidence intervals

Table 154: Errors in the Switching task for body fat groups

			Low BF		Normal BF		High BF			
		n	No. (95 % CI)	n	No. (95 % CI)	n	No. (95 % CI)	p	η^2
Errors	♂	4	3.44 (1.35-5.53)	73	3.70 (3.21- 4.19)	13	4.10 (2.94-5.25)	.79	.005
	♀	2	5.13 (1.24-9.00)	21	4.20 (3.00- 5.40)	15	3.95 (2.53- 5.37)	.84	.010
	Total	6	4.00 (2.16-5.84)	94	3.81 (3.35- 4.28)	28	4.02 (3.17-4.87)	.91	.002

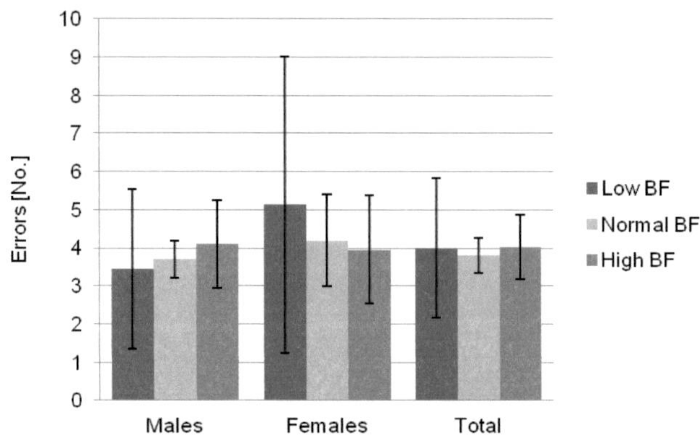

Figure 149: Errors during the Switching task for body fat groups; error bars represent 95% confidence intervals

P300

Body fat (%) does not show to have any effect on P300 amplitudes (table 155 & figure 150) and latencies (table 156 & figure 151) in the Flanker, Switching and Oddball task.

Table 155:   Mean P300 amplitudes for body fat groups during Flanker, Switching and Oddball task

		Low BF		Normal BF		High BF			
		n	µV (95 % CI)	n	µV (95 % CI)	n	µV (95 % CI)	p	η^2
Flan- ker	♂	4	5.90 (4.57- 7.23)	71	4.73 (4.41- 5.05)	14	4.33 (3.62- 5.04)	.12	.048
	♀	2	5.06 (2.89- 7.23)	20	4.26 (3.58- 4.95)	10	4.54 (3.57- 5.51)	.73	.021
	Tot.	6	5.62 (4.51-6.73)	91	4.63 (4.34-4.91)	24	4.42 (3.86-4.97)	.16	.030
Swit- ching	♂	3	3.45 (2.36- 4.53)	48	3.81 (3.53- 4.08)	10	3.88 (3.28-4.47)	.78	.008
	♀	2	4.45 (2.66- 6.24)	10	3.82 (3.02- 4.62)	10	4.71 (3.91- 5.51)	.27	.127
	Tot.	5	3.85 (2.94-4.76)	58	3.81 (3.54-4.08)	20	4.29 (3.84-4.75)	.19	.041
Odd- ball	♂	4	7.80 (5.40-10.20)	56	5.73 (5.09- 6.37)	12	5.34 (3.95- 6.73)	.21	.045
	♀	2	5.58 (2.25 - 8.91)	19	6.59 (5.50 - 7.67)	8	6.21 (4.55 - 7.88)	.81	.016
	Tot.	6	7.06 (5.13-8.99)	75	5.95 (5.40-6.49)	20	5.69 (4.63-6.75)	.47	.015

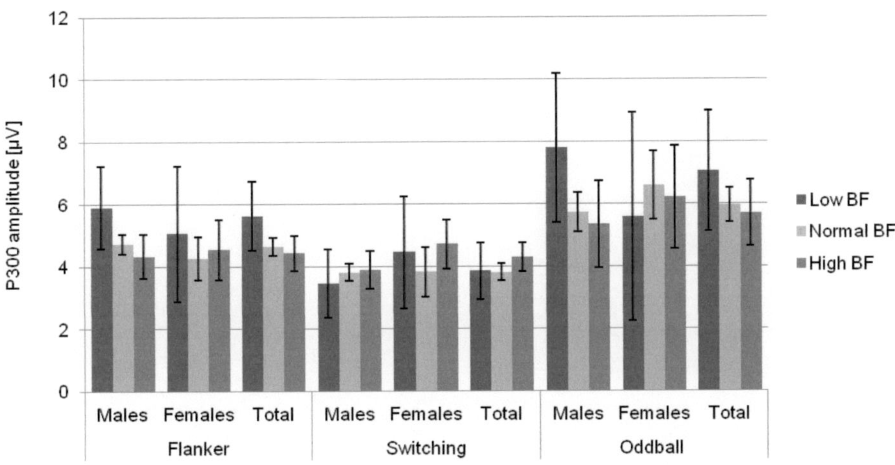

Figure 150: Mean P300 amplitudes during Flanker, Switching and Oddball task for body fat groups; error bars represent 95% confidence intervals

Table 156:   Mean P300 latencies for body fat groups during Flanker, Switching and Oddball task

		\multicolumn{2}{c}{Low BF}		Normal BF		High BF			
		n	ms (95 % CI)	n	ms (95 % CI)	n	ms (95 % CI)	p	η^2
Flan-ker	♂	4	367.87 (337.78- 397.96)	58	391.36 (383.46- 399.26)	11	403.01 (384.87- 421.16)	.14	.055
	♀	2	394.56 (310.10- 479.03)	17	393.10 (364.13- 422.07)	8	386.22 (343.99- 428.45)	.96	.003
	Tot.	6	376.77 (345.29-408.25)	75	391.75 (382.85-400.66)	19	395.94 (378.25-413.63)	.58	.011
Swit-ching	♂	1	343.95 (248.85- 439.04)	28	378.87 (360.90- 396.84)	6	417.35 (378.53- 456.17)	.14	.114
	♀	1	337.76 (233.84- 441.68)	7	381.85 (342.57- 421.12)	7	403.10 (363.82- 442.38)	.41	.138
	Tot.	2	340.85 (275.97-405.73)	35	379.46 (363.96-394.97)	13	409.68 (384.23-435.12)	.06	.116
Odd-ball	♂	4	410.35 (330.60- 490.10)	53	435.05 (413.14- 456.96)	11	442.64 (394.55- 490.73)	.79	.007
	♀	2	445.83 (370.02- 521.65)	17	388.93 (362.92- 414.93)	7	378.50 (337.97-419.02)	.28	.104
	Tot.	6	422.18 (360.70-483.66)	70	423.85 (405.85-441.85)	18	417.69 (382.20-453.19)	.95	.001

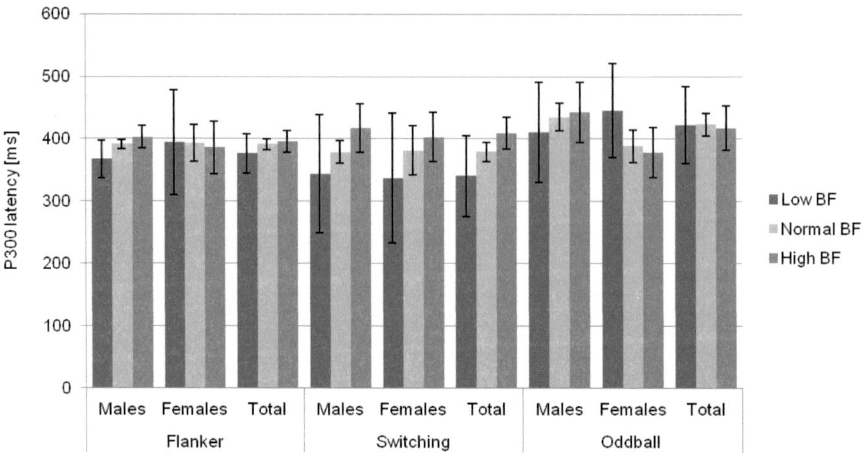

Figure 151: Mean P300 latencies during Flanker, Switching and Oddball task for body fat groups; error bars represent 95% confidence intervals

N100 and P200

There was a significant condition x body fat interaction for N100 amplitudes during the Flanker task (F [2, 120] = 3.67, p = .03, η^2 = .058) indicating larger amplitudes for normal body fat in the speed (-1.84 µV; 95 % CI -2.05 - -1.64) compared to accuracy condition (-1.69 µV; 95 % CI -1.90 - -1.49) but larger amplitudes in the accuracy compared to speed condition for low and high body fat groups. For P200 amplitude during the Flanker task, a main effect for females was observed (F [2, 120] = 2.09, p = .13, η^2 = .034; males: F [2, 87] = .27, p = .77, η^2 = .006; females: F [2, 30] = 6.55, p = <.01, η^2 = .304). Post-hoc analyses detect a significant difference between normal and high body fat groups (p = <.01). Another effect for females was observed for P200 amplitudes during the Switching task (F [2, 83] = .76, p = .47, η^2 = .018; males: F [2, 61] = .61, p = .54, η^2 = .020; females: F [2, 19] = 4.00, p = .04, η^2 = .296). Post-hoc analyses detect a significant difference between normal and high body fat groups (p = .04). Mean amplitudes are given in table 157 and figure 152 and 153.

Table 157: Mean N100 and P200 amplitudes for body fat groups in the Flanker, Switching and Oddball task

				Low BF		Normal BF		High BF			
			n	µV (95 % CI)	n	µV (95 % CI)	n	µV (95 % CI)	p	η^2
Flanker	N 100	♂	5	-1.53 (-2.31- -.74)	71	-1.77 (-1.98- -1.56)	14	-1.33 (-1.78- -.86)	.22	034
		♀	2	-1.40 (-2.70- -.11)	20	-1.75 (-2.16- -1.35)	11	-2.26 (-2.81- -1.70)	.25	.087
		Tot.	7	-1.49 (-2.17- -.81)	91	-1.77 (-1.96- -1.58)	25	-1.74 (-2.09- -1.38)	.74	.005
	P 200	♂	5	2.84 (1.51- 4.17)	71	3.04 (2.69- 3.39)	14	2.74 (1.94- 3.53)	.77	.006
		♀	2	4.82	20	3.04	11	5.08	<.01	.304

				Low BF		Normal BF		High BF			
				(2,58- 7.07)		(2.33- 3.75)		(4.12- 6.04)			
		Tot.	7	3.40	91	3.04	25	3.77	.13	.034
				(2.21-4.60)		(2.71-3.37)		(3.14-4.40)			
Switching	P 200	♂	3	4.28	49	4.06	12	3.56	.54	.020
				(2.58- 5.97)		(3.64- 4.48)		(2.72- 4.40)			
		♀	2	3.83	10	3.64	10	5.55	.04	.296
				(1.53- 6.12)		(2.62- 4.67)		(4.52- 6.58)			
		Tot.	5	4.10	59	3.99	22	4.47	.47	.018
				(2.72-5.48)		(3.59-4.39)		(3.81-5.12)			
Oddball	N 100	♂	4	-3.10	56	-4.72	12	-4.71	.36	.029
				(-5.28- -.93)		(-5.30- -4.14)		(-5.97- -3.46)			
		♀	2	-6.91	19	-5.06	9	-6.06	.27	.093
				(-9.74- -4.08)		(-5.98- -4.14)		(-7.40- -4.73)			
		Tot.	6	-4.37	75	-4.81	21	-5.29	.56	.012
				(-6.12- -2.63)		(-5.30- -4.31)		(-6.22- -4.36)			
	P 200	♂	4	2.37	56	3.09	12	2.85	.74	.009
				(.42- 4.32)		(2.57- 3.61)		(1.73- 3.98)			
		♀	2	2.18	19	2.48	9	3.36	.64	.032
				(-1.36- 5.71)		(1.34- 3.63)		(1.70- 5.03)			
		Tot.	6	2.31	75	2.94	21	3.07	.73	.006
				(.62-3.99)		(2.46-3.42)		(2.17-3.97)			

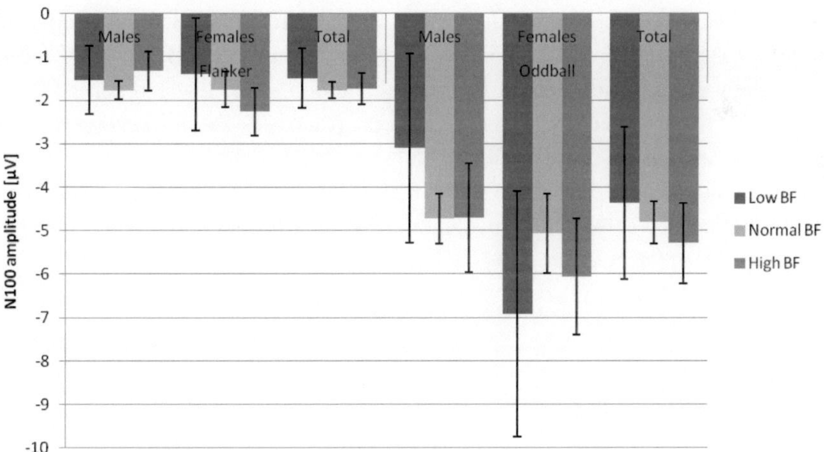

Figure 152: Mean N100 amplitudes during Flanker and Oddball task for body fat groups; error bars represent 95% confidence intervals

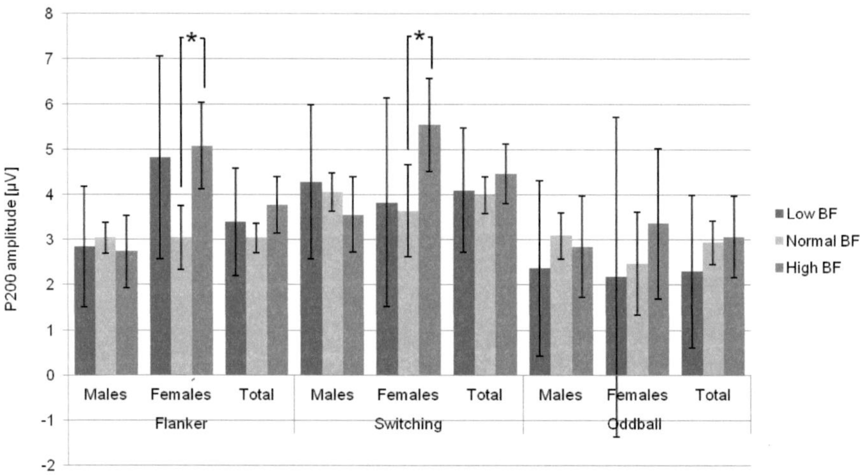

Figure 153: Mean P200 amplitudes during Flanker, Switching and Oddball task for body fat groups; error bars represent 95% confidence intervals

For P200 latency during the Flanker task, there was a congruency x body fat interaction (F [2, 90] = 6.79, p = <.01, η^2 = .131) indicating differences in latency between congruent and incongruent trials for low and high body fat groups but no differences for the normal body fat group. For P200 latency during the Switching task, there was a condition x body fat interaction (F [2, 76] = 5.79, p = <.01, η^2 = .132) indicating longer latencies for the speed condition in the low and normal fat groups but shorter latencies in the speed condition for the high fat group. For P200 latency during the Oddball task, a main effect was observed (F [2, 68] = 3.50, p = .04, η^2 = .093; males: F [2, 52] = 2.99, p = .06, η^2 = .103; females: F [2, 13] = .52, p = .61, η^2 = .074). Post-hoc analyses revealed a significant difference in P200 latency between the low and normal fat group (p = .03) with faster latencies for the low fat group. Mean latencies are presented in table 158 and figure 154 and 155.

Table 158: Mean N100 and P200 latencies for body fat groups in the Flanker, Switching and Oddball task

			Low BF		Normal BF		High BF			
			n	ms (95 % CI)	n	ms (95 % CI)	n	ms (95 % CI)	p	η^2
Flanker	N 100	♂	1	115.43 (86.92- 143.94)	30	140.33 (135.13- 145.54)	1	155.47 (126.96- 183.98)	.13	.131
		♀	-	-	8	141.06 (132.20- 149.93)	6	140.50 (130.26- 150.73)	.93	.001
		Tot.	1	115.43 (88.94- 141.92)	38	140.49 (136.19- 144.78)	7	142.63 (132.62- 152.65)	.16	.081
	P 200	♂	4	229.20 (201.69- 256.71)	57	223.57 (216.28- 230.86)	8	211.99 (192.53- 231.44)	.48	.022
		♀	2	233.11 (193.01- 273.21)	11	233.15 (216.05- 250.25)	11	219.21 (202.11- 236.31)	.47	.069
		Tot.	6	230.50 (208.40- 252.61)	68	225.12 (218.55-231.69)	19	216.17 (203.75- 228.59)	.37	.022

Task	Cond.	Sex	N	Low BF Mean (CI)	N	Normal BF Mean (CI)	N	High BF Mean (CI)	p	η²
Swit-ching	P 200	♂	3	198.76 (176.98- 220.55)	45	188.30 (182.68-193.93)	9	188.13 (175.55- 200.71)	.65	.016
		♀	2	208.20 (175.10- 241.31)	10	192.09 (177.29- 206.90)	10	184.47 (169.67- 199.28)	.38	.096
		Tot.	5	202.54 (185.16-219.91)	55	188.99 (183.75-194.23)	19	186.21 (177.29- 195.12)	.25	.035
Oddball	N 100	♂	3	169.79 (149.74- 189.85)	52	170.04 (165.23- 174.86)	10	167.19 (156.20- 179.17)	.89	.004
		♀	2	175.00 (157.52- 192.48)	18	176.74 (170.91- 182.56)	8	175.98 (167.24- 184.72)	.98	.002
		Tot.	5	171.88 (157.58- 186.17)	70	171.76 (167.94- 175.58)	18	171.09 (163.56-178.63)	.99	.000
	P 200	♂	3	232.29 (209.56- 255.03)	45	260.76 (254.89- 266.64)	7	260.94 (246.05- 275.82)	.06	.103
		♀	1	241.41 (196.41- 286.40)	10	263.28 (249.05-277.51)	10	259.38 (239.25- 279.50)	.61	.074
		Tot.	4	234.57 (215.14-254.00)	55	261.22 (255.98-266.46)	12	260.29 (249.07- 271.51)	.04	.093

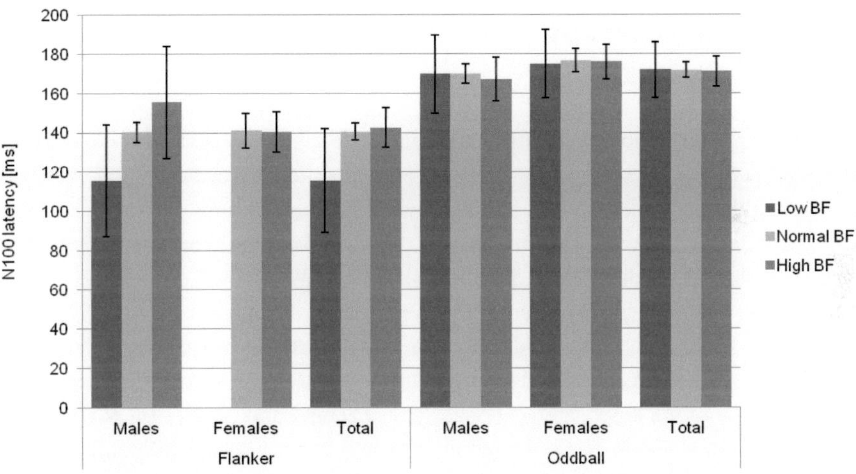

Figure 154: Mean N100 latencies during Flanker and Oddball task for body fat groups; error bars represent 95% confidence intervals

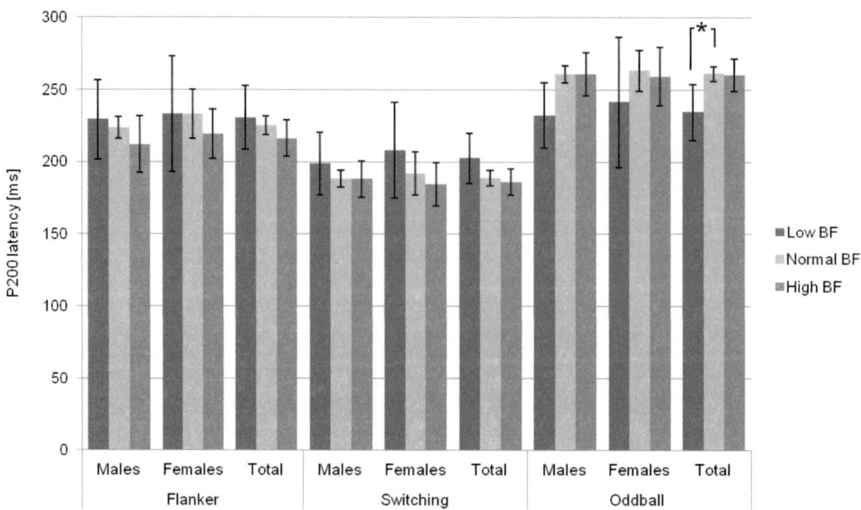

Figure 155: Mean P200 latencies during Flanker, Switching and Oddball task for body fat groups; error bars represent 95% confidence intervals

4.4.4.3 Discussion

Body composition is an important health-related factor that is hypothesized to modulate or affect the relationship between physical activity, fitness and cognition. Overweight and obesity have developed to a worldwide epidemic with still increasing prevalence rates and rank among the most important risk factors for several physiological (e.g. Kopelman, 2007) and, to a lesser extent, also psychological problems such as body dissatisfaction (e.g. Wardle & Cooke, 2005). Furthermore, recent studies have also shown that persons with a poor body composition have an impaired cognitive performance. In 2011, a study could provide evidence that increased body fat percentage in healthy older adults is negatively related to cognition as well as brain structure, which was indicated by a reduced volume of the hippocampus (Isaac, Sim, Zheng, Zagorodnov, Tai & Chee, 2011). The hippocampus is a brain region in the temporal lobe that has a relevant influence on memory. In children, there is also a negative correlation between body fat percentage and cognitive performance whereas physical fitness is positively correlated with cognition in this age group (Davis & Cooper, 2011; Donnelly & Lambourne, 2011). In young adults, there is a lack of studies assessing the relationship between body composition and cognitive performance. Furthermore, it is not yet determined how physical activity, fitness, and body composition are associated. One can basically assume that physical activity, besides nutritional factors and genetic disposition, has a positive and protective influence on the development of overweight. However, the extent of this influence is still unknown and there are contradictory results in literature. A study by Rowlands, Ingledew and Eston (2000) found evidence for a relationship between physical activity and body composition whereas Ortega, Tresaco, Ruiz, Moreno, Martin-Matillas et al. (2007) found no association. Further research of these interdependencies may shed light on the question of whether body composition is an independent influence factor on cognition or whether this relationship is affected by other parameters such as physical activity and fitness.

In the current study, no main effects of body composition and BMI were observed for behavioral measures. The high BMI group had a tendency for more errors in the Switching task. Males with a normal BMI had significantly larger P300 amplitudes during the Flanker task than males with a high BMI. Faster P300 latencies were further observed for females with a normal BMI in the Switching task, but the high BMI group consisted of only three people and thus, this result should not be interpreted. The same problem occurs for N100 latencies in males. Normal weighted participants also had faster latencies but the group sizes were very imbalanced with only one man in the overweight group.

Regarding P200 latencies during the Switching task, overweight compared to normal weighted participants had an improved processing speed as indicated by faster latencies. In contrast to this, participants with a low body fat percentage had significantly lower P200 latencies during the Oddball task.

Females with a high amount of body fat further had significantly larger P200 amplitudes during the Flanker and Switching task whereas no effects were observed for males. This might indicate that women profit from a higher body fat percentage with regard to their attention performance. These findings suggest that a higher body fat percentage might be beneficial for cognitive performance in the current sample, especially among females. This is unexpected and in contrast to findings in children. However, it could be explained by the fact that young females compared to children and men generally have a higher body fat amount due to hormonal influences and that a certain amount of body fat is essential for a good health which also comprises cognition.

4.4.5 Overall Fitness

In this chapter, the relationship between overall fitness and cognitive performance is described. The independent variable is an index built from the endurance, strength, coordination and body composition groups used in the chapters 4.4.1 - 4.4.4. Poor performance groups were coded with 0, good performance groups were coded with 1. The number of participants per group is given in table 159.

VO_{2max}: Poor (0), Good (1) Performance at IAT: Poor (0), Good (1)	Endurance (Mean: 0-1)	
Isometric strength: Poor (0), Good (1) Dynamic strength: Poor (0), Good (1)	Strength (Mean: 0-1)	Overall Fitness (Sum: 0-4)
Static Posturomed: Poor (0), Good (1) Dynamic Posturomed: Poor (0), Good (1) Balancing backwards: Poor (0), Good (1) Jumping side to side: Poor (0), Good (1)	Coordination (Mean: 0-1)	Low: ≤ 2 Medium: 2.25 - 2.75 High: ≥ 3
BMI: Under-/Overweight (0), Normal weight (1) Body fat: Low/High (0), Normal (1)	Body comp. (Mean: 0-1)	

Table 159:   Number of participants per BMI group

	Low n (%)	Medium n (%)	High n (%)	Total n
♂	25 (33)	23 (30)	28 (37)	76
♀	16 (46)	6 (17)	13 (37)	35
Total	41 (37)	29 (26)	41 (37)	111

RQ 12: Is overall fitness related to cognitive performance in young adulthood?

4.4.5.1 Results

Behavioral Measures

No effects were observed for behavioral measures (table 160 & 161, figure 156 & 157).

Table 160:   Mean response times for overall fitness groups in the Flanker, Switching and Oddball task

		Low fitness n	Low fitness ms (95 % CI)	Medium fitness n	Medium fitness ms (95 % CI)	High fitness n	High fitness ms (95 % CI)	p	η^2
Flan-ker	♂	23	496.20 (479.78 - 512.63)	21	494.26 (477.07 - 511.45)	24	481.30 (465.22 - 497.38)	.38	.030
	♀	12	516.32 (486.90 - 545.74)	6	512.69 (471.08 - 554.29)	11	509.04 (478.31 - 539.77)	.94	.005
	Tot.	35	503.10 (488.60 - 517.61)	27	498.36 (481.84 - 514.87)	35	490.02 (475.51 - 504.52)	.44	.017
Swit-ching	♂	22	704.30 (672.00 - 736.60)	20	691.17 (657.29 - 725.04)	23	697.13 (665.54 - 728.72)	.85	.005
	♀	12	703.31 (641.48 - 765.14)	6	709.26 (621.82 - 796.70)	11	708.17 (643.59 - 772.75)	.99	.001
	Tot.	34	703.95 (675.36 - 732.54)	26	695.34 (662.65 - 728.04)	34	700.70 (672.11 - 729.29)	.93	.002
Odd-ball	♂	23	495.50 (457.19 - 533.80)	20	517.07 (475.99 - 558.14)	23	502.12 (463.82 - 540.42)	.74	.010
	♀	13	541.29 (481.78 - 600.80)	6	500.01 (412.41 - 587.61)	13	550.12 (490.61 - 609.64)	.62	.032
	Tot.	36	512.03 (480.02 - 544.05)	26	513.13 (475.46 - 550.80)	36	519.46 (487.44 - 551.47)	.94	.001

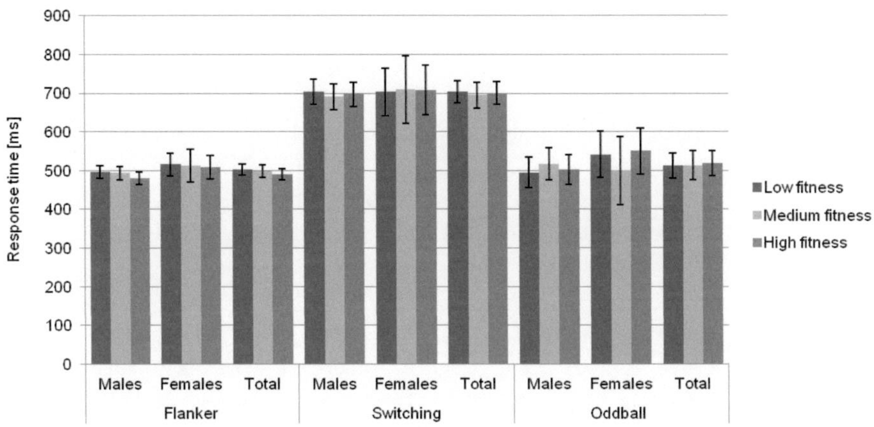

Figure 156: Mean response times during Flanker, Switching and Oddball task for overall fitness groups; error bars represent 95% confidence intervals

Table 161: Errors in the Switching task for overall fitness groups

		\multicolumn{2}{c}{Low fitness}		\multicolumn{2}{c}{Medium fitness}		\multicolumn{2}{c}{High fitness}			
		n	No. (95 % CI)	n	No. (95 % CI)	n	No. (95 % CI)	p	η^2
Errors	♂	22	4.14 (3.19 - 5.08)	21	3.54 (2.57 - 4.50)	23	4.07 (3.14 - 4.99)	.63	.015
	♀	13	3.90 (2.49 - 5.32)	6	5.54 (3.46 - 7.62)	12	3.46 (1.99 - 4.93)	.25	.094
	Tot.	35	4.05 (3.27 - 4.83)	27	3.98 (3.10 - 4.87)	35	3.86 (3.08 - 4.64)	.94	.001

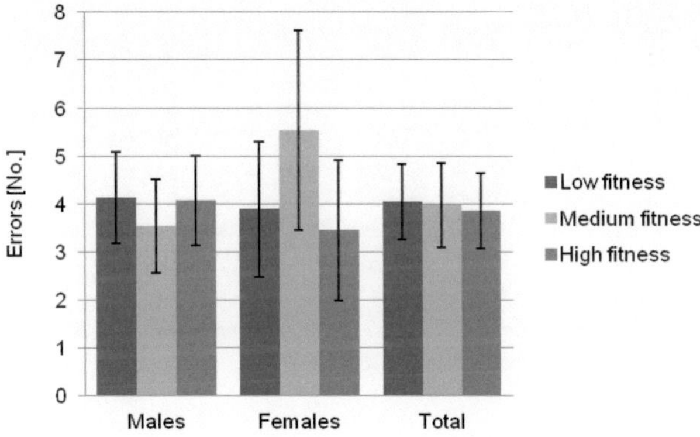

Figure 157: Errors during the Switching task for overall fitness groups; error bars represent 95% confidence intervals

P300

For P300 latencies (table 163 & figure 159), a main effect was observed for males during the Switching task (F [2, 35] = 2.33, p = .11, η^2 = .117; males: F [2, 23] = 3.95, p = .03, η^2 = .256; females: F [2, 9] = .32, p = .74, η^2 = .066). Post-hoc analyses detect a significant difference between low and medium fitness groups (p = .05). No effects were found for P300 amplitudes (table 162 & figure 160).

Table 162:   Mean P300 amplitudes for overall fitness groups during Flanker, Switching and Oddball task

		Low fitness		Medium fitness		High fitness			
		n	μV (95 % CI)	n	μV (95 % CI)	n	μV (95 % CI)	p	η^2
Flan-ker	♂	23	4.46 (3.90 - 5.02)	20	5.29 (4.69 - 5.89)	23	4.50 (3.94 - 5.06)	.09	.075
	♀	9	4.79 (3.67 - 5.91)	6	4.52 (3.15 - 5.89)	11	4.25 (3.24 - 5.26)	.76	.023
	Tot.	32	4.55 (4.05 - 5.05)	26	5.11 (4.56 - 5.67)	34	4.42 (3.94 - 4.90)	.15	.041
Swit-ching	♂	16	3.7 (3.22 - 4.20)	13	3.95 (3.41 - 4.50)	17	3.78 (3.31 - 4.25)	.79	.011
	♀	8	4.76 (3.79 - 5.72)	6	4.38 (3.27 - 5.50)	4	3.55 (2.19 - 4.92)	.34	.136
	Tot.	24	4.06 (3.61 - 4.50)	19	4.09 (3.59 - 4.59)	21	3.74 (3.26 - 4.21)	.51	.022
Odd-ball	♂	19	5.09 (3.97 - 6.22)	18	6.41 (5.26 - 7.57)	16	5.38 (4.15 - 6.61)	.24	.055
	♀	8	6.09 (4.36 - 7.81)	5	5.12 (2.94 - 7.30)	12	6.90 (5.49 - 8.31)	.37	.087
	Tot.	27	5.39 (4.45 - 6.33)	23	6.13 (5.12 - 7.15)	28	6.03 (5.11 - 6.95)	.49	.019

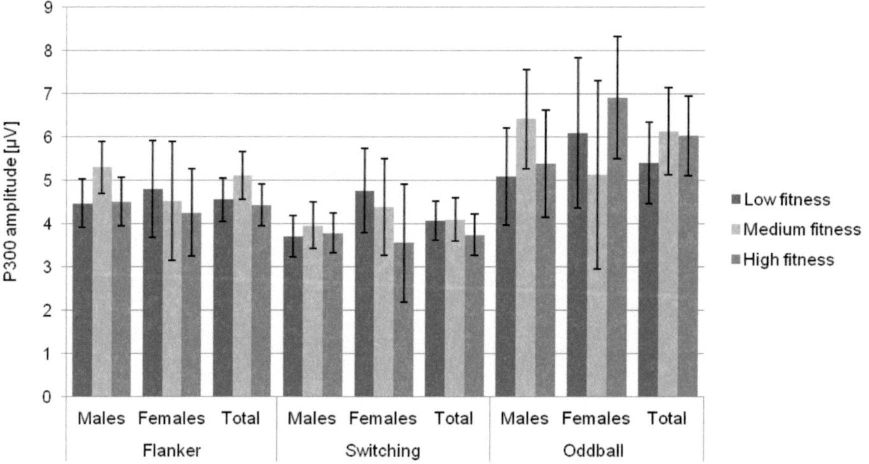

Figure 158: Mean P300 amplitudes during Flanker, Switching and Oddball task for overall fitness groups; error bars represent 95% confidence intervals

Table 163:   Mean P300 latencies for overall fitness groups during Flanker, Switching and Oddball task

		Low fitness		Medium fitness		High fitness			
		n	ms (95 % CI)	n	ms (95 % CI)	n	ms (95 % CI)	p	η^2
Flan-ker	♂	17	395.49 (380.59 - 410.40)	17	386.28 (371.38 - 401.18)	19	391.73 (377.63 - 405.83)	.68	.015
	♀	8	399.16 (358.51 - 439.82)	5	387.58 (327.16 - 430.01)	9	417.77 (379.44 - 456.10)	.45	.081
	Tot.	25	396.67 (381.18 - 412.15)	22	384.53 (368.03 - 401.04)	28	400.10 (385.47 - 414.73)	.35	.028
Swit-ching	♂	7	403.61 (373.49 - 433.72)	9	353.28 (326.72 - 379.84)	10	392.09 (366.89 - 417.29)	.03	.256

		n			n			n		p	η^2
	♀	5	395.83 (341.95 - 449.71)		4	387.00 (326.76 - 447.24)		3	365.10 (295.55 - 434.66)	.74	.066
	Tot.	12	400.37 (375.18 - 425.56)		13	363.65 (339.45 - 387.85)		13	385.86 (361.66 - 410.06)	.11	.117
Odd-ball	♂	16	424.27 (382.26 - 466.28)		18	448.51 (408.90 - 488.12)		15	445.14 (401.75 - 488.53)	.67	.017
	♀	6	391.58 (342.99 - 440.17)		5	412.76 (359.53 - 465.99)		12	383.23 (348.86 - 417.59)	.63	.045
	Tot.	22	415.35 (382.13 - 448.57)		23	440.74 (408.25 - 473.23)		27	417.62 (387.63 - 447.61)	.48	.021

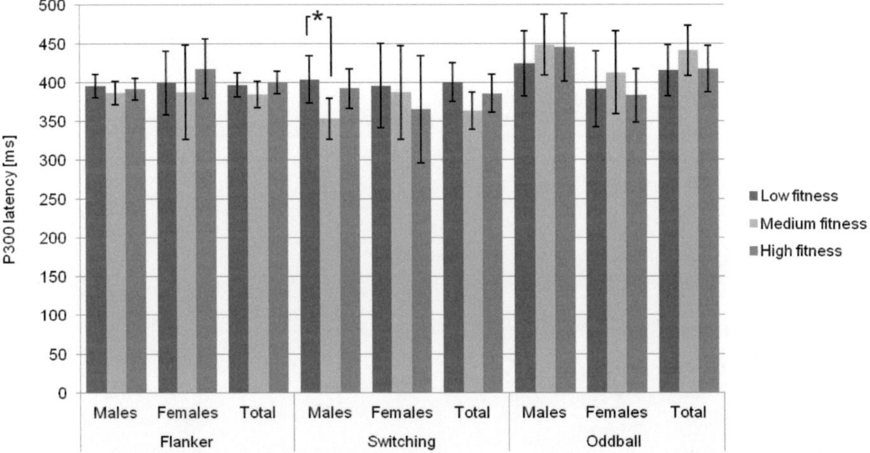

Figure 159: Mean P300 latencies during Flanker, Switching and Oddball task for overall fitness groups; error bars represent 95% confidence intervals

N100 and P200

For P200 amplitude during the Flanker task, a main effect was observed (F [2, 91] = 3.40, p = .04, η^2 = .069; males: F [2, 64] = 2.29, p = .11, η^2 = .067; females: F [2, 24] = 2.48, p = .11, η^2 = .171). Post-hoc analyses detect a marginally significant difference between low and medium fitness groups (p = .06). For N100 latency during the Flanker task, an effect for males was found (F [2, 28] = 3.11, p = .06, η^2 = .182; males: F [2, 17] = 4.04, p = .04, η^2 = .322; females: F [2, 8] = .02, p = .98, η^2 = .004). Post-hoc analyses detect a significant difference between low and high fitness groups (p = .03). N100 and P200 amplitudes are given in table 164 and figure 160 and 161, latencies are presented in table 165 and figure 162 and 163.

Table 164: Mean N100 and P200 amplitudes for overall fitness groups in the Flanker, Switching and Oddball task

			Low fitness		Medium fitness		High fitness			
			n	µV (95 % CI)	n	µV (95 % CI)	n	µV (95 % CI)	p	η^2
Flan-ker	N 100	♂	23	-1.51 (-1.84 - -1.17)	21	-1.55 (-1.90 - -1.20)	23	-1.58 (-1.91 - -1.25)	.95	.002

			N	Low fitness	N	Medium fitness	N	High fitness		
		♀	10	-2.27 (-2.83 - -1.70)	6	-2.22 (-2.95 - -1.49)	11	-1.41 (-1.94 - -.87)	.07	.205
		Tot.	33	-1.74 (-2.03 - -1.44)	27	-1.70 (-2.03 - -1.38)	34	-1.52 (-1.81 - -1.23)	.55	.013
	P 200	♂	23	2.73 (2.11 - 3.36)	21	3.51 (2.86 - 4.16)	23	2.61 (1.99 - 3.24)	.11	.067
		♀	10	3.23 (2.12 - 4.34)	6	5.15 (3.72 - 6.58)	11	3.68 (2.62 - 4.73)	.11	.171
		Tot.	33	2.88 (2.33 - 3.44)	27	3.87 (3.26 - 4.49)	34	2.96 (2.41 - 3.50)	.04	.069
Swit-ching	P 200	♂	17	3.37 (2.67 - 4.06)	14	4.64 (3.87 - 5.40)	17	4.01 (3.32 - 4.70)	.06	.121
		♀	8	4.59 (3.21 - 5.97)	6	4.81 (3.22 - 6.40)	4	4.23 (2.28 - 6.18)	.89	.016
		Tot.	25	3.76 (3.14 - 4.37)	20	4.69 (4.00 - 5.38)	21	4.05 (3.38 - 4.72)	.13	.062
Odd-ball	N 100	♂	19	-4.79 (-5.86 - -3.71)	18	-4.71 (-5.81 - -3.60)	16	-4.28 (-5.45 - -3.10)	.79	.009
		♀	9	-4.68 (-6.04 - -3.33)	5	-6.90 (-8.72 - -5.09)	12	-5.18 (-6.36 - -4.01)	.14	.156
		Tot.	28	-4.75 (-5.60 - -3.91)	23	-5.19 (-6.12 - -4.25)	28	-4.66 (-5.51 - -3.82)	.69	.010
	P 200	♂	19	2.56 (1.63 - 3.50)	18	3.50 (2.53 - 4.46)	16	3.23 (2.21 - 4.25)	.36	.040
		♀	9	2.91 (1.17 - 4.64)	5	3.42 (1.09 - 5.75)	12	2.63 (1.12 - 4.13)	.84	.015
		Tot.	28	2.67 (1.86 - 3.49)	23	3.48 (2.58 - 4.38)	28	2.97 (2.16 - 3.79)	.42	.023

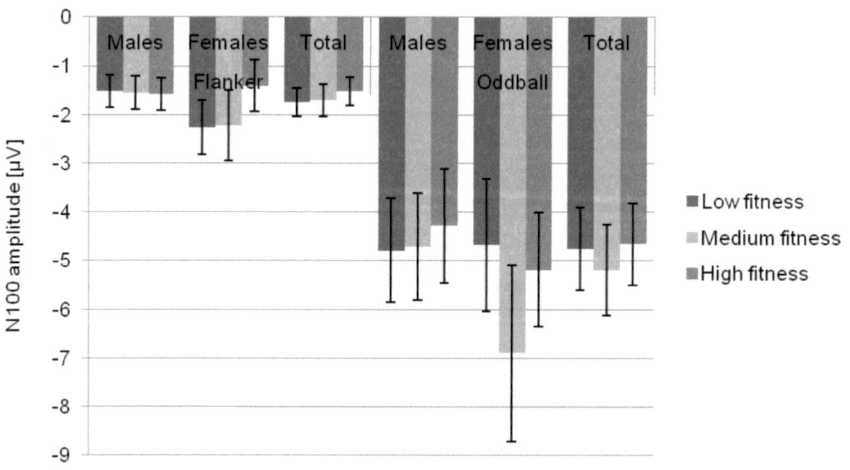

Figure 160: Mean N100 amplitudes during Flanker and Oddball task for overall fitness groups; error bars represent 95% confidence intervals

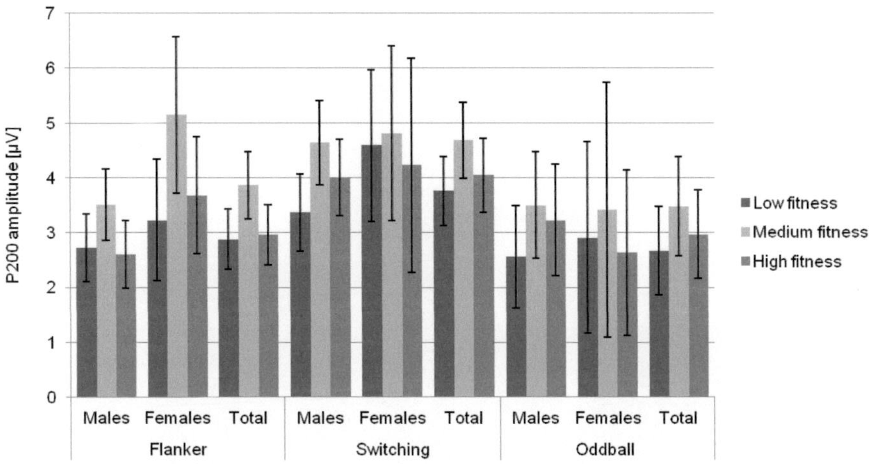

Figure 161: Mean P200 amplitudes during Flanker, Switching and Oddball task for overall fitness groups; error bars represent 95% confidence intervals

Table 165: Mean N100 and P200 latencies for overall fitness groups in the Flanker, Switching and Oddball task

			\multicolumn{2}{c}{Low fitness}		\multicolumn{2}{c}{Medium fitness}		\multicolumn{2}{c}{High fitness}		
			n	ms (95 % CI)	n	ms (95 % CI)	n	ms (95 % CI)	p	η^2
Flan-ker	N 100	♂	5	153.71 (140.51 - 166.92)	5	139.84 (126.64 - 153.05)	10	131.93 (122.60 - 141.27)	.04	.322
		♀	6	140.33 (132.44 - 148.23)	3	141.15 (129.98 - 152.31)	2	139.84 (126.17 - 153.52)	.98	.004
		Tot.	11	146.41 (138.59 - 154.23)	8	140.33 (131.16 - 149.50)	12	133.25 (125.77 - 140.74)	.06	.182
	P 200	♂	15	219.66 (204.69 - 234.64)	18	230.28 (216.61 - 243.95)	17	221.53 (207.46 - 235.60)	.52	.027
		♀	6	216.34 (190.73 - 241.95)	6	227.25 (201.64 - 252.86)	7	228.99 (205.28 - 252.70)	.72	.040
		Tot.	21	218.71 (206.32 - 231.11)	24	229.53 (217.93 - 241.12)	24	223.71 (212.11 - 235.30)	.45	.024
Swit-ching	P 200	♂	12	190.95 (179.86 - 202.04)	13	189.42 (178.77 - 200.08)	16	184.89 (175.28 - 194.49)	.68	.020
		♀	8	180.62 (162.60 - 198.63)	6	190.79 (169.99 - 211.59)	4	197.71 (172.23 - 223.18)	.49	.091
		Tot.	20	186.82 (177.68 - 195.95)	19	189.85 (180.48 - 199.23)	20	187.45 (178.32 - 196.59)	.89	.004
Odd-ball	N 100	♂	17	174.08 (165.30 - 182.86)	17	160.98 (152.20 - 169.76)	14	170.54 (160.86 - 180.21)	.10	.096
		♀	8	173.54 (164.37 - 182.70)	4	176.95 (163.99 - 189.91)	12	176.63 (169.15 - 184.11)	.84	.016
		Tot.	25	173.91 (167.31 - 180.50)	21	164.03 (156.83 - 171.22)	26	173.35 (166.88 - 179.82)	.09	.068
	P 200	♂	11	256.68 (243.72 - 269.63)	16	264.11 (253.37 - 274.85)	12	253.78 (241.37 - 266.18)	.42	.047

♀	5	261.72 (240.11 - 283.33)	3	270.05 (242.15 - 297.95)	6	257.03 (237.30 - 276.76)	.71	.060
Tot.	16	258.25 (247.80 - 268.71)	19	265.05 (255.45 - 274.65)	18	254.86 (245.00 - 264.72)	.33	.044

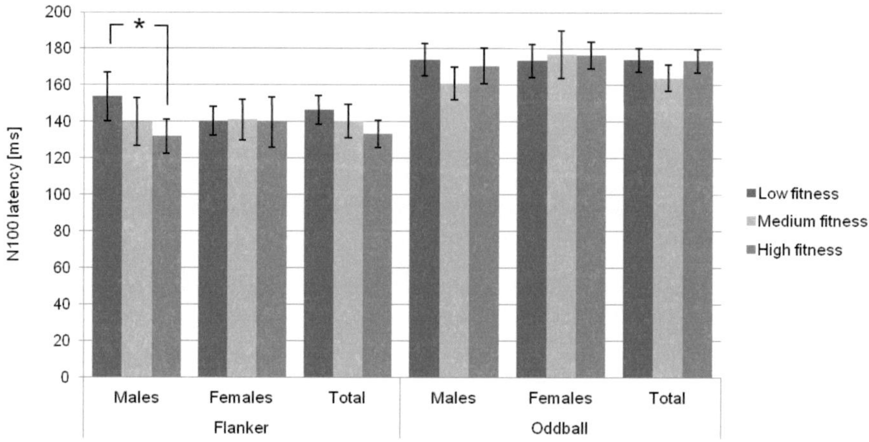

Figure 162: Mean N100 latencies during Flanker and Oddball task for overall fitness groups; error bars represent 95% confidence intervals

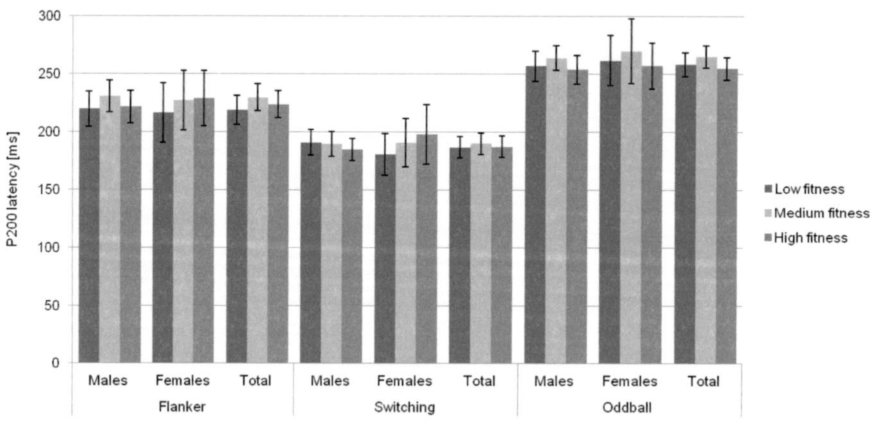

Figure 163: Mean P200 latencies during Flanker, Switching and Oddball task for overall fitness groups; error bars represent 95% confidence intervals

4.4.5.2 Discussion

Overall physical fitness was not related to behavioral cognitive measures. Regarding errors in the Switching task, males with a medium fitness tended to have fewer errors than the low and high fitness group, but females with a medium fitness had a tendency for the highest number of errors. There was also no effect on P300 component except for P300 latencies during the Switching task. The medium fitness group had faster latencies than the low fitness group. This effect was only significant for males and might indicate that young male adults with a moderate overall fitness level have the best effects on cognitive processing speed than compared to persons with a lower and even higher level. For the Flanker task, there was also a tendency for the medium fitness group having the fastest P300 latencies. This is also approved by the marginally significant finding on P200 amplitude in the Flanker task. The medium fitness group revealed the highest amplitudes which might indicate a better attention modulation. For the Switching and Oddball task, there was also a tendency for larger P200 amplitudes for the medium compared to the low and high fitness groups. With regard to Flanker N100 latencies, participants with a high fitness level showed the fastest cognitive processing speed.

It can be concluded that the overall fitness level is not a strong predictor for cognitive performance in young adults since only few significant effects were found. However, persons with a medium and high fitness level revealed some improved neurophysiologic functions compared to persons with lower fitness.

4.4.6 Summary

The results of this chapter on the relationship between physical fitness and cognition variables are summarized in the following table 166. In addition, tables 167 to 169 provide an overview of effect sizes for all analyses.

Table 166: Summary of results on fitness and cognition

Hypothesis	Summary of Results
RQ 8: Is aerobic endurance related to cognitive performance in young adulthood?	<u>Significant findings:</u> Faster response times during homogenous trials of the Switching task but longer response times during heterogeneous trials for the high compared to the poor VO_{2max} group. Participants with a good VO_{2max} had significantly fewer errors in the Switching task. Participants with a good VO_{2max} and performance at IAT revealed faster N100 latencies for the accuracy condition in the Flanker task, but longer latencies during the speed condition.
	<u>Trends:</u> Tendency for faster response times for the poor IAT performance group. Tendency for larger P300 amplitudes but longer P300 latencies across the cognitive tasks for VO_{2max} group.
	<u>Gender-specific:</u> Higher amplitudes for Flanker P200 component for females with good VO_{2max} compared to females with poor VO_{2max}. Males with good VO_{2max} revealed faster P200 latencies in the Flanker task than those with a poor VO_{2max}. Generally larger effect sizes for females (VO_{2max}: mean $\eta^2 = .038$; performance at IAT: mean $\eta^2 = .034$) than for males (VO_{2max}: mean $\eta^2 = .014$; performance at IAT: mean

	$\eta^2 = .015$).
	→ Research question can be mostly answered in the negative. Aerobic endurance is only slightly related to selected measures of cognitive performance in young adulthood and only for special conditions of the tasks. The overall mean effect sizes are small (VO_{2max}: $\eta^2 = .009$; performance at IAT: $\eta^2 = .007$).
RQ 9: Is maximal strength related to cognitive performance in young adulthood?	<u>Significant findings:</u> Good isometric strength was significantly related to longer P200 latencies during the Flanker task. Good dynamic strength group had significantly faster P300 latencies in the Oddball task.
	<u>Trends:</u> Females with good isometric strength had a tendency for longer P300 latencies in the Switching task. Males of the high strength group had a tendency for faster P300 latencies in the Oddball task.
	<u>Gender-specific:</u> Females with a good maximal isometric strength had significantly faster response times in the Flanker task. Males with a poor dynamic strength responded significantly faster in the Flanker task than males with a good dynamic strength. A good maximal isometric strength was beneficial for P300 latencies during the Switching (significant only for males) and Oddball task (significant only for females). Male participants with a good dynamic strength had faster P300 latencies during the Oddball task. Females with a good dynamic strength had smaller P300 amplitudes in the Switching task. For P200 amplitudes during the Flanker and Oddball task, females with a high isometric strength had significantly larger amplitudes. Good isometric strength was related to longer P200 latencies during the Flanker task for males. Females with good dynamic strength had faster P200 latencies during the Oddball task. Generally larger effect sizes for females (Isometric strength: mean $\eta^2 = .066$; dynamic strength: mean $\eta^2 = .048$) than for males (Isometric strength: mean $\eta^2 = .016$; dynamic strength: mean $\eta^2 = .024$).
	→ Research question can be mostly answered in the negative. Strength is only slightly related to selected measures of cognitive performance in young adulthood and the effects are mostly gender-specific. The overall mean effect sizes are small (Isometric strength: $\eta^2 = .009$; dynamic strength: $\eta^2 = .019$).
RQ 10: Is coordinative performance related to cognitive performance in young adulthood?	<u>Significant findings:</u> High jumping performance group consistently had significantly higher P300 amplitudes during the Flanker, Switching and Oddball task. Significantly higher amplitudes and faster latencies for P200 component during the Switching task and for P200 latencies during the Oddball task for the good jumping performance group. Longer response times during the Flanker task for the good jumping performance group. Good balancing performance group had larger P300 amplitudes in the homogenous condition of the Switching task but smaller amplitudes in the heterogeneous condition.
	<u>Trends:</u> P300 latencies were faster by trend for participants with a good jumping performance. Tendency for larger P200 amplitudes during Flanker and Oddball task for good jumping performance. Tendency for larger P300 amplitudes during all cognitive tasks for good static as well as dynamic Posturomed performance groups. Tendency for longer P300 latencies for the good static and dynamic Posturomed performance group. Tendency for shorter N100 latencies during the Flanker and the Switching task for males with a good dynamic Posturomed performance.
	<u>Gender-specific:</u> Significantly longer P300 latencies for males with a good dynamic

Posturomed performance in the Flanker task. Males with a good dynamic Posturomed performance had significantly longer P200 latencies during the Flanker task and smaller N100 amplitudes in the Oddball task. Females with a good dynamic performance responded significantly faster during the Flanker task. Females with a good balancing performance had a significantly longer P200 latency during the Switching task. Generally larger effect sizes for females (Static Posturomed performance: mean η^2 = .041; dynamic Posturomed performance: mean η^2 = .031; balancing performance: mean η^2 = .037; jumping performance: mean η^2 = .045) than for males (Static Posturomed performance: mean η^2 = .014; dynamic Posturomed performance: mean η^2 = .023; balancing performance: mean η^2 = .009; jumping performance: mean η^2 = .028).

→ Research question can be partly answered in the affirmative. Coordinative performance, especially performance in the jumping side-to-side task, is related to selected measures of cognitive performance in young adulthood. A higher jumping performance is mostly associated with a better cognitive performance, whereas higher coordinative performances in the other measures are partly negatively associated with cognitive performance. Overall mean effect sizes are small (Static Posturomed performance: η^2 = .011; dynamic Posturomed performance: η^2 = .016; balancing performance: η^2 = .005; jumping performance: η^2 = .025).

RQ 11: Is body fat percentage related to cognitive performance in young adulthood?

Significant findings: Faster P200 latencies during Switching task for overweight compared to normal weighted participants. Participants with low body fat percentage had significantly lower P200 latencies during the Oddball task.

Trends: Tendency for more errors in the Switching task for high BMI group.

Gender-specific: Males with a normal BMI had significantly larger P300 amplitudes during the Flanker task than males with a high BMI. Faster P300 latencies for females with normal BMI in the Switching task. Females with a high amount of body fat had significantly larger P200 amplitudes during the Flanker and Switching task. Generally larger effect sizes for females (BMI: mean η^2 = .051; body fat percentage: mean η^2 = .080) than for males (BMI: mean η^2 = .019; body fat percentage: mean η^2 = .037).

→ Research question can be mostly answered in the negative. Most effects observed only for P200 component but no effects on P300. The overall mean effect sizes are small (BMI: η^2 = .012; body fat percentage: η^2 = .029).

RQ 12: Is overall fitness related to cognitive performance in young adulthood?

Significant findings: Marginally significant larger P200 amplitudes in the Flanker task for medium fitness group. Participants with a high fitness level showed the fastest Flanker N100 latencies.

Trends: Tendency for males with a medium fitness for fewer errors in Switching task. Tendency for females with a medium fitness for the highest number of errors. Tendency for the medium fitness group for fastest P300 latencies in the Flanker task. Tendency for larger P200 amplitudes for the medium compared to the low and high fitness groups during Switching and Oddball task.

Gender-specific: Males in the medium fitness group had faster P300 latencies during Switching task than low fitness group. Generally larger effect size for females (mean η^2 = .067) than for males (mean η^2 = .062).

→ Research question can be partly answered in the affirmative. Overall fitness is slightly related to selected measures of cognitive performance in young adulthood.

> However, most effects are gender-specific. The overall mean effect size is small (η^2 = .038).

The largest effect sizes for the independent variables on cognitive performance with regard to the total study sample were found for overall fitness, body fat percentage and jumping performance (table 167). These factors explain the highest variance in cognitive performance (overall fitness: 3.8 %; body fat: 2.9 %; jumping performance: 2.5 %). When looking at the dependent variables, the largest effects of physical fitness were found on N100 latencies (mean explained variance: 2.4 %), P300 latencies (mean explained variance: 2.2 %) and P200 latencies (mean explained variance: 2.2 %).

For the male subsample (table 168), the largest effects on cognitive performance were found for overall fitness (explained variance: 6.2 %), body fat percentage (explained variance: 3.7 %) and jumping performance (explained variance: 2.8 %). The largest effects of physical fitness variables were observed for N100 latencies (explained variance: 4.5 %) and P300 latencies (explained variance: 3.4 %).

For the female subsample (table 169), the effect sizes were consistently larger than those for males. However, due to the small number of female participants in this study, most effects were not significant. The largest effects on cognitive performance were found for body fat percentage (explained variance: 8.0 %), the overall fitness (explained variance: 6.7 %) and maximal isometric strength (explained variance: 6.6 %). The largest effects of physical fitness variables were observed for P200 latencies (explained variance: 6.4 %) and P200 amplitudes (explained variance: 6.3 %).

Table 167: Summary of effect sizes (η^2) for all cognitive and physical fitness measures for the total sample (only main effects, interactions are not displayed)

		VO2max [η^2]	Perform. at IAT [η^2]	Max. isom. strength [η^2]	Max. dyn. strength [η^2]	Stat. Posturom. [η^2]	Dyn. Posturom. [η^2]	Balanc. [η^2]	Jump. [η^2]	BMI [η^2]	Body fat [η^2]	Overall Fitness [η^2]	Total (Mean η^2)
Resp. Times	Flanker	.005	.001	.003	.012	.000	.030	.000	**.034***	.002	.027	.017	.012
	Switching	.006	.020	.001	.014	.001	.014	.001	.003	.000	.021	.002	.008
	Oddball	.004	.009	.007	.004	.001	.010	.000	.000	.005	.003	.001	.004
Errors	Switching	**.038***	.015	.001	.003	.003	.000	.001	.001	.009	.002	.001	.007
P300 ampl.	Flanker	.002	.008	.000	.015	.013	.009	.007	**.032***	.025	.030	.041	.017
	Switching	.003	.007	.004	.005	.025	.027	.009	**.148***	.022	.041	.022	.028
	Oddball	.017	.024	.000	.018	.020	.025	.001	**.041***	.006	.015	.019	.017
P300 lat.	Flanker	.000	.003	.005	.016	.004	.015	.012	.028	.010	.011	.028	.012
	Switching	.017	.006	.032	.034	.006	.041	.001	.023	.019	.116	.117	.037
	Oddball	.010	.000	.016	**.077***	.000	.019	.000	.039	.000	.001	.021	.017
N100 ampl.	Flanker	.000	.001	.000	.002	.003	.007	.011	.004	.005	.005	.013	.005
	Oddball	.003	.007	.001	.019	.000	.029	.000	.000	.016	.012	.010	.009
N100 lat.	Flanker	.002	.000	.008	.004	.017	.024	.003	.006	.015	.081	.182	.031
	Oddball	.030	.000	.001	.033	.000	.042	.003	.006	.000	.000	.068	.017
P200 ampl.	Flanker	.011	.014	.004	.007	.003	.000	.000	.009	.001	.034	**.069***	.014
	Switching	.015	.013	.003	.004	.000	.002	.000	**.063***	.016	.018	.062	.018
	Oddball	.002	.001	.022	.004	.001	.000	.003	.008	.001	.006	.023	.006
P200 lat.	Flanker	.009	.008	**.056***	.015	**.070***	.020	.002	.001	.023	.022	.024	.023
	Switching	.000	.001	.004	.040	.022	.000	.022	.000	**.051***	.035	.004	.016
	Oddball	.001	.000	.007	.048	.023	.002	.021	**.059***	.004	**.093***	.044	.027
Total (Mean η^2)		.009	.007	.009	.019	.011	.016	.005	.025	.012	.029	.038	

Table 168:   Summary of effect sizes (η^2) for all cognitive and physical fitness measures for males (only main effects, interactions are not displayed)

		VO2max [η^2]	Perform. at IAT [η^2]	Max. isom. strength [η^2]	Max. dyn. strength [η^2]	Stat. Posturom. [η^2]	Dyn. Posturom. [η^2]	Balanc. [η^2]	Jump. [η^2]	BMI [η^2]	Body fat [η^2]	Overall Fitness [η^2]	Total (Mean η^2)
Resp. Times	Flanker	.000	.000	.012	**.042***	.000	.006	.002	**.046***	.002	.026	.030	.015
	Switching	.007	.007	.000	.004	.012	.008	.000	.014	.000	.022	.005	.007
	Oddball	.023	.021	.001	.001	.002	.001	.013	.003	.003	.029	.010	.010
Errors	Switching	**.044***	.022	.000	.007	.006	.001	.006	.000	.001	.005	.015	.010
P300 ampl.	Flanker	.008	.012	.000	.006	.002	.001	.020	**.052***	**.051***	.048	.075	.025
	Switching	.012	.004	.007	.006	.000	.011	.012	**.195***	.002	.008	.011	.024
	Oddball	.017	.029	.004	.033	.002	.015	.000	.033	.023	.045	.055	.023
P300 lat.	Flanker	.001	.001	.003	.026	.000	**.085***	.016	.016	.021	.055	.015	.022
	Switching	.002	.010	**.156***	.068	.004	.031	.011	.006	.002	.114	**.256***	.060
	Oddball	.022	.011	.002	**.092***	.000	.033	.001	.046	.002	.007	.017	.021
N100 ampl.	Flanker	.008	.006	.002	.020	.010	.008	.004	.000	.023	.034	.002	.011
	Oddball	.006	.020	.003	**.070***	.008	**.067***	.000	.000	.001	.029	.009	.019
N100 lat.	Flanker	.017	.012	.011	.012	.021	.054	.034	.002	**.164***	.131	**.322***	.071
	Oddball	.039	.004	.002	.016	.000	.052	.001	.013	.000	.004	.096	.021
P200 ampl.	Flanker	.012	.001	.001	.000	.009	.005	.012	.003	.007	.006	.067	.011
	Switching	.035	.036	.025	.001	.002	.003	.001	.046	.034	.020	.121	.029
	Oddball	.002	.016	.004	.003	.012	.019	.002	.015	.026	.009	.040	.013
P200 lat.	Flanker	.013	**.056***	**.066***	.013	**.110***	.047	.001	.000	.018	.022	.027	.034
	Switching	.010	.024	.005	.052	.018	.000	.000	.002	.004	.016	.020	.014
	Oddball	.001	.001	.018	.016	.056	.008	.048	.068	.001	.103	.047	.033
Total (Mean η^2)		.014	.015	.016	.024	.014	.023	.009	.028	.019	.037	.062	

Table 169: Summary of effect sizes (η^2) for all cognitive and physical fitness measures for females (only main effects, interactions are not displayed)

		VO2max [η^2]	Perform. at IAT [η^2]	Max. isom. strength [η^2]	Max. dyn. strength [η^2]	Stat. Posturom. [η^2]	Dyn. Posturom. [η^2]	Balanc. [η^2]	Jum p. [η^2]	BMI [η^2]	Body fat [η^2]	Overall Fitness [η^2]	Total (Mean η^2)
Resp. Times	Flanker	.035	.006	**.151***	.023	.013	**.127***	.000	.010	.008	.008	.005	.035
	Switching	.005	.064	.003	.049	.071	.029	.012	.004	.003	.097	.001	.031
	Oddball	.020	.000	.019	.008	.002	.036	.020	.020	.009	.026	.032	.017
Errors	Switching	.029	.005	.002	.000	.000	.000	.002	.014	.040	.010	.094	.018
P300 ampl.	Flanker	.003	.001	.004	.046	.115	.067	.000	.007	.000	.021	.023	.026
	Switching	.124	.019	.000	**.194***	.180	.075	.002	.085	.093	.127	.136	.094
	Oddball	.008	.009	.071	.007	.101	.062	.007	.053	.011	.016	.087	.039
P300 lat.	Flanker	.000	.042	.004	.008	.025	.005	.011	.061	.001	.003	.081	.022
	Switching	.248	.000	.090	.000	.012	.066	.026	.102	**.276***	.138	.066	.093
	Oddball	.004	.023	**.201***	.000	.071	.000	.060	.005	.000	.104	.045	.047
N100 ampl.	Flanker	.074	.007	.015	.062	.008	.008	.097	.055	.024	.087	.205	.058
	Oddball	.001	.001	.002	.015	.039	.000	.004	.007	.100	.093	.156	.038
N100 lat.	Flanker	.070	.128	.002	.008	.008	.000	.107	.256	.079	.001	.004	.060
	Oddball	.020	.013	.001	.058	.047	.035	.002	.014	.001	.002	.016	.019
P200 ampl.	Flanker	.013	**.127***	**.139***	.030	.084	.014	.012	.027	.007	**.304***	.171	.084
	Switching	.000	.003	.051	.042	.000	.004	.012	.119	.001	**.296***	.016	.049
	Oddball	.003	.028	**.354***	.015	.004	.069	.044	.002	.047	.032	.015	.056
P200 lat.	Flanker	.001	.121	.029	.015	.004	.006	.053	.001	.046	.069	.040	.035
	Switching	.051	.070	.139	.018	.029	.001	**.263***	.010	**.255***	.096	.091	.093
	Oddball	.044	.013	.050	**.363***	.009	.013	.009	.047	.022	.074	.060	.064
Total (Mean η^2)		.038	.034	.066	.048	.041	.031	.037	.045	.051	.080	.067	

Chapter 5

Summary and Conclusion

5.1 Major Findings

The purpose of this study was to determine the interdependences of physical (in-) activity, fitness, and cognitive performance in young adults. To answer this question, several physical activity and fitness factors were assessed in a study sample of 152 young adults and the relationships between those factors and cognitive performance were analyzed.

With regard to physical activity variables, the most significant effects were found for general sports participation, participation in competitive sports and frequency of exercise.

For the first factor, athletetes compared to nonathletes showed a better cognitive performance as assessed by faster response times during the Oddball task, larger P300 amplitudes in the Oddball task and faster P300 latencies in the Flanker task. It may be supposed that general physical (in-) activity as assessed by the simple question "Do you regularly exercise?" not only reflects the existence or abstinence of regular sport activity, but potentially also separates people by their general attitude referring to a healthy and active way of life. This generally active and healthy way of life is certainly linked to many other factors such as nutrition and sedentary lifestyle besides physical activity. All these factors underlying the answer to the question stated above are in sum of great importance for explaining the difference in cognitive performance between active and inactive people. Regarding the second factor, non-competitive athletes had an improved cognition compared to competitive athletes and nonathletes as indicated by faster response times during the Oddball, faster P300 latencies during the Flanker task and faster P200 latencies during the Flanker task. Thus, regular sports activity seems to be beneficial for cognitive processing speed and response preparation speed in young adulthood. But, based on the results from this study, sports activity should be performed without high efforts for practicing and without pressure and expectations coming along with challenging situations and competitions.

The last factor showed that frequency of sports activity might play a role underlying the relationships between sports activity and cognitive performance in young adults. There is evidence for larger P300 amplitudes during the Flanker and Oddball task as well as faster P30 latencies in the Oddball task for athletes who exercise $\geq$ 4 sessions per week. Participants with fewer sessions had an impaired cognitive performance compared to the $\geq$ 4 sessions group. These findings might indicate that a continuous and frequently performed sports activity per week is more important for exercise-induced effects on cognitive performance in this age group than exercise duration or even intensity.

With regard to fitness measures, coordination under time pressure as measured by the jumping side-to-side task revealed the most significant and strongest associations with cognitive performance. Participants with a good jumping performance had larger P300 amplitudes and thus an enhanced attention performance during all cognitive tasks. They further had larger P200 amplitudes in the

Switching and faster P200 latencies in the Oddball task. However, there was a negative association since participants with good jumping performance had an impaired response speed in the Flanker task. Nevertheless, these findings might indicate that sports associated with high motor control might be particularly valuable in promoting cognition and brain health in young adulthood. There is a lack of research on the relationship between coordination and cognition in young adults but some evidence for older adults suggesting that coordination training enhances cognition. It is hypothesized that coordination training leads to functional and structural changes and adaptation in selected brain regions which might be in turn beneficial for cognitive functioning.

Moreover, body fat percentage and the overall fitness level had higher mean effects sizes than other fitness variables across all analyses. However, the results were inconsistent since there was no clear tendency for a beneficial effect on cognitive performance for one subgroup.

Interestingly, the cardiovascular fitness hypothesis could not be confirmed in this study. Cardiovascular fitness as assessed by VO_{2max} and performance at individual anaerobic threshold was not related to cognitive performance in this sample. However, first analyses on heart rate variability data, which were conducted during the present study but not included in this thesis, indicate that heart rate variability is related to faster response time and larger amplitude for males but not females. These findings might indicate that vagally mediated cardiac control is related to cognitive performance in young adults independent of aerobic endurance predictors. However, these analyses need further verification.

Importantly, it must be noted that the results on cognitive performance are partly inconsistent as most significant effects were determined only for selected tasks (e.g. only for the Oddball but not for the Flanker task) or only for specific electrode sites, conditions or trials (e.g. only for P300 amplitudes at Cz electrode under speed instruction of the Flanker task). However, this inconsistency is in line with many previous studies in which effects of physical activity and fitness on cognition could also be determined only for particular conditions of a certain paradigm or only for selected cognitive tasks.

In addition, this study detected gender-specific differences in the relationships between physical activity, fitness and cognition. Over all cognitive performance measures, there were stronger relations to physical activity and fitness measures (as indicated by mean η^2) observed for females compared to males. Due to the imbalanced sample sizes with only 45 females compared to 107 males, most effects for females became not significant although the effect sizes were large.

5.2   Limitations and Strengths of the Study

There are several limitations of this study that need to be discussed in this chapter. The major limitation is certainly the cross-sectional design, which does not allow for drawing any conclusions about the causality of the results. It is plausible that the participants included in this study differ on many factors such as genetics, personality characteristics and lifestyle factors. These factors could lead to the observed relationships between physical activity, fitness and neuroelectric and behavioral cognitive measures. However, the results may be used as a basis for designing an intervention study in young adults with an experimental design.

A second limitation is the homogenous sample of young and healthy students at a University. Students were further not representatively selected, but could voluntarily get in touch with study investigators in order to become participants, provided that they fulfilled the inclusion criteria. Related to the young age and high educational level of participants, the amount of sports and habitual physical activity was very high compared to older adults or probably also to participants with a lower educational level. The same might apply to the level of physical fitness measures including body composition. Furthermore, there was an imbalance of athletes (n = 119) and nonathletes (n = 33) as well as of males (n = 107) and females (n = 45) in the study sample. This is due to the fact that participants were mainly recruited from the campus of a technical University with a larger number of male compared to female students. Furthermore, it should be considered that physically active and high-fit adults and also those with a high cognitive performance probably had a bigger interest in participating in this study than inactive, low-fit and/or adults with a poor cognitive performance. This might also bias the results to a substantial extent.

With regard to the study methods, one limitation is the assessment of physical activity which was done using a questionnaire. Self-reported physical activity is assumed to be higher than objectively and more accurately measured physical activity using accelerometers (see also chapter 2.2.1). However, in most previous studies on this topic in young adults, self-reported instead of objectively measured physical activity was also used. The amount of sports and habitual physical was estimated from different items of the questionnaire using a formula which comprises bias. Nevertheless, the physical activity questionnaire offered the chance to assess other relevant parameters such as type of sport, intensity, frequency and duration of exercise sessions as well as sport history and was therefore used in this study. During the measurements and tests of the participants in the laboratories, potential interfering variables were controlled (i.e. by closing doors, keeping room temperature constant using ventilators) or minimized (i.e. by instructing participants not to eat for at least 2 hours prior to testing). However, it was not possible to control for all interfering variables that might have affected the present findings (i.e. time of day, time of the year, lack of sleep).

The cognitive tasks were selected as they are widely accepted and were often used in comparable studies on the relationship between physical activity, fitness and cognition in young adults (see tables 3-5). However, the tests turned out to be too easy for the study sample regarding response accuracy. Participants made too little errors, especially in the Flanker and Oddball task so that an analysis of results in terms of mean differences was impossible. Analyses of response times for all cognitive tasks and behavioral measures of the Switching task but not Flanker or Oddball task could be conducted. Another limitation is that there is no subdivision of P300 component into P3a and P3b in the Oddball task. Instead, only the P300 component obtained from the target stimulus could be analyzed which is comparable to the P3b component.

The strength of the study is the sample size which is larger than in most previously published studies on this topic (see tables 3 - 5). Moreover, an extensive physical performance testing battery was used to accurately assess physical fitness. In addition, this study not only assessed response times and P300 event-related brain potential but also P200 and N100 components.

5.3   Recommendations for Future Research

Given that young adults are on the peak of their cognitive performance, there is suggested to be only little range for exercise-related improvement in contrast to elderly or very young children. Due to this fact, Kamijo and Takeda (2010, p. 311) proposed that cognitive paradigms should be selected carefully to avoid an underestimation of exercise-induced effects. In most studies, effortful tasks and trials which require higher cognitive functions are therefore used.

However, this study could show that beneficial effects can also be investigated by using a relatively simple Oddball task. Nevertheless, it is doubtlessly recommended to use more complex and demanding tasks instead of easier ones when working with young adults. An innovative approach might also be not only to limit cognitive performance testing to standardized paradigms but to also think about new techniques such as driving simulators for example. In the current study, this technique was also used. Participants were asked to perform a lane change task which involves driving on a virtual lane and changing lanes as instructed by road signs. While completing the lane change task, participants were instructed to perform secondary n-back tasks.

Figure 164: Participant performing the tests in the driving simulator

In this study, no effects could be found for behavioral data such as errors in these tasks. But it would be interesting to test if event-related brain potentials assessed during such a task would be affected by physical activity or fitness measures. However, such an analysis was not possible in this study.

Future research should continue in examining the effects of general sports participation, types, durations, frequencies and intensities of exercise as well as fitness parameters on cognitive performance in young adults. Larger sample sizes, especially for females, are needed in order to ensure a gender-specific data analysis. Further, recruitment of participants should not only be limited to the university setting as done so far in most studies. Additionally, the use of objective techniques to assess physical activity habits and also fitness status is stringently required. However, since cross-sectional designs do not allow for detection of causally determined associations, intervention and longitudinal study designs are increasingly needed.

This study could not test any potential mechanisms that might underlie the effects of activity and fitness on cognition. Since possible mechanisms and adaptations of physical activity or fitness on brain structure and executive functions have been described in literature, potentially underlying mechanisms can only be suggested from the current findings. Therefore, more basic research is needed to investigate the potential mechanisms that might underlie exercise-induced effects on cognitive performance in young adults. It would be further very interesting to know in which way benefits on cognitive performance as assessed by standardized paradigms are related to cognitive performance and behavior in real life.

5.4 Conclusion

This study provides further evidence that physical activity and fitness might be related to cognitive performance in young adulthood.

However, the cross-sectional study design does not allow for the interpretation of any causality based on the data. This means that regular physical activity or a good coordinative performance might be beneficial for cognition on the one hand. On the other hand, young adults with a higher cognitive performance might be more likely to regularly engage in physical activity and to perform sports that enhance coordinative skills. If the first option turned out to be true, physical activity and fitness would lead to adaptations in cognition, but in the second option, cognitive skills would be the underlying factor for a person to the decide whether to be physically active or not. Whether physical activity might be the cause or effect of/on cognitive performance in young adulthood can only be answered by true experimental design studies.

It can be concluded that especially the general participation in regular physical activity, the frequency of sports activity and coordinative skills performance are associated with higher cognitive performance and central information processing in young adults.

References

Åberg, M. A. I., Pedersen, N. L., Torén, K., Svartengren, M., Bäckstrand, B., Johnsson, T., Cooper-Kuhn, C. M., Åberg, N. D., Nilsson, M. & Kuhn, H. G. (2009). Cardiovascular fitness is associated with cognition in young adulthood. *Proceedings of the National Academy of Sciences of the United States of America, 106* (49), 20906-20911.

Ainsworth, B. E., Haskell, W. L., Herrmann, S. D., Meckes, N., Bassett, D. R., Tudor-Locke, C., Greer, J. L., Vezina, J., Whitt-Glover, M. C. & Leon, A. S. (2011). Compendium of physical activities: A second update of codes and MET values. *Medicine and Science in Sports and Exercise, 43*, 1575-1581.

Alvarez, J. A. & Emory, E. (2006). Executive function and the frontal lobes: A meta-analytic review. *Neuropsychology Review, 16* (1), 17-42.

Angevaren, M., Aufdemkampe, G., Verhaar, H. J. J., Aleman, A. & Vanhees, L. (2008). Physical activity and enhanced fitness to improve cognitive function in older people without known cognitive impairment (Review of the Cochrane Collaboration). *Cochrane Database of Systematic Reviews,* Issue 3: CD005381. doi: 10.1002/14651858.CD005381.pub3.

Baddeley, A. D. (2003). Working memory: looking back and looking forward. *Nature Reviews Neuroscience, 4*, 829-839.

Baddeley, A. D. & Hitch, G. (1974). Working memory. In G. H. Bower (Ed.), *The psychology of learning and motivation* (Vol. 8; pp. 47-89). New York: Academic Press.

Bassett, D. R. & Howley, E. T. (2000). Limiting factors for maximum oxygen uptake and determinants of endurance performance. *Medicine and Science in Sports and Exercise, 32* (1), 70-84.

Bear, M. F., Connors, B. W. & Paradiso, M. A. (2007). *Neuroscience. Exploring the Brain (3rd edition)*. Baltimore: Lipincott Williams & Wilkins.

Blair, S. N., Kohl, H. W., Gordon, N. F. & Paffenbarger, R. S. (1992). How much physical activity is good for health? *Annual Review of Public Health, 13*, 99-126.

Bös, K. (1987). *Handbuch Sportmotorischer Tests*. Göttingen: Hogrefe Verlag.

Bös, K. & Mechling, H. (1983). *Dimensionen Sportmotorischer Leistung*. Schorndorf: Hofmann Verlag.

Bös, K., Schlenker, L., Büsch, D., Lämmle, L., Müller, H., Oberger, J., Seidel, I. & Tittlbach, S. (2009). *Deutscher Motorik-Test 6-18 (DMT 6-18)*. Hamburg: Czwalina Verlag.

Borg, G. A. (1982). Psychophysical bases of perceived exertion. *Medicine and Science in Sports and Exercise, 14* (5), 377-381.

Borkenau, P. & Ostendorf, F. (1993). *NEO-Fünf-Faktoren Inventar (NEO-FFI) nach Costa und McCrae. Handanweisung.* Göttingen: Hogrefe.

Botvinick, M. M., Cohen, J. D. & Carter, C. S. (2004). Conflict monitoring and anterior cingulate cortex: an update. *Trends in Cognitive Sciences, 8* (12), 539-546.

Bouchard, C., Blair, S. N. & Haskell, W. L. (2012). Why study physical activity and health? In C. Bouchard, S. N. Blair & W. L. Haskell (Eds.), *Physical Activity and Health* (2nd edition; pp. 3-20). Champaign: Human Kinetics.

Bouchard, C., Shephard, R. J. & Stevens, T. (1994). *Physical activity, fitness and health: international proceedings and consensus statement.* Champaign: Human Kinetics.

Braver, T. S., Cohen, J. D., Nystrom, L. E., Jonides, J., Smith, E. E. & Noll, D. C. (1997). A parametric study of prefrontal cortex involvement in human working memory. *Neuroimage, 5*, 49-62.

Brickman, A. M., Zimmerman, M. E., Paul, R. H., Grieve, S. M., Tate, D. F., Cohen, R. A., Williams, L. M., Clark, C. R. & Gordon, E. (2006). Regional white matter and neuropsychological functioning across the adult lifespan. *Biological Psychiatry, 60* (5), 444-453.

Burdette, J. H., Laurienti, P. J., Espeland, M. A., Morgan, A., Telesford, Q., Vechlekar, C. D., Hayasaka, S., Jennings, J. M., Katula, J. A., Kraft, R. A. & Rejeski, W. J. (2010). Using network science to evaluate exercise-associated brain changes in older adults. *Frontiers of Aging and Neuroscience, 7* (2), 23. doi: 10.3389/fnagi.2010.00023.

Burkhalter, T. M. & Hillman, C. H. (2011). A narrative review of physical activity, nutrition, and obesity to cognition and scholastic performance across the human lifespan. *Advances in Nutrition, 2*, 201S-206S.

Burpee, R. H. & Stroll, W. (1936). Measuring reaction time of athletes. *Research Quarterly for Exercise and Sport, 7*, 110-118.

Cacioppo, J. T. & Hawkley, L. C. (2009). Perceived social isolation and cognition. *Trends in Cognitive Sciences, 13* (10), 447-454.

Canadian Society for Exercise Physiology (1992). Revision of the Physical Activity Readiness Questionnaire (PAR-Q). *Canadian Journal of Sport Sciences*, 17, 338-345.

Caspersen, C. J., Powell, K. E. & Christenson, G. M. (1985). Physical activity, exercise, and physical fitness: Definitions and distinctions for health-related research. *Public Health Reports, 100* (2), 126-131.

Cassilhas, R. C., Viana, V. A., Grassmann, V., Santos, R. T., Santos, R. F., Tufik, S. & Mello, M. T. (2007). The impact of resistance exercise on the cognitive function of the elderly. *Medicine and Science in Sports and Exercise, 39* (8), 1401-1407.

Chaddock, L., Pontifex, M. B., Hillman, C. H. & Kramer, A. F. (2011). A review of the relation of aerobic fitness and physical activity to brain structure and function in children. *Journal of the International Neuropsychological Society, 17*, 975-985.

Chan, R. C. K., Shum, D., Toulopoulou, T. & Chen, E. Y. H. (2008). Assessment of executive functions: Review of instruments and identification of critical issues. *Archives of Clinical Neuropsychology, 23* (2), 201-216.

Chang, Y. K., Labban, J. D., Gapin, J. I. & Etnier, J. L. (2012). The effects of acute exercise on cognitive performance: A meta-analysis. *Brain Research, 1453*, 87-101.

Cohen, J. D., Perlstein, W. M., Braver, T. S., Nystrom, L. E., Noll, D. C., Jonides, J. & Smith, E. E. (1997). Temporal dynamics of brain activation during a working memory task. *Nature, 386*, 604-608.

Coles, K. & Tomporowski, P. D. (2008). Effects of acute exercise on executive processing, short-term and log-term memory. *Journal of Sports Sciences, 26* (3), 333-344.

Colcombe, S. & Kramer, A. F. (2003). Fitness effects on the cognitive function of older adults: A meta-analytic study. *Psychological Science, 14* (2), 125-130.

Conn, V. S., Hafdahl, A. R., Cooper, P. S., Brown, L. M. & Lusk, S. L. (2009). Meta-analysis of workplace physical activity interventions. *American Journal of Preventive Medicine, 37*, 330-339.

Conroy, M. A. & Polich, J. (2007). Normative variation of P3a and P3b from a large sample. *Journal of Psychophysiology, 21* (1), 22-32.

Copeland, J. & Heggie, L. (2008). IGF-1 and IGFBP-3 during continuous and interval exercise. *International Journal of Sports Medicine, 29*, 182-187.

Cotman, C. W. & Berchtold, N. C. (2002). Exercise: a behavioral intervention to enhance brain health and plasticity. *Trends in Neurosciences*, 25, 295-301.

Craik, F. I. M. & Bialystok, E. (2006). Cognition through the lifespan: mechanisms of change. *Trends in Cognitive Sciences, 10* (3), 131-138.

Curlik, D. M. & Shors, T. J. (2013). Training your brain: Do mental and physical (MAP) training enhance cognition through the process of neurogenesis in the hippocampus? *Neuropharmacology, 64*, 506-514.

Curtis, M. A., Kam, M. & Faull, R. L. (2011). Neurogenesis in humans. *European Journal of Neuroscience, 33* (6), 1170-1174.

Davelaar, E. J. (2013). When the ignored gets bound: sequential effects in the flanker task. *Frontiers in Psychology, 3*, 552. doi: 10.3389/fpsyg.2012.00552.

Davelaar, E. J. & Stevens, J. (2009). Sequential dependencies in the Eriksen flanker task: A direct comparison of two competing accounts. *Psychonomic Bulletin & Review, 16*, 121-126.

Davidson, M. C., Amso, D., Anderson, L. C. & Diamond, A. (2006). Development of cognitive control and executive functions from 4 to 13 years: Evidence from manipulations of memory, inhibition and task switching. *Neuropsychologia, 44* (11), 2037-2078.

Davis, C. L. & Cooper, S. (2011). Fitness, fatness, cognition, behavior, and academic achievement among overweight children: do cross-sectional associations correspond to exercise trial outcomes? *Preventive Medicine, 1* (52) Suppl 1, 65-9.

Davis, C. L. & Lambourne, K. (2009). Exercise and cognition in children. In T. McMorris, P. Tomporowski & M. Audiffren (Eds.), *Exercise and Cognitive Function* (pp. 249-267). Hoboken: Wiley-Blackwell.

Dempster, P. & Aitkens, S. (1995). A new air displacement method for the determination of human body composition. *Medicine and Science in Sports and Exercise, 27*, 1692-1697.

Donnelly, J. E. & Lambourne, K. (2011). Classroom-based physical activity, cognition, and academic achievement. *Preventive Medicine, 52*, S36-S42.

Duncan, C. C., Barry, R. J., Connolly, J. F., Fischer, C., Michie, P. T., Näätänen, R., Polich, J., Reinvang, I. & Van Petten, C. (2009). Event-related potentials in clinical research: Guidelines for eliciting, recording, and quantifying mismatch negativity, P300, and N400. *Clinical Neurophysiology, 120*, 1883-1908.

Dustman, R. E., Emmerson, R. Y., Ruhling, R. O., Shearer, D. E., Steinhaus, L. A., Johnson, S. C., Bonekat, H. W. & Shigeoka, J. W. (1990). Age and fitness effects on EEG, ERPs, visual sensitivity, and cognition. *Neurobiology of Aging, 11*, 193-200.

Engelmann, J. (2012). Untersuchung des Zusammenhangs von EEG basierten ereigniskorrelierten Potentialen und physischer und kognitiver Fitness bei jungen Studierenden. Unpublished diploma thesis at the Department of Informatics at the Karlsruhe Institute of Technology (KIT).

Erickson, K. I. (2012). Physical activity and brain functions. In C. Bouchard, S. N. Blair & W. L. Haskell (Eds.), *Physical Activity and Health* (2nd edition; pp. 317-330). Champaign: Human Kinetics.

Erickson, K. I., Prakash, R. S., Voss, M. W., Chaddock, L., Heo, S., McLaren, M., Pence, B. D., Martin, S. A., Vieira, V. J., Woods, J. A., McAuley, E. & Kramer, A. F. (2010). Brain-derived neurotrophic factor is associated with age-related decline in hippocampal volume. *Journal of Neuroscience, 30* (15), 5368-5375.

Eriksen, B. A. & Eriksen, C. W. (1974). Effects of noise letters upon the identification of a target letter in nonsearch task. *Perception & Psychophysics, 16*, 143-149.

Etnier, J. L., Nowell, P. M., Landers, D. M. & Sibley, B. A. (2006). A meta-regression to examine the relationship between aerobic fitness and cognitive performance. *Brain Research Reviews, 52*, 119-130.

Euston, D. R., Gruber, A. J. & McNaughton, B. L. (2012). The role of medial prefrontal cortex in memory and decision making. *Neuron, 76* (6), 1057-1070.

Eysenck, M. W., Derakshan, N., Santos, R. & Calvo, M. G. (2007). Anxiety and cognitive performance: attentional control theory. *Emotion, 7* (2), 336-353.

Farmer, J., Zhao, X., Van Praag, H., Wodtke, K., Gage, F. H. & Christie, B. R. (2004). Effects of voluntary exercise on synaptic plasticity and gene expression in the dentate gyrus of adult male Sprague-Dawley rats in vivo. *Neuroscience, 124* (1), 71-79.

Faude, O., Kindermann, W. & Meyer, T. (2009). Lactate threshold concepts. How valid are they? *Sports Medicine, 36* (6), 469-490.

Fedewa, A. L. & Ahn, S. (2011). The effects of physical activity and physical fitness on children's achievement and cognitive outcomes: A meta-analysis. *Research Quarterly for Exercise and Sport, 82* (3), 521-535.

Ferris, L. T., Williams, J. S. & Shen, C. L. (2007). The effect of acute exercise on serum brain-derived neurotrophic factor levels and cognitive function. *Medicine and Science in Sports and Exercise, 39*, 728-734.

Fields, D. A., Goran, M. I. & McCrory, M. A. (2002). Body-composition assessment via air-displacement plethysmography in adults and children: a review. *American Journal of Clinical Nutrition, 75*, 453-467.

Finger, J. D., Tylleskär, T., Lampert, T. & Mensink, G. B. (2012). Physical activity patterns and socioeconomic position: the German National Health Interview and Examination Survey 1998 (GNHIES98). *BMC Public Health, 12*, 1079. doi: 10.1186/1471-2458-12-1079.

Frensch, P. A. (2006). Kognition. In J. Funke & P. A. Frensch (Eds.), *Handbuch der allgemeinen Psychologie - Kognition* (pp. 19-28). Göttingen: Hogrefe Verlag.

Garber, C. E., Blissmer, B., Deschenes, M. R., Franklin, B. A., Lamonte, M. J., Lee, I. M., Nieman, D. C. & Swain, D. P. (2011). Quantity and quality of exercise for developing and maintaining cardiorespiratory, musculoskeletal, and neuromotor fitness in apparently healthy adults: Guidance for prescribing exercise. ACSM Position Stand. *Medicine and Science in Sports and Exercise, 43* (7), 1334-1359.

Gazzaniga, M. S., Ivry, R. B. & Mangun, G. R. (2009). *Cognitive Neuroscience. The Biology of the Mind (3rd edition)*. New York: W.W. Norton & Company, Inc.

Gläscher, J., Adolphs, R., Damasio, H., Bechara, A., Rudrauf, D., Calamia, M., Paul, L. & Tranel, D. (2012). Lesion mapping of cognitive control and value-based decision making in the prefrontal cortex. *Proceedings of the National Academy of Sciences of the United States of America, 109* (36), 14681-14686.

Goekint, M., De Pauw, K., Roelands, B., Njemini, R., Bautmans, I., Mets, T. & Meeusen, R. (2010). Strength training does not influence serum brain-derived neurotrophic factor. *European Journal of Applied Physiology, 110* (2), 285-293.

Going, S. B. (2005). Hydrodensitometry and air displacement plethysmography. In S. B. Heymsfield, T. G. Lohman, Z. Wang & S. B. Going (Eds.), *Human Body Composition* (2nd Ed.; pp. 17-33). Champaign: Human Kinetics.

Gomez-Pinilla, F., Vaynman, S. & Ying, Z. (2008). Brain-derived neurotrophic factor functions as a metabotrophin to mediate the effects of exercise on cognition. *European Journal of Neuroscience, 28* (11), 2278-2287.

Green, H. J. (2012). Skeletal muscle adaptation to regular physical activity. In C. Bouchard, S. N. Blair & W. L. Haskell (Eds.), *Physical Activity and Health* (2nd edition; pp. 121-148). Champaign: Human Kinetics.

Griesbach, G. S., Hovda, D. A. & Gomez-Pinilla, F. (2009). Exercise-induced improvement in cognitive performance after traumatic brain injury in rats is dependent on BDNF activation. *Brain Research, 1288*, 105-115.

Güllich, A. & Schmidtbleicher, D. (1999). Struktur der Kraftfähigkeiten und ihrer Trainingsmethoden. *Deutsche Zeitschrift für Sportmedizin, 50* (7/8), 223-234.

Guiney, H. & Machado, L. (2013). Benefits of regular aerobic exercise for executive functioning in healthy populations. *Psychonomic Bulletin and Review, 20*, 73-86.

Haapala, E. A. (2013). Cardiorespiratory fitness and motor skills in relation to cognition and academic performance in children - A review. *Journal of Human Kinetics, 36*, 55-68.

Hamilton, M. & Owen, N. (2012). Sedentary behavior and inactivity physiology. In C. Bouchard, S .N. Blair & W. L. Haskell (Eds.), *Physical Activity and Health* (2nd edition; pp. 53-68). Champaign: Human Kinetics.

Hansen, A. L., Johnsen, B. H., Sollers III, J. J., Stenvik, K. & Thayer, J. F. (2004). Heart rate variability and its relation to prefrontal cognitive function: the effects of training and detraining. *European Journal of Applied Physiology, 93*, 263-272.

Hansen, A. L., Johnsen, B. H. & Thayer, J. F. (2003). Vagal influence on working memory and attention. *International Journal of Psychophysiology, 48* (3), 263-274.

Hardman, A. E. (2012). Acute responses to physical activity and exercise. In C. Bouchard, S. N. Blair & W. L. Haskell (Eds.), *Physical Activity and Health* (2nd edition; pp. 87-102). Champaign: Human Kinetics.

Haskell, W. L. (2012). Dose-response issues in physical activity, fitness, and health. In C. Bouchard, S. N. Blair & W. L. Haskell (Eds.), *Physical Activity and Health* (2nd edition; pp. 345-358). Champaign: Human Kinetics.

Helmerhorst, H. J., Brage, S., Warren, J., Besson, H. & Ekelund, U. (2012). A systematic review of reliability and objective criterion-related validity of physical activity questionnaires. *International Journal of Behavioral Nutrition and Physical Activity, 9,* 103. doi: 10.1186/1479-5868-9-103.

Herrmann, C. S. & Knight, R. T. (2001). Mechanisms of human attention: event-related potentials and oscillations. *Neuroscience and Biobehavioral Reviews, 25* (6), 465-476.

Heyward, V. H. (1998). *Advanced Fitness Assessment & Exercise Prescription* (3rd Edition). Champaign: Human Kinetics.

Hillman, C. H., Buck, S. M., Themanson J. R., Pontifex, M. B. & Castelli, D. M. (2009). Aerobic fitness and cognitive development: Event-related potential and task performance indices of executive control in preadolescent children. *Development Psychology, 45* (1), 114-129.

Hillman, C. H., Castelli, D. M. & Buck, S. M. (2005). Aerobic fitness and neurocognitive function in healthy preadolescent adults. *Medicine and Science in Sports and Exercise, 37* (11), 1967-1974.

Hillman, C. H., Erickson, K. I. & Kramer, A. F. (2008). Be smart, exercise your heart: exercise effects on brain and cognition. *Nature Reviews. Neuroscience*, 9, 58-65.

Hillman, C. H., Kamijo, K. & Scudder, M. (2011). A review of chronic and acute physical activity participation on neuroelectric measures of brain health and cognition during childhood. *Preventive Medicine, 52*, S21-S28.

Hillman, C. H., Kramer, A. F., Belopolsky, A. V. & Smith, D. P. (2006). A cross-sectional examination of age and physical activity on performance and event-related brain potentials in a task switching paradigm. *International Journal of Psychophysiology, 59*, 30-39.

Hillman, C. H., Motl, R. W., Pontifex, M. B., Posthuma, D., Stubbe, J. H., Boomsma, D. I. & de Geus, E. J. (2006). Physical activity and cognitive function in a cross-section of younger and older community-dwelling individuals. *Health Psychology, 25* (6), 678-687.

Hillman, C. H., Snook, E. M. & Jerome, G. J. (2003). Acute cardiovascular exercise and executive control function. *International Journal of Psychophysiology, 48*, 307-314.

Hillman, C. H., Weiss, E. P., Hagberg, J. M. & Hatfield, B. D. (2002). The relationship of age and cardiovascular fitness to cognitive and motor processes. *Psychophysiology, 39*, 303-312.

Hirtz, P. (1985). *Koordinative Fähigkeiten im Schulsport.* Berlin: Sportverlag.

Hirtz, P. (1994). Koordinative Fähigkeiten. In G. Schnabel, D. Harre & A. Borde (Eds.), *Trainingswissenschaft* (pp. 137-146). Berlin: Sportverlag.

Hirtz, P., Hotz, A. & Ludwig, G. (2003). *Bewegungsgefühl*. Schorndorf: Hofmann Verlag.

Hohmann, A., Lames, M. & Letzelter, M. (2003). *Einführung in die Trainingswissenschaft* (3. Aufl.). Wiebelsheim: Limpert Verlag.

Hollmann, W. & Strüder, H. K. (2009). *Sportmedizin. Grundlagen für körperliche Aktivität, Training und Präventivmedizin* (5. Aufl.). Stuttgart: Schattauer Verlag.

Hopkins, M. E., Davis, F. C., Vantieghem, M. R., Whalen, P. J. & Bucci, D. J. (2012). Differential effects of acute and regular physical exercise on cognition and affect. *Neuroscience, 215*, 59-68.

Howley, E. T. (2001). Type of activity: resistance, aerobic and leisure versus occupational physical activity. *Medicine and Science in Sports and Exercise, 33* (6), 364-369.

Howley, E. T. (2012). Metabolic, cardiovascular, and respiratory responses to physical activity. In C. Bouchard, S. N. Blair & W. L. Haskell (Eds.), *Physical Activity and Health* (2nd edition; pp. 71-86). Champaign: Human Kinetics.

Hughes, C. & Graham, A. (2002). Measuring executive functions in childhood: Problems and solutions. *Child and Adolescent Mental Health, 7*, 131-142.

Isaac, V., Sim, S., Zheng, H., Zagorodnov, V., Tai, E. S. & Chee, M. (2011). Adverse associations between visceral adiposity, brain structure, and cognitive performance in healthy elderly. *Frontiers in Aging Neuroscience, 3* (12). doi: 10.3389/fnagi.2011.00012.

Iwadate, M., Mori, A., Ashizuka, T., Takayose, M. & Ozawa, T. (2005). Long-term physical exercise and somatosensory event-related potentials. *Experimental Brain Research, 160* (4), 528-532.

Janssen, I. & LeBlanc, A. G. (2010). Systematic review of the health benefits of physical activity and fitness in school-aged children and youth. *International Journal of Behavioral Nutrition and Physical Activity, 7*, 40. doi: 10.1186/1479-5868-7-40.

Jeneson, A. & Squire, L. R. (2012). Working memory, long-term memory, and medial temporal lobe function. *Learning & Memory, 19* (1), 15-25.

Jurado, M. B. & Rosselli, M. (2007). The exclusive nature of executive functions: A review of our current understanding. *Neuropsychology Review, 17*, 213-233.

Kamijo, K., Nishihira, Y., Hatta, A., Kaneda, T., Wasaka, T., Kida, T. & Kuroiwa, K. (2004). Differential influences of exercise intensity on information processing in the central nervous system. *European Journal of Applied Physiology, 92*, 305-311.

Kamijo, K., Nishihira, Y., Higashiura, T. & Kuroiwa, K. (2007). The interactive effect of exercise intensity and task difficulty on human cognitive processing. *International Journal of Psychophysiology, 65*, 114-121.

Kamijo, K., O'Leary, K. C., Pontifex, M. B., Themanson, J. R. & Hillman, C. H. (2010). The relation of aerobic fitness to neuroelectric indices of cognitive and motor task preparation. *Psychophysiology, 47*, 814-821.

Kamijo, K. & Takeda, Y. (2009). General physical activity levels influence positive and negative priming effects in young adults. *Clinical Neurophysiology, 120*, 511-519.

Kamijo, K. & Takeda, Y. (2010). Regular physical activity improves executive function during task switching in young adults. *International Journal of Psychophysiology, 75*, 304-311.

Kane, M. J., Conway, A. R., Miura, T. K. & Colflesh, G. J. (2007). Working memory, attention control, and the N-back task: a question of construct validity. *Journal of Experimental Psychology. Learning, Memory and Cognition, 33* (3), 615-622.

Karch, S. & Mulert, C. (2010). Cognition. In C. Mulert & L. Lemieux (Eds.), *EEG-fMRI. Physiological Basis, Technique, and Applications* (pp. 419-450). Heidelberg: Springer Verlag.

Katzmarzyk, P. T. (2012). Physical activity and fitness with age, sex, and ethnic differences. In C. Bouchard, S. N. Blair & W. L. Haskell (Eds.), *Physical Activity and Health* (2nd edition; pp. 39-51). Champaign: Human Kinetics.

Kayser, D. (2003). Fitness. In P. Röthig & R. Prohl (Eds.), *Sportwissenschaftliches Lexikon* (7., völlig neu bearb. Aufl.; p. 200). Schorndorf: Hofmann Verlag.

Kempermann, G., Kuhn, H. G. & Gage, F. H. (1997). More hippocampal neurons in adult mice living in an enriched environment. *Nature, 386*, 493-495.

Kenney, W. L., Wilmore, J. H. & Costill, D. L. (2012). *Physiology of Sport and Exercise* (5th edition). Champaign: Human Kinetics.

Keul, J., Simon, G., Berg, A., Dickhuth, H.-H., Goerttler, I. & Kübel, R. (1979). Bestimmung der individuellen anaeroben Schwelle zur Leistungsbewertung und Trainingsgestaltung. *Deutsche Zeitschrift für Sportmedizin, 30* (7), 212-218.

Key, A. P. F., Dove, G. O. & Maguire, M. J. (2005). Linking brainwaves to the brain: An ERP primer. *Developmental Neuropsychology, 27* (2), 183-215.

Knab, A. M., Nieman, D. C., Sha, W., Broman-Fulks, J. J. & Canu, W. H. (2012). Exercise frequency is related to psychopathology but not neurocognitive function. *Medicine and Science in Sports and Exercise, 44* (7), 1395-1400.

Knaepen, K., Goekint, M., Heyman, E. M. & Meeusen, R. (2010). Neuroplasticity - Exercise-induced response of peripheral brain-derived neurotrophic factor. *Sports Medicine, 40* (9), 765-801.

Kolb, B., Mychasiuk, R., Muhammad, A., Li, Y., Frost, D. O. & Gibb, R. (2012). Experience and the developing prefrontal cortex. *Proceedings of the National Academy of Sciences of the United States of America, 109* (2), 17186-17193.

Kopelman, P. (2007). Health risks associated with overweight and obesity. *Obesity Reviews, 8* (Suppl. 1), 13-17.

Kramer, A. F. & Hillman, C. H. (2006). Aging, physical activity, and neurocognitive function. In E. O. Acevedo & P. Ekkekakis (Eds.), *Psychobiology of Physical Activity* (pp. 45-59). Champaign: Human Kinetics.

Krug, S., Jordan, S., Mensink, G. B. M., Müters, S., Finger, J. D. & Lampert, T. (2013). Körperliche Aktivität. Ergebnisse der Studie zur Gesundheit Erwachsener in Deutschland (DEGS1). *Bundesgesundheitsblatt - Gesundheitsforschung - Gesundheitsschutz 56* (5/6), 765-771.

Lämmle, L., Tittlbach, S., Oberger, J., Worth, A. & Bös, K. (2010). A two-level model of motor performance ability. *Journal of Exercise Science & Fitness, 8* (1), 41-49.

Lambourne, K. & Tomporowksi, P. (2010). The effect of exercise-induced arousal on cognitive task performance: A meta-regression analysis. *Brain Research, 1341*, 12-24.

Lampert, T., Mensink, G. B. M. & Müters, S. (2012). Körperlich-sportliche Aktivität bei Erwachsenen in Deutschland. Ergebnisse der Studie „Gesundheit in Deutschland aktuell 2009". *Bundesgesundheitsblatt-Gesundheitsforschung-Gesundheitsschutz, 55* (1), 102-110.

Lardon, M. T. & Polich, J. (1996). EEG changes from long-term physical exercise. *Biological Psychology, 44* (1), 19-30.

Lee, I.-M. (2012). Physical activity, fitness, and cancer. In C. Bouchard, S. N. Blair & W. L. Haskell (Eds.), *Physical Activity and Health* (2nd edition; pp. 231-244). Champaign: Human Kinetics.

Levine, B. D. (2008). VO_{2max}: what do we know, and what do we still need to know? *Journal of Physiology, 586* (1), 25-34.

Linden, D. E. J. (2007). The working memory networks of the human brain. *The Neuroscientist, 13* (3), 257-267.

Liu-Ambrose, T. & Donaldson, M. G. (2009). Exercise and cognition in older adults: is there a role for resistance training programmes? *British Journal of Sports Medicine, 43* (1), 25-27.

Liu-Ambrose, T., Nagamatsu, L. S., Graf, P., Beattie, B. L., Ashe, M. C. & Handy, T. C. (2010). Resistance training and executive functions. A 12-month randomized controlled trial. *Archives of Internal Medicine, 170* (2), 170-178.

Mader, A., Liesen, H., Heck, H., Phillipi, H., Rost, R., Schürch, P. & Hollmann, W. (1976). Zur Beurteilung der sportartspezifischen Ausdauerleistungsfähigkeit im Labor. *Sportarzt und Sportmedizin, 27* (4) 80-88; (5) 109-112.

Magnié, M.-N., Bermon, S., Martin, F., Madany-Lounis, M., Suisse, G., Muhammad, W. & Dolisi, C. (2000). P300, N400, aerobic fitness, and maximal aerobic exercise. *Psychophysiology, 37*, 369-377.

Maher, B. (2008). Poll results: look who's doping. *Nature, 452*, 674-675.

Malina, R. M. (2005). Variation in body composition associated with sex and ethnicity. In S. B. Heymsfield, T. G. Lohman, Z. Wang & S. B. Going (Eds.), *Human Body Composition* (2nd Ed.; pp. 271-298). Champaign: Human Kinetics.

McDowell, K., Kerick, S. E., Santa Maria, D. L. & Hatfield, B. D. (2003). Aging, physical activity, and cognitive processing: an examination of P300. *Neurobiology of Aging, 24*, 597-606.

McMorris, T., Sproule, J., Turner, A. & Hale, B. J. (2011). Acute, intermediate intensity exercise, and speed and accuracy in working memory tasks: A meta-analytical comparison of effects. *Physiology & Behavior, 102*, 421-428.

Meeusen, R. (2006). Physical activity and neurotransmitter release. In E. O. Acevedo & P. Ekkekakis (Eds.), *Psychobiology of Physical Activity* (pp. 129-143). Champaign: Human Kinetics.

Memmert, D. & Weikgenannt, J. (2006). Zum Einfluss sportlicher Aktivität auf die Konzentrations-leistung im Kindesalter. *Spektrum der Sportwissenschaft, 18* (2), 77-99.

Meyer, D. E. & Kieras, D. E. (1997). A computational theory of executive cognitive processes and multi-task performance: Part 1. Basic mechanisms. *Psychological Review, 104*, 3-65.

Middelbeek, R. J. & Goodyear, L. J. (2012). Physical activity, fitness, and diabetes mellitus. In C. Bouchard, S. N. Blair & W. L. Haskell (Eds.), *Physical Activity and Health* (2nd edition; pp. 215-229). Champaign: Human Kinetics.

Miller, E. K. (2000). The prefrontal cortex and cognitive control. *Nature Reviews Neuroscience, 1* (1), 59-65.

Mitchell, J. H., Haskell, W. L, Snell, P. & Van Camp, S. P. (2005). Task Force 8: Classification of Sports. *Journal of the American College of Cardiology, 45* (8), 1364-1367.

Monsell, S. (2003). Task switching. *Trends in Cognitive Sciences, 7* (3), 134-140.

Neumaier, A. (1999). *Koordinatives Anforderungsprofil und Koordinationstraining*. Köln: Strauß Verlag.

Nocon, M., Hiemann, T., Müller-Riemenschneider, F., Thalau, F., Roll, S. & Willich, S. N. (2008). Association of physical activity with all-cause and cardiovascular mortality: a systematic re-view and meta-analysis. *European Journal of Cardiovascular Prevention & Rehabilitation, 15* (3), 239-246.

Norman, D. A. & Shallice, T. (1986). Attention to action: Willed and automatic control of behavior. In R. J. Davidson, G. E. Schwartz & D. Sharipo (Eds.), *Consciousness and self-regulation: Vol. 4. Advances in research and theory* (pp. 1-18). New York: Plenum Press.

Ode, J. J., Pivarnik, J. M., Reeves, M. J. & Knous, J. L. (2007). Body mass index as a predictor of percent fat in college athletes and nonathletes. *Medicine and Science in Sports and Exercise, 39* (3), 403-409.

Oja, P., Laukkanen, R., Pasanen, M., Tyry, T. & Vuori, I. (1991). A 2-km walking test for assessing the cardiorespiratory fitness of healthy adults. *International Journal of Sports Medicine, 12* (4), 356-362.

Oja, P. & Titze, S. (2011). Physical activity recommendations for public health: development and policy context. *EPMA Journal, 2* (3), 253-259.

Opper, E., Oberger, J., Worth, A., Woll, A. & Bös, K. (2008). Wie motorisch leistungsfähig sind aktive Kinder und Jugendliche in Deutschland? *Motorik, 31* (2), 60-73.

Orszag, P. R. & Ellis, P. (2007). The challenge of rising health care costs - A view from the congressional budget office. *New England Journal of Medicine, 357* (18), 1793-1795.

Ortega, B. F., Tresaco, B., Ruiz, J. R., Moreno, L. A., Martin-Matillas, M., Mesa, J. L. Warnberg, J., Bueno, M., Tercedor, P., Gutiérrez, A., Castillo, M. J. & the AVENA Study Group (2007). Cardiorespiratory fitness and sedentary activities are associated with adiposity in adolescents. *Obesity, 15* (6), 1589-1599.

Pate, R. R. (1988). The evolving definition of fitness. *Quest, 40,* 174-179.

Pereira, A. C., Huddleston, D. E., Brickman, A. M., Sosunov, A. A., Hen, R., McKhann, G. M., Sloan, R., Gage, F. H., Brown, T. R. & Small, S. A. (2007). An in vivo correlate of exercise-induced neurogenesis in the adult dentate gyrus. *Proceedings of the National Academy of Sciences of the United States of America, 104* (13), 5638-5643.

Pesce, C. (2009). An integrated approach to the effect of acute and chronic exercise on cognition: the linked role of individual and task constraints. In T. McMorris, P. Tomporowski & M. Audiffren (Eds.), *Exercise and Cognitive Function* (pp. 213-226). Hoboken: Wiley-Blackwell.

Peters, M., Laeng, B., Latham, K., Jackson, M., Zaiyouna, R. & Richardson, C. (1995). A redrawn Vandenberg and Kuse Mental Rotations Test: Different versions and factors that affect performance. *Brain and Cognition, 28,* 39-58.

Polich, J. (2007). Updating P300: An integrative theory of P3a and P3b. *Clinical Neurophysiology, 118,* 2128-2148.

Polich, J. & Kok, A. (1995). Cognitive and biological determinants of P300: an integrative review. *Biological Psychology, 41* (2), 103-146.

Polich, J. & Lardon, M. T. (1997). P300 and long-term physical exercise. *Electroencephalography and Clinical Neurophysiology, 103* (4), 493-498.

Pontifex, M. B., Hillman, C. H. & Polich, J. (2009). Age, physical fitness, and attention: P3a and P3b. *Psychophysiology, 46*, 379-387.

President's Council on Physical Fitness and Sports (1971). *Physical Fitness Research Digest.* Washington D.C. (Series 1, Number 1).

Raglin, J. S. & Wilson, G. S. (2012). Exercise and its effects on mental health. In C. Bouchard, S. N. Blair & W. L. Haskell (Eds.), *Physical Activity and Health* (2nd edition; pp. 331-342). Champaign: Human Kinetics.

Rasberry, C. N., Lee, S. M., Robin, L., Laris, B. A., Russell, L. A., Coyle, K. K. & Nihiser, A. J. (2011). The association between school-based physical activity, including physical education, and academic performance: A systematic review of the literature. *Preventive Medicine, 52*, S10-S20.

Rasmussen, P., Brassard, P., Adser, H., Pedersen, M. V., Leick, L., Hart, E., Secher, N. H., Pedersen, B. K. & Pilegaard, H. (2009). Evidence for a release of brain derived neurotrophic factor from the brain during exercise. *Experimental Physiology, 94*, 1062-1069.

Ratey, J. J. & Loehr, J. E. (2011). The positive impact of physical activity on cognition during adulthood: a review of underlying mechanisms, evidence and recommendations. *Reviews in the Neurosciences, 22* (2), 171-185.

Rheinberg, F., Vollmeyer, R. & Burns, B. D. (2001). QCM. A questionnaire to assess current motivation in learning situations. *Diagnostica, 47*, 57-66.

Rindermann, H., Baumeister, A. E. & Gröper, A. (2013). Cognitive abilities of Emirati and German engineering university students. *Journal of Biosocial Science.* doi:10.1017/S0021932013000266.

Ross, R. & Janssen, I. (2012). Physical activity, fitness and obesity. In C. Bouchard, S. N. Blair & W. L. Haskell (Eds.), *Physical Activity and Health* (2nd edition; pp. 197-214). Champaign: Human Kinetics.

Roth, K. (1982). *Strukturanalyse koordinativer Fähigkeiten.* Bad Homburg: Limpert Verlag.

Rowlands, A. V., Ingledew, D. K. & Eston, R. G. (2000) The effect of type of physical activity measure on the relationship between body fatness and habitual physical activity in children: a meta analysis. *Annals of Human Biology, 27* (5), 479-497.

Rütten, A., Abu-Omar, K., Lampert, T. & Ziese, T. (2005). *Körperliche Aktivität. Gesundheitsberichterstattung des Bundes. Heft 26.* Berlin: Robert Koch-Institut.

Sahakian, B. & Morein-Zamir, S. (2007). Professor's little helper. *Nature, 450*, 1157-1159.

Sardinha, L. B. & Teixeira, P. J. (2005). Measuring adiposity and fat distribution in relation to health. In S. B. Heymsfield, T. G. Lohman, Z. Wang & S. B. Going (Eds.), *Human Body Composition* (2nd Ed.; pp. 177-201). Champaign: Human Kinetics.

Schiffer, T., Schulte, S., Hollmann, W., Bloch, W. & Strüder, H. (2009). Effects of strength and endurance training on brain-derived neurotrophic factor and insulin-like growth factor 1 in humans. *Hormone and Metabolic Research, 41*, 250-254.

Schönfeld, R. (2008). *Alters- und Geschlechtsdifferenzen in der Raumkognition - auch eine Frage der Lösungsstrategien.* Dissertation, Fakultät für Naturwissenschaften, Otto-von-Guericke-Universität Magdeburg.

Schuenke, M., Schulte, E. & Schumacher, U. (2010). *Thieme Atlas of Anatomy. Head and Neuroanatomy.* Stuttgart: Thieme Verlag.

Schousboe, K., Willemsen, G., Kyvik, K. O., Mortensen, J., Boomsma, D. I., Cornes, B. K., Davis, C. J., Fagnani, C., Hjelmborg, J., Kaprio, J., De Lange, M., Luciano, M., Martin, N. G., Pedersen, N., Pietiläinen, K. H., Rissanen, A., Saarni, S., Sørensen, T. I., Van Baal, G. C. & Harris, J. R. (2003). Sex differences in heritability of BMI: a comparative study of results from twin studies in eight countries. *Twin Research, 6* (5), 409-421.

Scisco, J. L., Leynes, P. A. & Kang, J. (2008). Cardiovascular fitness and executive control during task-switching: An ERP study. *International Journal of Psychophysiology, 69*, 52-60.

Seifert, T., Brassard, P., Wissenberg, M., Rasmussen, P., Nordby, P., Stallknecht, B., Adser, H., Jakobsen, A., Pilegaard, A., Nielsen, H. & Sechr, N. (2010). Endurance training enhances BDNF release from the human brain. *American Journal of Physiology - Regulatory, Integrative and Comparative Physiology, 298*, 372-377.

Seiferth, N. Y., Thienel, R. & Kircher, T. (2007). Exekutive Funktionen. In F. Scheider & G. R. Fink (Eds.), *Funktionelle MRT in Psychiatrie und Neurologie* (pp. 266-277). Heidelberg: Springer Verlag.

Shen, W., St-Onge, M. P., Wang, Z. & Heymsfield, S. B. (2005). Study of body composition: An overview. In S. B. Heymsfield, T. G. Lohman, Z. Wang & S. B. Going (Eds.), *Human Body Composition* (2nd Ed.; pp. 3-16). Champaign: Human Kinetics.

Shephard, R. J. (2003). Limits to the measurement of habitual physical activity by questionnaires. *British Journal of Sports Medicine, 37* (3), 197-206.

Sherwood, D. E. & Selder, D. J. (1979). Cardiorespiratory health, reaction time and aging. *Medicine and Science in Sports, 11*, 186-189.

Sibley, B. A. & Etnier, J. L. (2003). The relationship between physical activity and cognition in children: A meta-analysis. *Pediatric Exercise Science, 15*, 243-256.

Singh, A., Uijtdewilligen, L., Twisk, J. W. R., van Mechelen, W. & Chinapaw, M. J. M. (2012). Physical activity and performance at school. A systematic review of the literature including a methodological quality assessment. *Archives of Pediatrics & Adolescent Medicine, 166* (1), 49-55.

Siri, W. E. (1956). The gross composition of the body. *Advances in Biological and Medical Physics, 4*, 239-280.

Slusher, H. S. (1964). Personality and intelligence characteristics of selected high-school athletes and nonathletes. *Research Quarterly, 35* (4), 539-545.

Smith, P. J., Blumenthal, J. A., Hoffman, B. M., Cooper, H., Strauman, T. A., Welsh-Bohmer, K., Browndyke, J. N. & Sherwood, A. (2010). Aerobic exercise and neurocognitive performance: a meta-analytic review of randomized controlled trials. *Psychosomatic Medicine, 72* (3), 239-252.

Solso, R. L. (2001). *Cognitive Psychology* (6th edition). Needham Heights: Allyn & Bacon.

Spirduso, W. W. (1975). Reaction and movement time as a function of age and physical activity level. *Journal of Gerontology, 30*, 435-440.

Stangl, W. (1991). Der Freizeit-Interessen-Test (FIT). *Zeitschrift für Differentielle und Diagnostische Psychologie 4*, 231-244.

Stegmann, H., Kindermann, W. & Schnabel, A. (1981). Lactate kinetics and individual anaerobic threshold. *International Journal of Sports Medicine, 2* (3), 160-165.

Stroth, S., Hille, K., Spitzer, M. & Reinhardt, R. (2009). Aerobic endurance exercise benefits memory and affect in young adults. *Neuropsychological Rehabilitation, 19* (2), 223-243.

Stroth, S., Reinhardt, R. K., Thöne, J., Hille, K., Schneider, M., Härtel, S., Weidemann, W., Bös, K. & Spitzer, M. (2010). Impact of aerobic exercise training on cognitive functions and affect associated to the COMT polymorphism in young adults. *Neurobiology of Learning and Memory, 94*, 364-372.

Stuss, D. T., Alexander, M. P., Floden, D., Binns, M. A., Levine, B., McInstosh, A. R., Rajah, M. N. & Hevenor, S. J. (2002). Fractionation and localization of distinct frontal lobe processes: Evidence from focal lesions in humans. In D. T. Stuss & R. T. Knight (Eds.), *Principles of frontal lobe function* (pp. 392-407). New York: Oxford University Press.

Squire, L. R., Stark, C. E. L. & Clark, R. E. (2004). The medial temporal lobe. *Annual Review of Neuroscience, 27*, 279-306.

Tang, S. W., Chu, E., Hui, T., Helmeste, D. & Law, C. (2008). Influence of exercise on serum brain-derived neurotrophic factor concentrations in healthy human subjects. *Neuroscience Letters, 431*, 62-65.

Teplan, M. (2002). Fundamentals of EEG measurement. *Measurement Science Review, 2* (2), 1-11.

Themanson, J. R. & Hillman, C. H. (2006). Cardiorespiratory fitness and acute aerobic exercise effects on neuroelectric and behavioral measures of action monitoring. *Neuroscience, 141*, 757-767.

Themanson, J. R., Pontifex, M. B. & Hillman, C. H. (2008). Fitness and action monitoring: Evidence for improved cognitive flexibility in young adults. *Neuroscience, 157*, 319-328.

Thomas, A. G., Dennis, A., Bandettini, P. A. & Johansen-Berg, H. (2012). The effects of aerobic activity on brain structure. *Frontiers in Psychology, 3* (86). doi: 10.3389/fpsyg.2012.00086.

Tomporowski, P. D. (2003). Effects of acute bouts of exercise on cognition. *Acta Psychologica, 112*, 297-324.

Tomporowski, P. D., Lambourne, K. & Okumura, M. S. (2011). Physical activity interventions and children's mental function: an introduction and overview. *Preventive Medicine, 52*, S3-S9.

Trejo, J. L., Carro, E. & Torres-Aleman, I. (2001). Circulating insulin-like growth factor 1 mediates exercise induced increases in the number of neurons in the adult hippocampus. *The Journal of Neuroscience, 21,* 1628-1634.

Troiano, R. P., Berrigan, D., Dodd, K. D., Mâsse, L. C., Tilert, T. & McDowell, M. (2008). Physical activity in the United States measured by accelerometer. *Medicine and Science in Sports and Exercise, 40* (1), 181-188.

Van Petten, C. (2004). Relationship between hippocampal volume and memory ability in healthy individuals across the lifespan: review and meta-analysis. *Neuropsychologia, 42*, 1394-1413.

Van Praag, H. (2006). Exercise, neurogenesis, and learning in rodents. In E. O. Acevedo & P. Ekkekakis (Eds.), *Psychobiology of Physical Activity* (pp. 61-73). Champaign: Human Kinetics.

Van Praag, H., Kempermann, G. & Gage, F. H. (1999). Running increases cell profileration and neurogenesis in the adult mouse dentate gyrus. *Nature Neuroscience, 2*, 266-270.

Vaynman, S., Ying, Z. & Gomez-Pinilla, F. (2003). Interplay between brain-derived neurotrophic factor and signal transduction modulators in the regulation of the effects of exercise on synaptic-plasticity. *Neuroscience, 122*, 647-657.

Voelcker-Rehage, C., Godde, B. & Staudinger, U. M. (2011). Cardiovascular and coordination training differentially improve cognitive performance and neural processing in older adults. *Frontiers in Human Neuroscience, 5* (26). doi: 10.3389/fnhum.2011.00026.

Voelcker-Rehage, C., Godde, B. & Staudinger, U. M. (2010). Physical and motor fitness are both related to cognition in old age. *The European Journal of Neuroscience, 31*, 167-176.

Voelcker-Rehage, C. & Windisch, C. (2013). Structural and functional brain changes related to different types of physical activity across the lifespan. *Neuroscience and Biobehavioral Reviews.* doi: 10.1016/j.neubiorev.2013.01.028.

Vogel, T., Brechat, P.-H., Leprêtre, P. M., Kaltenbach, G., Berthel, M. & Lonsdorfer, J. (2009). Health benefits of physical activity in older patients: a review. *International Journal of Clinical Practice, 63* (2), 303-320.

Voss, M. W., Nagamatsu, L. S., Liu-Ambrose, T. & Kramer, A. F. (2011). Exercise, brain, and cognition across the life span. *Journal of Applied Physiology, 111* (5), 1505-1513.

Wagner, T. D. & Smith, E. E. (2003). Neuroimaging studies of working memory: A meta analysis. *Behavioral Neuroscience, 3,* 241-253.

Wardle, J. & Cooke, L. (2005). The impact of obesity on psychological well-being. *Best Practice & Research Clinical Endocrinology & Metabolism, 19* (3), 421-440.

Weinstein, A. M., Voss, M. W., Prakash, R. S., Chaddock, l., Szabo, A., White, S. M., Wojcicki, T. R., Mailey, E., McAuley, E., Kramer, A. F. & Erickson, K. I. (2011). The association between aerobic fitness and executive function is mediated by prefrontal cortex volume. *Brain, Behavior, and Immunity, 26* (5), 811-819.

Weiß, R. H. (2006). *Grundintelligenztest Skala 2- Revision. CFT 20-R.* Göttingen: Hogrefe Verlag.

Werner, C. & Jansen, C. P. (2010). *Koordiniere Dich schlau - Auswirkung von anaerobem Koordinationstraining auf kognitive Fähigkeiten.* Unpublished B. A. thesis at the Institute of Sport and Sport Science, Karlsruhe Institute of Technology.

Woll, A. (2004). Diagnose körperlich-sportlicher Aktivität im Erwachsenenalter. *Zeitschrift für Sportpsychologie, 11* (2), 54-70.

Woll, A., Kurth, B. M., Opper, E., Worth, A. & Bös, K. (2011). The 'Motorik-Modul' (MoMo): physical fitness and physical activity in German children and adolescents. *European Journal of Pediatrics, 170* (9), 1129-1142.

World Health Organiziation (1968). *Meeting of investigators on exercise tests in relation to cardiovascular function.* Geneva: World Health Organization.

World Health Organization (2010). *Global recommendations on Physical Activity for Health.* Geneva: World Health Organization.

Zimmerman, M. E., Brickman, A. M., Paul, R. H., Grieve, S. M., Tate, D. F., Gunstad, J., Cohen, R. A., Aloia, M. S., Williams, L. M., Clark, C. R., Whitford, T. J. & Gordon, E. (2006). The relationship between frontal gray matter volume and cognition varies across the healthy adult lifespan. *The American Journal of Geriatric Psychiatry, 14* (10), 823-833.

Zelazo, P. D., Craik, F. I. M. & Booth, L. (2004). Executive function across the life span. *Acta Psychologica, 115* (2-3), 167-183.

Zoladz, J., Pilc, A., Majerczak, J., Grandys, M., Zapart-Bukowska, J. & Duda, K. (2008). Endurance training increases plasma brain-derived neurotrophic factor concentration in young healthy men. *Journal of Physiology and Pharmacology, 59,* 119-132.

List of Figures

List of Tables

Appendix

Chronological sequence of measurements

Testing day I (CSL)	
General	Information about testing procedure
Motivation & personality	- Questionnaire on current motivation QCM (Rheinberg, Vollmeyer & Burns, 2001) - Questionnaire on attitude towards virtual agents - Personality inventory NEO-FFI (Borkenau & Ostendorf, 1993)
Cognition & Driving simulation	- Computerized cognitive testing battery: Oddball-, modified Flanker-, and Switching-Task, with 16 channel-EEG recording - Driving simulator: Lane change task with two secondary tasks (2-back task and visual search task) and interaction task (communication with virtual co-driver system and multiple-choice memory task) - Task load assessment tool (NASA-TLX) after different tasks

Testing day II (IfSS)	
General	- Information about testing procedure - Written informed consent to perform a graded cycle ergometer test
Leisure time	- Questionnaire on habitual and work-related physical activity, and recreational activities - Leisure Interest Inventory (Stangl, 1991)
Anthropometry & Body Comp	- Anthropometric measurements: height, waist and hip circumference - Body composition measurements: Air-displacement plethysmography (BOD POD Gold Standard® System; Life Measurement Inc., USA), Bioelectrical impedance spectroscopy (Body Composition Monitor including Fluid Management Tool; Fresenius Medical Care, Germany), Bioelectrical impedance analysis (Tanita BC-545 Innerscan Body Composition Monitor; Tanita Europe B.V., The Netherlands)
Body temperature	- Body surface temperature measurement using a thermal imaging infrared camera (ThermaCAM® PM545, Flir Systems, Inc., USA) before and after endurance test - Body core temperature measurement via heat-flux using two double-sensors located at the forehead and the manubrium sterni (Health Lab, Koralewski Industrie, Germany)

Heart rate variability	11 minutes heart rate variability measurement (rest condition) using Polar RS 800 heart rate monitor (Polar Electro Oy, Finland), and a three lead ECG (Health Lab, Koralewski Industrie, Germany)
Endurance	Graded cycle ergometer test to volitional exhaustion. During the tests, participants were fitted with a mobile cardiopulmonary exercise testing device, two double-sensors located at the forehead and the manubrium sterni to assess body temperature via heat-flux, Polar RS 800 heart rate monitor, and a three lead ECG

Testing day III (IfSS)	
General	Information about testing procedure
Mental rotation & Intelligence	- Mental rotation test MRT-B (Peters, Laeng, Latham, Jackson, Zaiyouna & Richardson, 1995) & Strategy and experience questionnaire (Schönfeld, 2008) - Intelligence test CFT 20-R (Weiß, 2006)
Coordination	Static and dynamic one-legged-stabilization tasks on the Posturomed (Posturomed 202; Haider Bioswing GmbH, Germany)
Strength	- Short practice on leg press and warm-up phase on treadmill - Isometric lower limb maximal strength test (leg press) - Dynamic lower limb strength test (Counter-movement jump)
Coordination	- Backward balancing and jumping side-to-side tests (Bös, Schlenker, Büsch, Lämmle, Müller et al., 2009) - Backward medicine ball throw test (Hirtz, Hotz & Ludwig, 2003)

Acknowledgment

Mein besonderer Dank gilt meinem Betreuer Prof. Dr. Klaus Bös für seine Förderung in den vergangenen Jahren, seine fachlichen Ratschläge und die vielen gewinnbringenden Gespräche nicht zuletzt auch auf persönlicher Ebene. Prof. Dr. Alexander Woll danke ich für die Erstellung des Zweitgutachtens. Weiterhin möchte ich mich bei dem Initiator der Studie, Dr. Alexander Stahn vom Zentrum für Weltraummedizin der Charité Berlin bedanken, der durch seine inhaltliche Betreuung einen Beitrag zum Gelingen der Arbeit leistete und immer ein offenes Ohr für Fragen und Probleme hatte. Ein großes Dankeschön richte ich zudem an Prof. Dr. Tanja Schultz, Dominic Heger und Felix Putze vom Cognitive Systems Lab am KIT für die freundliche Kooperation, den konstruktiven Austausch und die tatkräftige Unterstützung der Studie.

Desweiteren möchte ich der Forschungsförderung des KIT für die finanzielle Unterstützung der Studie sowie der leistungsdiagnostischen Abteilung und dem BioMotion Center am Institut für Sport und Sportwissenschaft für die Bereitstellung der Testgeräte und –räumlichkeiten danken; ebenso auch allen Studierenden, die mich bei der Studienorganisation und Datenerhebung unterstützt haben. Erwähnt sei hier Sabrina Benzinger, die mir mit ihrem unermüdlichen Einsatz stets eine wichtige Hilfe war. All meinen Kolleginnen und Kollegen möchte ich für die nette Arbeitsatmosphäre und viele schöne Gespräche danken, insbesondere Dr. Susanne Krug und Claire Kutzner für ihre Freundschaften auch nach unserer gemeinsamen Zeit am Institut sowie Steffen Schmidt, Dr. Sascha Härtel und Birte von Haaren für die gute Zusammenarbeit und den verständnisvollen Austausch.

Von ganzem Herzen möchte ich mich abschließend bei meinen Eltern und meiner Schwester für ihre immerwährende Unterstützung bedanken sowie bei meinem Ehemann Raphael für seinen Rückhalt und sein Vertrauen in mich.

Janina Krell-Rösch